P9-EDT-973

The Genetic Basis of
Plant Physiological Processes

The Genetic Basis of Plant Physiological Processes

JOHN KING

University of Saskatchewan

New York   Oxford
OXFORD UNIVERSITY PRESS
1991

Oxford University Press

Oxford   New York   Toronto
Delhi   Bombay   Calcutta   Madras   Karachi
Petaling Jaya   Singapore   Hong Kong   Tokyo
Nairobi   Dar es Salaam   Cape Town
Melbourne   Auckland

and associated companies in
Berlin   Ibadan

Copyright © 1991 by Oxford University Press, Inc.

Published by Oxford University Press, Inc.,
200 Madison Avenue, New York, New York 10016

Oxford is a registered trademark of Oxford University Press

All rights reserved. No part of this publication may be reproduced.
stored in a retrieval system, or transmitted, in any form or by any
means, electronic, mechanical, photocopying, recording, or otherwise,
without the prior permission of Oxford University Press.

Library of Congress Cataloging-in-Publication Data
King, John, 1938–
The genetic basis of plant physiological processes / John King.
p. cm.   Includes bibliographical references and index.
ISBN 0-19-504857-1
1. Plant physiology — Genetic aspects — Case studies.
2. Plant mutation — Case studies.   I. Title.
QK711.2.K57   1991
581.1 — dc20   91-11129   72 984

9 8 7 6 5 4 3 2 1

Printed in the United States of America
on acid-free paper

How odd it is
that anyone should not see
that all observation must be
for or against some view
if it is to be of any service.

C. Darwin

AUGUSTANA UNIVERSITY COLLEGE
LIBRARY

Preface

Plant physiology can be divided into such areas as water relations, mineral nutrition, metabolism, photosynthesis, and growth and development. The purpose of studies within these broad areas is to provide insights into the ways in which plants function. A wide variety of approaches and techniques can be used to investigate these plant physiological processes among which are those involving the use of variant genetic material.

An increasing number of examples exist where genetic (especially single gene) characters are being investigated from the point of view of their biochemistry and physiology or where specific mutants and variants are being sought to study particular plant biochemical and physiological processes. Sometimes mention is made of these studies in textbooks of plant physiology; often only in a fragmentary way as discoveries made through the use of mutants and variants have contributed to the knowledge of plant function arrived at by techniques other than those of genetics. There are increasing numbers of cases, however, where a genetical approach is central or significantly helpful to the understanding of a physiological problem. An increasing array of genetically variant plant material already exists which is being used to dissect physiological processes. Genetics also represents a powerful tool with which to generate new variant plants or cells to investigate particular plant functions. In addition, advances in molecular biology are making it increasingly necessary that students of plant physiology become more aware of genetics and how it can be used as a tool in their studies (one might note that the opposite is also true, that is, students of genetics will need to become more familiar with the physiology of the plant material they are studying if they hope to employ molecular biology techniques to modify plant functions).

The primary objective of this book is to illustrate in some detail, through the use of Case Studies, examples where genetics has contributed significantly to the investigation of physiological phenomena. The intention is not to provide a complete account of how mutants and variants have been used in plant physiology. The latter approach would be a task of encyclopedic proportions and would defeat the main purpose which is to demonstrate to students of plant physiology how genetically variant material has been or could be used to advantage in physiological investigations. To achieve this aim, a small number of examples has been selected, in as many of the areas of plant physiology as possible, to illustrate how mutants and variants are being used in research. In this way, it is hoped that the imaginations of some readers might be stimulated sufficiently for them to contemplate putting the powerful tools of genetics to work in their own research at some point in the future.

Thus, the book might be described as a companion to an elementary text in plant physiology. Certainly, it is essential for the reader to have already a background in plant physiology, although some basic physiological information is provided throughout the text. The references for each Case Study have also been kept to a minimum. Only what were regarded as those publications necessary to explain the main discoveries under each topic have been cited. The reader is encouraged to use these references as a starting point for further reading in areas of particular interest.

The book is divided into two parts, each of which has three chapters. Each chapter begins with a general introduction incorporated into which is some background information about those aspects of plant physiology pertinent to the discussions in the Case Studies in the chapter. Most Case Studies take the form of a very brief introduction followed by a discussion of the isolation and characterization of the mutants being considered. The Case Study then usually concludes with a few examples of the ways in which the mutants have been used to answer physiological questions. In other cases, existing genotypes rather than newly generated mutants are discussed; here the format used may be different.

Cast Studies in Part I deals with plant nutrition and metabolism. Thus, Chapter 1 deals with water relations and mineral nutrition; Chapter 2 with mutants in metabolism; and Chapter 3 with mutants affecting the chloroplast and photosynthesis. Part II encompasses the vast area of plant growth and development within which Chapter 4 covers examples of mutants that affect vegetative growth and development; Chapter 5 deals with mutants in flower and fruit development; and Chapter 6 with mutants affecting seed dormancy and development.

Although reference to the references in each chapter will show that mutants and variants have been used in plant physiological and biochemical investigations for several decades it is, nevertheless, the case that there has been an enormous acceleration of research activity in this area since about 1980. Because of the contemporary nature of the field it is difficult to find comprehensive reference works to which the reader can be directed for further study. However, there are two recent, edited books that I found to be of great assistance in the preparation of Case Studies in several chapters: *A Genetic Approach to Plant Biochemistry*, edited by A.D. Blonstein and P.J. King; *Developmental Mutants in Higher Plants*, edited by H. Thomas and D. Grierson. In the main, these two collections of reviews confine themselves, as I have done, to mutants and variants of higher plants or to those derived from cultures of higher plant cells. To do otherwise would involve another vast and increasing body of literature that, although of great importance, is often of questionable applicability to higher plants. Such a compendium would only serve to confuse the relatively uninformed reader and defeat the primary purpose of arousing the interest of the novice in the field.

I should like to acknowledge the help, advice and positive criticism of the following people: Dr. I.M. Sussex for his critical reading of the manuscript; Dr. A.D.M. Glass and the Department of Botany, University of British Columbia, Vancouver, where I spent a productive sabbatical leave in 1987 - 1988, during which much progress was made with the manuscript; Myrna K. King, who not only helped with the preparation of the text but also provided essential succor at difficult points in the writing; Evelyn Peters and Joan Ryan for their help in manuscript preparation; and Gordon Holtslander for the preparation of figures and tables, and for his inestimable expertise with desk-top publishing programs.

J.K.
Saskatoon

June 1991

Contents

*The Genetic Basis of
Plant Physiological Processes*

I. Mutants Affecting Plant Nutrition and Metabolism

Included under the general heading of *Plant Nutrition and Metabolism* are the acquisition of water and nutrients from the environment, the biosynthetic pathways of the plant, the distribution of substances throughout the plant body, and photosynthesis. In some of these areas, such as in the case of photosynthesis, many mutants have been isolated because of the wide array of visual phenotypes that are manifested when aspects of the photosynthetic pathway are disrupted. In most other cases, fewer mutants or variants have been isolated that can be used for plant physiological investigations.

In *CHAPTER 1*, examples are given of some mutants that affect plant-water relations and mineral nutrition. Thus, the *"wilty"* mutants of tomato are being used to investigate the control of the stomatal pore; cultivars and variants resistant to environmental salt stress or to anaerobic conditions are helping to uncover the processes plants have developed to combat stresses they encounter within their natural habitats; and the *chloronerva* mutants of tomato are proving to be a valuable tool in the development of an understanding of the distribution and processing of the essential nutrient, iron, within the plant.

CHAPTER 2 is devoted to an array of mutants and variants with modifications to their cellular metabolism generated both as whole plants and by tissue culture techniques. These range from the classic examples of the anthocyanin-deficient mutants of maize, used to unravel the intricacies of the control of a metabolic pathway as well as in the discovery of transposons, to the cases of the generation of auxotrophic mutants and their uses in plant biochemistry and cell biology, to those mutants and variants with resistance either to natural metabolites (e.g., amino acids) and their analogues or to herbicides, classes of plants and cells that are useful in probing the nature and control of a number of plant enzymes.

Presented in *CHAPTER 3* are only some of the classes of photosynthetic mutants. These are chosen to illustrate the different ways photosynthesis can be investigated using genetic techniques. Thus, among those mutants affecting the light reactions of photosynthesis are the chlorophyll b-deficient mutants, the *virescent* mutants, the *high chlorophyll fluorescence (hcf)* mutants, and the chloroplast lipid mutants, all of which have an effect on some structural components of the chloroplast. Following those are two cases of mutants that affect the dark reactions of photosynthesis, namely, mutations affecting the enzyme ribulose-1,5-bisphosphate carboxylase (Rubisco) and those in the photorespiratory pathway.

1. Mutants Affecting Water Relations and Mineral Nutrition

Many plant functions depend on the properties of water and on the mineral ions dissolved in it. The first topics to be dealt with, therefore, have been chosen to illustrate ways in which mutants have been used to study how modifications to the plant can affect plant - water - mineral ion relations. But first, some background information about the physiology of water and mineral ion acquisition and distribution through the plant and the central role of the stomate in these processes.

Most of the water absorbed by a plant through the roots is passed via the transpiration stream to the leaves and then through the stomates into the surrounding air. The carbon dioxide (CO_2) essential for photosynthesis, enters the plant through these same pores, the stomates. The dilemma faced by the plant is how to acquire as much CO_2 as possible from the atmosphere, in which it is extremely dilute, while, at the same time, retaining as much water as possible for essential functions. Thus, the performance and control of the stomate is a central issue in any understanding of plant - water relations (Salisbury and Ross, 1985).

Typical stomates (at least of dicots) consist of two kidney-shaped guard cells. Stomates open because the guard cells take up water (H_2O) and swell (Salisbury and Ross, 1985). What causes guard cells to take up water is complex and might be summarized in the following way: Two important feedback loops, one for CO_2 and the other for H_2O, appear to control stomatal action. Light promotes photosynthesis, which lowers CO_2 levels in the leaf (Fig.1-1, *left*); the response of the leaf is to cause more potassium (K^+) to move into the guard cells; water follows osmotically causing stomates to open. When more water leaves the plant than is being replaced from the roots, abscisic acid (ABA) is released from the mesophyll cells of the leaf, which leads to movement of K^+ out of the guard cells; water follows osmotically and the stomates close (Fig.1-1, *right*).

ABA, therefore, is assumed to be the messenger that causes stomates to close under water stress. Most investigations into the physiological roles of ABA have involved external applications of the growth substance or correlative studies in which developmental or physiological events are correlated with changes in endogenous ABA levels. Such studies have

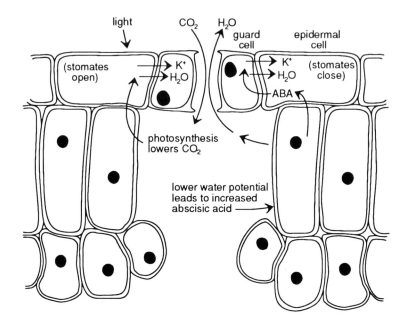

Figure 1-1. The control of stomatol action. (From *Plant Physiology*, Third Edition, by Frank B. Salisbury and Cleon W. Ross © 1985 by Wadsworth, Inc. Reprinted by permission of the publisher.)

serious limitations in that responses to externally applied ABA may not be the same as those produced by endogenous levels.

An alternative approach to investigations of the roles played by ABA would be to manipulate the endogenous levels of the growth substance by chemical or genetic means followed by observations of any resulting physiological changes. Mutants with reduced ABA levels are available including the *wilty* mutants of tomato, all of which have less ABA than the wild type. In some of these, the mutant stomates remain open in darkness, in wilted leaves, and even when the guard cells are plasmolyzed. They, therefore, provide a system with which to investigate the role of ABA in stomatal action, the subject of Case Study 1-1.

An important branch of environmental physiology deals with how plants and animals respond to environmental conditions, that are quite different from those that are optimal for the particular organism. Investigations into this field of stress physiology could enhance our understanding of what limits the distribution of plants in natural environments.

Several types of stress are known to bring about specific and reproducible patterns of gene expression in a number of plant species. These stresses include heat shock, anaerobiosis, oxygen free radicals, heavy metals, water stress, and chilling. The rapidity and repeatability of the changes in gene expression brought about by such stresses make these genetic systems ideal for examination of the factors and mechanisms involved in gene regulation (Matters and Scandalios, 1986).

A fully mature plant cell expresses an array of genes required for carrying out its normal functions. Yet, when severe changes in environmental conditions occur, the cell is capable of responding almost immediately by increasing or decreasing the expression of specific genes. It can be presumed that those genes whose expression is increased during times of stress are critical to the response of the organism to the environmental challenge. The identification of these genes and their products may help to understand the biological processes developed in plants to withstand stress and may also point the way to the identification of methods that could be used in agriculture to design stress-resistant crops. These genetic systems are also of interest because they can be used to study the molecular events that control quantitative gene expression. Thus, the way in which plants recognize and respond to signals to alter gene expression is important physiologically and in the understanding of gene regulation.

One of the natural or induced stresses that plants come in contact with in the environment is created by salts. In Case Study 1-2, the physiological responses to high salt environments are examined in suspensions of cultured cells derived from some of the wild halophytes. Comparisons are also made to the response of other salt-resistant cells and tissues taken from plant species of economic importance. At the same time it is recognized that each type of stress evokes its own specific responses within a plant and that, therefore, the observations made and conclusions drawn in the examples included here are only a sample of the many that could be made.

Higher plants are obligate aerobes. Thus, when roots become hypoxic, under conditions of high rainfall or poor drainage, for example, root growth stops and the roots eventually die. Only if the root cells receive oxygen by aerenchyma can the roots continue to function and grow. For example, crops that are flood tolerant, such as rice, are not so because of tolerance to hypoxia but because they can avoid it (Roberts et al., 1984b). Introducing flood tolerance into other crop species would be of value, but the prospect of doing so through the development of aerenchyma in these plants is negligible. The path of development leading to aerenchyma is long and involved. At present, there is no way to introduce all the traits for aerenchyma into a species that does not have them. In any case, in many crops the exposure to hypoxia is only infrequent and sporadic.

All plants can tolerate anoxia for longer periods than can, say, any vertebrate animal. The ability of plants to carry out a mainly ethanolic fermentation, rather than the exclusively lactic fermentation typical of higher animals, seems to be a key element of their tolerance. What is still puzzling is why some plants are more tolerant than others when all can carry out ethanolic fermentation (Roberts et al., 1984b). Understanding these differences may allow the identification of biochemical markers for tolerance of transient flooding, which can be used in crop selection.

Maize is a species of higher plants in which there has been an intensive investigation of the responses of roots to anaerobic environments. When different lines of maize are compared vast differences are usually found in their protein profiles under anaerobic conditions. There are extra proteins, missing proteins, electrophoretic variations, apparent size variations, and a great deal of variation in the balance among individual anaerobic proteins (Freeling and Bennett, 1985). About 20 polypeptides are induced specifically under anaerobic conditions, among which are those for alcohol dehydrogenase (ADH). Three ADH isozymes are induced in the roots subjected to anaerobiosis, which are formed by the random dimerization of polypeptides produced by two genes: *Adh1* and *Adh2*. Of these, *Adh1* is the only gene among all those activated by anaerobiosis whose function has been proved necessary for survival in near-anaerobic conditions such as total flooding. This proof was possible only because *Adh1*-deficient mutants were available. Case Study 1-3 is devoted to a discussion of the use of *Adh1* corn mutants to investigate the source of the resistance of certain roots to flooding tolerance.

A major component of the autotrophic capability of higher plants is the ability to "mine" inorganic compounds directly from the environment. This survival strategy, evolved by green plants and certain microorganisms, involves the absorption of mineral nutrients from the surrounding medium, usually the soil. The remarkable fact is that all living cells can absorb certain essential solutes so fast and over such long periods that their concentrations become much higher within the cells than in the external solution. This selectivity of solute transport applies not only to mineral ions but also to organic compounds, and is important evidence supporting the hypothesis that proteinaceous carriers in the membranes move solutes into cells. Such carriers, therefore, have a genetic origin and should be open to manipulation by mutation.

Among the studies of the mechanisms used by plants to accumulate ions from the environment has been the use of mutants either deficient in uptake capacity for specific ions or with some other genetic change that modifies accumulation or utilization. In Case Study 1-4, one aspect of the control of the acquisition of an essential mineral ion from the rhizosphere is described. The auxotrophic *chloronerva* mutant of tomato has been used to investigate the role of the metabolite nicotianamine in the chelation of iron (Fe) within cells. Iron is one of the most abundant, and yet least available, of the minerals in the soil, which are essential to plants.

Case Study 1-1.   The *wilty* Mutants of Tomato and
                   Stomatal Control

When plants develop water deficits they adapt by modifying their water relations in particular ways, their control of gas exchange through stomates, and their growth and development. The endogenous plant growth regulator (+)*cis,trans* abscisic acid (ABA) appears to play a key role in this adaptation to water stress because known adaptive changes can be initiated either by the imposition of water deficits directly or by treatment of plants with exogenous ABA.

The series of three nonallelic *wilty* mutants of tomato, *flacca* (*flc*), *sitiens* (*sit*), and *notabilis* (*not*), provides a well-characterized set of material to investigate the physiology of ABA in stress adaptation. These mutants have an increasing relative tendency to wilt compared with wild type. The wiltiness is the result of excessive transpiration induced by increased stomatal conductance. The three mutants cover a range of phenotypic expression of water stress from the relatively mild *not*, in which endogenous ABA concentrations seem to be between one-third and one-half those in the wild type, to the relatively severe *sit*, where ABA levels are less than 15 percent of the wild type (Jones et al., 1987).

Our first Case Study is concerned with the characterization of these tomato mutants and with their value in investigating the important topic of stress physiology, in particular, the link between water stress, the control of stomatal action, and the role of ABA in this control.

1-1.1   Transpiration Rates and Stomatal Control in the *wilty*
         Mutants

As already mentioned, most of the investigations into the physiological roles of ABA have been through its external application or through the correlation of developmental or physiological events with changes in endogenous ABA levels. The main problems with studies of this kind are, first, that responses of a plant to externally applied substances may not be the same as to endogenous levels of the same compound and, second, that correlated variations may not be the true cause of the response observed in a plant.

An alternative approach to the study of the physiology of ABA action would be to manipulate endogenous ABA levels chemically or genetically. There are no specific inhibitors of ABA biosynthesis available, which

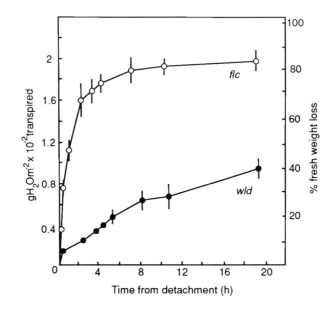

Figure 1-2 The water loss from detached leaves of wild-type *Lycopersicon esculentum* Mill. cv. Ailsa Craig and a *flacca* mutant. (Neill and Horgan, *J. Exp.Bot. 36*: 1222-1231, 1985; reproduced by permission of Oxford University Press.)

makes precise chemical manipulation impossible. Mutants with reduced ABA levels, however, are available, among which are the *wilty* tomato mutants that wilt as a result of excessive transpiration induced by abnormal stomatal behavior (Neill and Horgan, 1985). An example of this increased water loss from the stomates of a *wilty* mutant is given in Figure 1-2 where a comparison is made between the water loss from detached leaves of wild-type tomato and a *flc* mutant over a period of 20 hours. The stomates of wild-type leaves responded to rapid water loss by closing in contrast to the stomates of the *flc* mutant that did not close and in which rapid transpiration continued until the leaves became very dessicated (Fig. 1-2; Neill and Horgan, 1985).

When ABA was supplied to single leaves through cut petioles there was an 80 percent decrease in transpiration rate for the wild type and 50 percent for *flc* (Fig. 1-3).

Because the transpiration rate of *flc* in the presence of ABA was still as great as that of the wild-type control leaves, the results in Figure 1-3 do not provide a fully convincing demonstration of the role of ABA in stomatal action. They do demonstrate that the stomates of *flc* leaves can still respond to exogenous ABA by closing and have not lost their ability to do so.

A more convincing demonstration of the link between ABA and stomatal action was observed in experiments with a different *flc* mutant

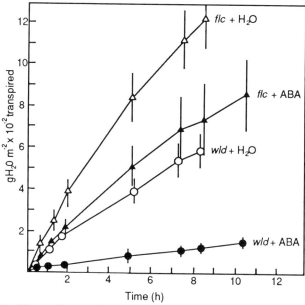

Figure 1-3. The effects of 0.1 mole m³ ABA (▲,●) or H₂O (△,○) on transpiration by detached leaves of *L. esculentum* cv. Ailsa Craig *wld* (○,●) and *flc* (△,▲). (Neill and Horgan, *J. Exp. Bot. 36*: 1222-1231, 1985; reproduced by permission of Oxford University Press.)

of the tomato cultivar, Rheinlands Rhum (Imber and Tal, 1970). Here, the water loss from detached leaves of the mutant and normal plants was compared at different concentrations of applied ABA (Fig. 1-4).

When *dl*-ABA was applied to seedlings, either by foliar spray or in root solution, a change in the growth habit of the mutant was observed after several days (Imber and Tal, 1970). The mutant plants looked like the normal ones in stem thickness and height, and in the size, shape, and turgidity of their leaves. Water loss in detached leaves from untreated mutant plants decreased proportionately with the hormone concentration when leaves were exposed to ABA solutions (Fig. 1-4), which suggested that normal stomatal behavior was dependent on the continuous presence of ABA. These conclusions were consistent with those made by others in which the concentrations of ABA in the *flc*, *sit*, and *not* mutants of Ailsa Craig tomato were shown to be much lower than in the wild type (Table 1-1; Neill and Horgan, 1985). The degree of wiltiness of these three *wilty* mutants was correlated with their basal ABA levels. As seen in Table 1-1, *not* had the highest level of basal ABA among the three mutants and *sit*, the lowest.

Although these investigations established a connection between ABA and stomatal behavior there still remained the need to show a correlation between endogenous ABA concentrations and stomatal responses at

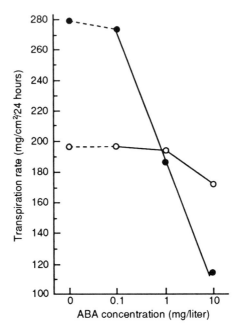

Figure 1-4 Water loss from detached leaves from mutant (●) and normal (○) plants treated with ABA by foliar spray. (Imber and Tal, *Science 169*: 592-593; Copyright 1970 by the AAAS.)

different water potentials. The *wilty* mutants proved useful here, also. Thus, potted wild-type Ailsa Craig tomato plants were allowed to dry out. Their water potential began to decrease soon after water was withheld and after 4 days had decreased from minus 0.25 to minus 1.1 MPa (Fig. 1-5A; Neill and Horgan, 1985). At the same time as the water potential of the wild-type Ailsa Craig leaves was decreasing, ABA concentrations in the same leaves increased from 1.0 nmol/g to between 8.0 and 14.0 nmol/g (Fig. 1-5A). Rewatering led to a rapid restoration of turgor and reduction in ABA concentration. Potted *flc* plants wilted very rapidly after watering was stopped and leaf water potential decreased from minus 0.3 to minus 1.4 MPa in only 1.5 days. On the other hand,

Table 1-1. Basal ABA concentrations in leaves of *L. esculentum* cv. Ailsa Craig wild type, *not*, *flc*, and *sit* mutants.

	Wild type	*not*	*flc*	*sit*
[ABA], nmol g^{-1} fr. wt.	1.52±0.5	0.75±0.01	0.39±0.08	0.23±0.05
Ratio	1.0	0.49	0.26	0.15

(Neill and Horgan, *J. Exp. Bot. 36*: 1222-1231, 1985; reproduced by permission of Oxford University Press.)

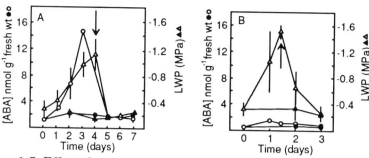

Figure 1-5. Effect of watering regime on ABA content (●,○) and leaf water potential (LWP) (▲,△) in potted plants of *L. esculentum* Mill. cv. Ailsa Craig (A) and the *flacca* mutant (B). ▲ ,●: well watered; △,○: water withheld from day 0. LWP: mean of seven determinations + SE, [ABA] in duplicate. Arrow indicates time of rewatering. (Neill and Horgan, *J. Exp. Bot. 36*: 1222-1231, 1985; reproduced by permission of Oxford University Press.)

ABA levels remained essentially constant (Fig. 1-5B). These experiments clearly established a close correlation between ABA and stomatal responses in the wild type and the mutant leaves.

The case for the involvement of ABA in stomatal control was strengthened further when the response of detached leaves of wild type and the three *wilty* genotypes of tomato to rapid stress was examined in greater detail. It was discovered that in the wild type after 2 hours, the levels of ABA had increased from about 1.2 to 1.6 nmol/g and continued to increase up to 3.4 nmol/g after 24 hours (Fig. 1-6). Leaves of the *not* mutant also responded by synthesizing some ABA, up to a level of about 1.3 nmol/g after 3 hours. However, no further increase occurred.

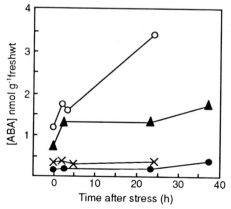

Figure 1-6. Effect of water stress on ABA levels in detached leaves of *L. esculentum* cv. Ailsa Craig *wld* (○), *not* (▲), *flc* (x), and *sit* (●). *wld* = wild type. (Neill and Horgan, *J. Exp. Bot. 36:* 1222-1231, 1985; reproduced by permission of Oxford University Press.)

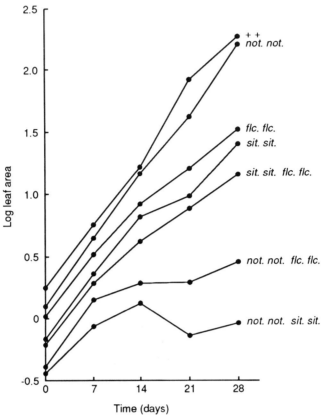

Figure 1-7. The effect of wilty mutant genes on growth rate. (Taylor and Tarr, *Theor. Appl. Genet. 68:* 115-119, 1984.)

Detached leaves of both *flc* and *sit* failed to respond to a rapid stress by synthesizing ABA (Fig. 1-6; Neill and Horgan, 1985).

The conclusions from these results are that wilting in the tomato mutants occurs because of excessive transpiration due to a failure of stomates to close when the water supply is limiting. The severity of the wilting phenotype can be correlated approximately with the concentrations of ABA in the tissues of the leaf and with the varying ability of mutant leaves to produce more ABA in response to rapid water stress. In regard to this last point, the ability to produce ABA rapidly may be more important than initial basal levels. ABA synthesis has also been found to be necessary for the maintenance of stomatal closure. Stomates reopen rapidly after ABA is removed from the transpiration stream and it may be that continued arrival of ABA at guard cells is required to maintain closure (Neill and Horgan, 1985). Mutants with restricted ABA content could also be useful then in these and other more detailed studies of stomatal action.

Finally, several hypotheses may be advanced to explain the small quantities of ABA present in the mutant tissue, the simplest of which is that ABA biosynthesis is blocked at the same or different steps in each mutant giving rise to variously leaky phenotypes. Data indicate that the mutants may be leaky in the order *not* greater than *flc* greater than *sit*. It has been demonstrated that the three mutants are nonallelic. Crosses between *not* and *sit* and *not* and *flc* lead to much more extreme phenotypes as could be predicted if the mutations affect two different steps in a metabolic pathway that are additive in effect. Combinations of *flc* and *sit*, however, were not additive. The double mutant grew nearly as well as either single mutant alone. This could be explained if the *flc* and *sit* mutations affected the same enzyme. Combining them resulted in only a small additional loss of enzyme activity (Fig. 1-7; Taylor and Tarr, 1984). The possibility that ABA precursors accumulate behind metabolic blocks makes these mutants of considerable value for biosynthetic studies (see Case Study 6-2).

1-1.2 The Use of the *wilty* Mutants in Physiological Studies

In addition to being useful in studying the relationship between stomatal action and ABA directly as already illustrated, the *wilty* mutants can also be exploited to investigate other physiological events hypothesized to be influenced by water stress. Three very brief examples follow where the use of these mutants has provided answers to certain physiological questions.

1-1.2.1 The Influence of Water Stress on Leaf Growth and Water Relations

Leaf expansion is known to be highly sensitive to water stress. Studies of the relationship between leaf growth and water stress, however, are often complicated by the need to alternate periods of stress with rewatering to maintain the viability of leaves.

In the *flc* mutant of tomato, low leaf water potential occurs daily despite a constant high soil water potential. Differences between *flc* and wild-type plants develop daily throughout growth under otherwise identical conditions, thus offering the opportunity to study shoot responses to a relatively constant degree of water stress.

Plants of wild type (*L. esculentum* Mill. cv. Rheinlands Rhum) and *flc* were sprayed with solutions of either 10 μM or 30 μM ABA daily. Leaf expansion in the wild type was not significantly affected by ABA (Table 1-2; Bradford, 1983). However, the leaf areas of *flc* plants were 44 percent and 65 percent, respectively, of that of wild-type plants and ABA treatment increased leaf area to more normal levels in *flc* (Table 1-2) as

Table 1-2 Final leaf areas of *L. esculentum* Mill. cv. Rheinlands Rhum (RR) and *flc* mutant plants as influenced by ABA sprays. Within each experiment, means (±SE) in a column followed by the same letter are not signifigantly different (P<0.05)

Treatment		Leaf Area (cm)
Exp. 1, 10 µM ABA	RR	$1,784 \pm 78a$
	RR + ABA	$1,904 \pm 94a$
	flc	$791 \pm 25b$
	flc + ABA	$1,268 \pm 31c$
Exp. 2, 30 µM ABA	RR	$1,230 \pm 36a$
	RR + ABA	$1,160 \pm 54a$
	flc	$794 \pm 24b$
	flc + ABA	$1,019 \pm 43c$

(Bradford, *Plant Physiol. 72*, 1983; reproduced by permission of the ASPP.)

it did leaf, shoot, and root dry weights (data not shown). Because it is known that ABA has the effect of decreasing transpiration by causing stomatal closure and also increasing the hydraulic conductance of mutants, such as *flc*, it may be concluded that ABA promotes leaf growth by its effect on leaf water balance.

It may also be that these mutants can be phenotypically reverted to wild-type by applications of growth substances and can be used to examine the influence of water stress on a wide range of growth phenomena because shoot and root as well as leaf growth were strongly affected by ABA additions (Bradford, 1983). Certainly, the clear-cut conclusion in the case of leaf expansion should provide encouragement to those looking for a suitable experimental system to investigate the influence of water stress on growth.

1-1.2.2   *Activity of the Ethylene Biosynthetic Pathway in Relation to ABA and Water Stress*

In a 1986 paper, Neill et al. reported observations by others that *flc* plants contained more indole acetic acid (IAA)-like growth substances than did wild-type tomato plants. The rate of ethylene evolution was also found to be higher in *flc* plants owing in part, it was suggested, to the increased auxin content. When *flc* plants were phenotypically reverted to a more wild-type state by foliar spraying with ABA solutions the rate of ethylene evolution was also reportedly reduced (Neill et al., 1986). A

Table 1-3. Ethylene production (nmol/g/hour) by detached leaves of *L. esculentum* cv. Ailsa Craig wild type and *flc* averaged over 8 hours after excision.

	Turgid		Wilted	
	+H$_2$O	+ABA	+H$_2$O	+ABA
wild type	0.026±0.004	0.021±0.006	0.07±0.01	0.068±0.03
Flc	0.025±0.005	0.028±0.003	0.065±0.014	0.11±0.02

(Neill et al., *J. Exp. Bot. 37*: 535-541, 1986; reproduced by permission of Oxford University Press.)

question that arose from such observations was whether a causal relationship existed between ABA content and stress-related ethylene production. A second more difficult question was whether a connection between endogenous ABA, IAA, and ethylene levels could be established and opened to investigation.

The concentration of ABA in *flc* leaves is only 20 to 40 percent of that found in wild-type leaves, therefore, it was conceivable that the higher rates of ethylene evolution reported for *flc* leaves were a consequence of this. The hypothesis could be tested by comparing the effect of ABA spraying on the production of ethylene by wild-type and *flc* leaves (Table 1-3; Neill et al., 1986).

The results in Table 1-3 indicate that there was little difference in ethylene production between *flc* and wild-type leaves. Pretreatment with ABA did not prevent stress-induced ethylene production; rather, production under stress was greater in ABA-treated *flc* plants. Thus, the suggestion that the low endogenous ABA content of *flc* plants would be reflected by increased ethylene levels was not borne out in these experiments. Nor did ABA pretreatment significantly reduce basal levels of ethylene. Thus, it cannot be said that the production of ethylene is necessarily regulated by ABA.

In this case, the availability of a mutant with reduced ABA content provided an opportunity to design experiments to determine whether there is a causal relationship between ABA content and stress-related ethylene production, a question difficult to answer definitively in plants with high endogenous levels of ABA. Any connection of ABA with IAA could have been investigated here also using this material and experimental protocol, but was not.

1-1.2.3   The Relationship Between Stomatal Conductance and Carbon Assimilation

Two important factors influence carbon assimilation rates of leaves: stomatal conductance of CO_2 and the capacity of the leaf mesophyll cells to assimilate CO_2. The former is determined by stomatal frequency and by reversible movements of the guard cells; the latter, by the amounts and activities of enzymes, light-harvesting structures, and electron transport components of the chloroplast. A close relationship has been found between stomatal conductance and photosynthetic capacity suggesting that they are coordinated but the mechanism of coordination is not understood (Bradford et al., 1983).

Some research groups have found that ABA can influence stomatal conductance without affecting photosynthetic capacity. Others have suggested a role for ABA in adjusting stomatal conductance in accordance with the prevailing photosynthetic rate (see Bradford et al., 1983, for a discussion of this work). One way to test the latter hypothesis directly would be to use the *wilty* mutants of tomato in which stomatal conductance can be influenced through the application of foliar sprays of ABA. At the same time assimilation rates (*A* in Table 1-4) could be measured to see if ABA might be involved in coordinating stomatal conductance (g_e) and photosynthetic capacity (Bradford et al., 1983).

Tests were devised to find out whether variation in g_e alone would alter mesophyll function. Therefore, given the high g_e of *flc* leaves, would mesophyll photosynthetic capacity increase, maintaining a constant calculated intercellular CO_2 partial pressure (C_i), or would photosynthetic

Table 1-4. The effect of 30 µM ABA spray on gas exchange parameters of *L. esculentum* Mill. cv. Rheinlands Ruhm (RR) wild type and *flc* mutant in light (1000 µE/m2/s). *A*, net CO_2 assimilation rate; g_e, epidermal (stomatal + cuticular) conductance to water vapor; *Ci*, calculated intercellular CO_2 partial pressure. Letters within columns followed by the same letter are not significantly different (P <0.05).

Treatment	A $\mu mol\ m^{-2}\ s^{-1}$	g_e $mol\ m^{-2}\ s^{-1}$	C_i μbar	$\partial A/\partial C_i$ $\mu mol\ m^{-2}\ s^{-1}$ μbar
RR control	21.5a	0.40a	235a	0.15a[1]
RR + ABA	18.4b	0.28b	220b	0.13a
flc + control	22.8a	0.81c	274c	0.12a
flc +ABA	19.1b	0.27b	220b	0.14a

(Bradford et al., *Plant Physiol. 72*: 245-250, 1983; reproduced by permission of the ASPP.)

capacity remain constant with consequent increase in C_i? The results in Table 1-4 indicate that the latter is the case. When g_e was changed by ABA treatment, A/C_i ratios were not affected significantly but C_i was lowered substantially. ABA had no effect on photosynthetic capacity.

Thus, the main role of ABA seems to be to override other influences when water deficits threaten such as enabling stomatal closure to occur despite the strong promotion of stomatal opening by light. Such a mechanism shifts priority toward water conservation at the expense of carbon assimilation, as evidenced by lower C_i values (Table 1-4).

Case Study 1-2.   The Responses of Plants and Their Cells to Environmental Salt Stress

Salinity is a form of water-related stress responsible for major crop losses worldwide, especially in semiarid and irrigated agriculture. In many major agricultural areas, salinity is regarded as the most pervasive problem of irrigated agricultural lands, a concern that can now be extended to other cultivated lands where salinity problems are also becoming manifest. The purpose of this Case Study is to present and discuss data from some studies in which both salt-sensitive and salt-tolerant plant materials were used to understand the basis of salt tolerance in higher plants. The eventual hope of these and other investigations is that tolerance, which clearly has a genetic basis, will be thoroughly understood and lend itself to genetic engineering into important crop species.

Plant scientists know little of the mechanisms that enable salt-tolerant plants to survive, and even thrive and reproduce, in saline environments that are fatal or severely inhibitory to salt-sensitive ones. However, plant physiologists have used a variety of strategies in an attempt to understand this important form of environmental adaptation. Experiments have been performed on salt-sensitive plants, on halophytic species, and on cells in culture. The use of cell culture methods to study salt tolerance offers several advantages among which is the ability to characterize at the cellular level the processes that are involved in salt tolerance. The relative lack of differentiation in cultured cells eliminates the complication arising from the morphological variability and the highly differentiated state of the various tissues of the whole plant. Growing plant cells on defined culture media also permits a relatively uniform treatment of these tissues with salt.

Selection and breeding for resistance to any environmental stress ultimately depends on two factors: (1) genetic variability with respect to resistance to stress and (2) exposure of genetically variable populations to the stress (Epstein et al., 1980). There are certainly large collections of germplasm that can be screened for salt tolerance. In addition, salt-tolerant wild relatives of crop species can be exploited as a genetic resource. The imposition of an appropriate stress, for instance to cells in culture that may be salt tolerant, is another matter, however. Saline soils and natural water supplies are extremely variable in both the kinds of salts present and their concentrations. Therefore, it is necessary to test whether selections made under one set of conditions, such as those in cell culture, are equally effective under a different saline regime (Epstein et al., 1980). Inferences drawn from experiments carried out under controlled conditions, such as those found in cell culture, must be viewed with a great deal of caution until more has been learned about the basis of stress tolerance in general and salt tolerance in particular.

1-2.1   Mechanisms of Salt Tolerance in Plant Cells

1-2.1.1   The adjustment of ions in cells subjected to salt stress

The mechanisms that confer salt tolerance to plant cells in culture and the molecular strategies by which cells survive in saline environments are by no means completely understood. Many cell lines have been derived from glycophytes that are "adapted' to grow in culture in the presence of salts, but only a few of these have proved to retain their tolerance over long periods in nonselective (salt-free) media, or have given rise to regenerated plants in which the expression of salt tolerance could be demonstrated (Daines and Gould, 1985).

An alternative to the use of cell lines derived from glycophytes to study salt tolerance is to isolate tissue cultures from halophytes in which salt tolerance is known to be based within the cells themselves. Understanding the spontaneous expression of salt tolerance in these cells could more clearly identify key cellular strategies involved in this trait than would an examination of tolerant cell lines isolated after lengthy periods of adaptation to saline conditions.

Several mechanisms could account for cellular salt tolerance including salt-tolerant enzymes, the exclusion from cells of the injurious ions, or their compartmentation within the cell. The last two of these strategies would also involve the correction of osmotic imbalances caused by any redistribution of ions across membranes.

A good example of the use of cell cultures derived from a halophyte to investigate the phenomenon of salt tolerance is that of the graminoid spike grass, *Distichlis spicata*. Expression of the considerable tolerance of cells from this species to salinized media requires no prior adaptation to salt and is immediately manifested in cells grown for several years in salt-free conditions and then exposed to a high salt medium (Daines and Gould, 1985).

A major consequence of life in a saline environment for any living organism is the need to accumulate osmotically active solutes to sustain

Table 1-5. The effect of saline media on cellular osmotic potential and sodium content of *Distichlis spicata*. Calli of *Distichlis spicata* (DS), previously grown in the absence of salt, were transfered to saline media and cultured for 2 weeks at which time tissue was removed for analysis. *DS+* refers to cultures grown continuously in media containing 200 µM salt.

Culture/Media salinity	Cellular osmotic potential (bars)	Sodium	
		ng/mg FW	ng/mg DW
DS / 0 mM	- 6.3	0.12	1.64
DS+ / 200 mM	-15.9	0.78	9.42
DS / 100 mM	-12.0	0.48	5.88
DS / 200 mM	-17.7	0.81	9.37
DS / 300 mM	-24.2	1.22	11.98
DS / 400 mM	-26.6	1.64	13.45

(Daines and Gould, *J. Plant Physiol 119*: 269-280, 1985; reproduced by permission of Gustav Fischer Verlag.)

favorable osmotic gradients so that cells can maintain turgor. In most halophytes, this process is twofold. First, cells take up salt and sequester it in their central vacuoles. Data in Table 1-5 illustrate the ability of the cells of *Distichlis* to accumulate sodium when grown in a high salt medium for 2 weeks and then analyzed. The data also support the idea that cells of *D. spicata* are sodium accumulators, not excluders, and that osmotic adjustment is partially achieved by sodium accumulation. Furthermore, the amount of sodium found in the cells reflected the level of sodium in the agar medium on which the callus tissue had been grown.

1-2.1.2 The Accumulation of Osmolytes in Cells Subjected to Salt Stress

In *Distichlis*, after the initial osmotic adjustment caused by the accumulation of sodium (Na^+), possibly in the vacuole, there is the accumulation of proline in the cytosol. A time course of proline accumulation was established for cultures with a first time inoculation into sodium chloride (NaCl) (Fig. 1-8). During the first 10 hours postinoculation, proline levels remained unchanged from initial control values. This was followed by a period of rapid proline accumulation beginning about 12 hours after inoculation, which coincided with the development of the greatest osmotic potential difference between *Distichlis* cells and their culture medium (Fig. 1-9).

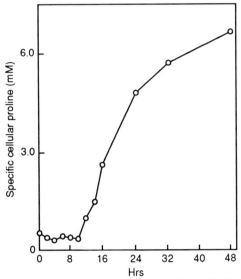

Figure 1-8. Time course of proline accumulation of *Distichlis spicata* cell suspensions inoculated from salt-free medium into medium containing 200 mm NaCl. (Daines and Gould, *J. Plant Physiol. 119*: 269-280, 1985; reproduced by permission of Gustav Fischer Verlag.)

The highest levels of proline accumulation were achieved at about 48 hours, just before the onset of cell division in the suspensions (Fig. 1-8). A convincing correlation between NaCl concentration in the medium

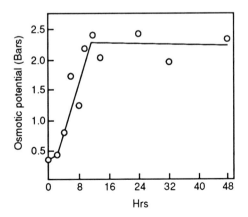

Figure 1-9. Time course demonstrating the change in osmotic potential plotted as the difference between *Distichlis* cells and their culture medium. Cells grown previously without salt were inoculated into medium containing 200 mM NaCl. (Daines and Gould, *J. Plant Physiol. 119*: 269-280, 1985; reproduced by permission of Gustav Fischer Verlag.)

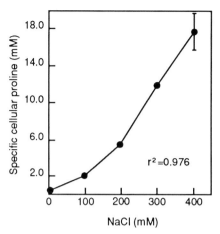

Figure 1-10. Correlation of the final concentration of proline accumulated in *Distichlis spicata* cells with the level of NaCl in the culture medium. Values were calculated as (proline/cell). (volume/cell)-1. (Daines and Gould, *J. Plant Physiol. 119*: 269-280, 1985; reproduced by permission of Gustav Fischer Verlag.)

and the final level of proline in cells was established by regression analysis (Fig. 1-10).

All of these data are consistent with a model in which vacuolar compartmentation of salt is followed by osmotic compensation in the cytoplasm by an organic solute, in this case, proline (Daines and Gould, 1985).

1-2.1.3   *Examples of osmolytes*

At this point, it may be beneficial to digress and discuss the involvement of substances, such as proline, in osmoregulation associated with salt tolerance.

Many organisms have learned to use a simple rule of chemistry to live in a world deficient in available water. They have evolved sophisticated mechanisms to balance their osmotic strength with that of their surroundings. A wide variety of metabolites have been tested in microorganisms as potential protectants. In *Escherichia coli*, where more than 150 compounds were tested, only the betaine series proved to be potent protectants, the most active being glycine and proline betaines and choline; proline was also active but at lower stress levels (Rudulier et al., 1984). One of the striking chemical features of such molecules (Fig. 1-11) is their extreme solubility in water (about 14 kg per kilogram of water). This, and other chemical characteristics, makes these compounds ideal as osmoprotectants.

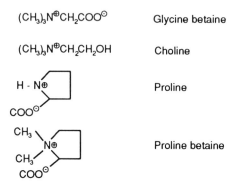

Figure 1-11. Structures of some osmoprotectants.

1-2.2 Evidence for Osmoregulation in Response to Salt Stress in Plant Species Other Than *Distichlis*

1-2.2.1 Responses to Salts

Thus, the evidence from studies using cells of the halophyte *D. spicata* suggested that a major strategy followed by plants with a cell-based mechanism of salt tolerance is, (1) to accumulate salt and sequester it

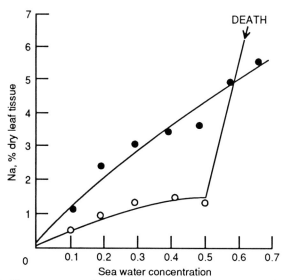

Figure 1-12. Percentages of sodium in dry leaf tissue as a function of the salinization of the nutrient solution with seawater salt mix. *Lycopersicon cheesmanii* (●); *L. esculentum* (○). (Rush and Epstein, *Plant Physiol.* 57: 162-166, 1976; reproduced by permission of the ASPP.)

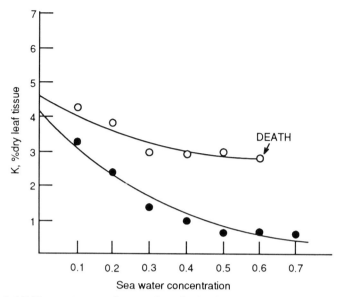

Figure 1-13 Percentages of potassium in dry leaf tissue as a function of the nutrient solution with seawater salt mix. *Lycopersicon cheesmanii* (●); *L. esculentum* (○). (Rush and Epstein, *Plant Physiol. 57*: 162-166, 1976; reproduced by permission of the ASPP.)

and, (2) to compensate for the inevitable increase in osmotic strength of the cytosol by synthesizing osmoregulators such as proline or other highly soluble metabolites. The question then arises as to whether plant species other than halophytes develop similar responses when selected for growth in high salt conditions. A few examples were chosen to illustrate that in plants, as opposed to microorganisms, the answer to this question is not yet entirely clear.

In the first example, an ecotype of the species *Lycopersicon cheesmanii* ssp. minor (Hook.) C.H. Mull. (salt-tolerant) from the Galapagos Islands was compared with *L. esculentum* Mill. cv. VF 36 (salt-sensitive). Consistently, leaf Na levels remained low in the salt-sensitive variety under conditions of increasing Na concentrations, up to a certain point. Beyond this point, usually at about half seawater concentration, leaf Na levels increased sharply and the plants died. In the *L. cheesmanii* ecotype, Na content in the leaves increased with each increase in the nutrient solution (Fig. 1-12; Rush and Epstein, 1976).

The sensitive tomato ecotype tended to exclude Na from the leaf tissue, to which it was toxic, whereas the tolerant ecotype freely accumulated Na in the leaf with no similar toxic effects. Also, K^+ levels in the sensitive VF 36 remained high in the tissue even when Na was present in the bathing solution at high concentrations. In *L. cheesmanii*, K^+ levels decreased sharply as the Na levels increased (Fig. 1-13),

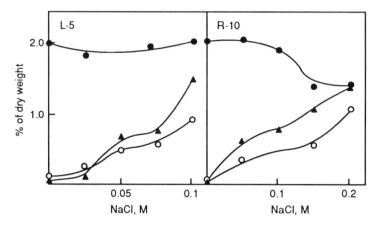

Figure 1-14   Accumulation of Na+ (○), K+ (●) and Cl- (▲) in L-5 (non-selected) and R-10 (selected) cells of *Citrus sinensis* L. Osbeck as a function of NaCl concentrations (note the different scales on the abscissae of the two graphs). (Ben-Hayyim and Kochba, *Plant Physiol.* 72: 685-690, 1983; reproduced by permission of the ASPP.)

suggesting that the *L. cheesmanii* ecotype compensates to an extent for the accumulation of Na in its leaf tissue by exchanging K^+ with the surrounding medium. The inability of the VF 36 to withstand salinity seemed to be linked to its limited efficiency in keeping Na in the leaf tissue below toxic levels. In contrast, the *L. cheesmanii* could accumulate large quantities of Na (up to 7 percent of dry leaf tissue) without toxic effect (Rush and Epstein, 1976).

Results with other species suggest that although some cell lines, such as the *L. cheesmanii* ecotype, may survive salinity by a general shift toward a halophytic mode of salt tolerance (i.e., salt accumulation), not all may do so. For example, in the case of a salt-selected line (R-10) of *Citrus sinensis*, survival most probably depended on the partial avoidance of NaCl.

To study the mechanism by which R-10 cells of *C. sinensis* tolerated NaCl in their medium, the concentrations of Na^+, K^+ and Cl^- in L-5 (salt-sensitive) and R-10 cells were determined. R-10 cells took up much less Na^+ and Cl^- compared with R-5 cells growing at the same external NaCl concentration (Fig. 1-14; Ben-Hayyim and Kochba, 1983). Not only did the R-10 cells take up much less Na^+ and Cl^- compared with L-5 but the amount of K^+ in L-5 cells was unaffected by NaCl, whereas in R-10, the concentration of K^+ decreased as NaCl increased. Thus, the R-10 cell line acted like a true genetic variant that tolerates NaCl by partially excluding it from the cell rather than by developing a mechanism of salt tolerance as in the previous case in tomato. The R-10 did resemble *L. cheesmanii*, however, by apparently compensating for the uptake of Na^+ by exchanging K^+ with the medium.

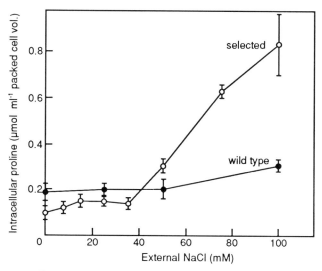

Figure 1-15. Proline accumulation in wild-type and NaCl-selected *Nicotiana* cells as a function of NaCl concentration in the medium. Proline was measured 7 days after inoculation. The selected cells had been grown in 150 mM NaCl medium then transferred gradually over 7 days to standard medium before inoculation into the experimental media. (Watad et al., *Plant Physiol. 73*: 624-629, 1983; reproduced by permission of the ASPP.)

1-2.2.2   *The Presence of Osmolytes in Salt-tolerant Lines*

From the limited number of cases described in the preceding section it would seem that examples do exist in which the accumulation of NaCl is a strategy used by some species other than halophytes to accommodate saline growth conditions. The question then arises as to whether the formation of osmolytes, the second phase of the strategy of salt tolerance in some halophytes, also occurs in those species in which salt accumulation forms the first phase of salt tolerance.

In one series of studies, an NaCl-resistant line was developed from suspension-cultured cells of *Nicotiana tabacum/gossii* by stepwise increases in the NaCl concentration in the growth medium from 85 to 400 mM. These cells and wild-type cells were grown at various concentrations of NaCl and proline in the cells measured 7 days after inoculation (Fig. 1-15; Watad et al., 1983).

Above a threshold of 35 mM NaCl, proline concentration in the selected cells rose almost linearly with external NaCl. On the other hand, in wild-type cells, the proline concentration rose only slightly (Fig. 1-15). The trait allowing the observed change in proline was stable through many generations in the absence of NaCl and no other amino acid was affected. The accumulation of proline was also fully reversible, as might be expected if there was a causal link between salt tolerance and proline accumulation (Fig. 1-16).

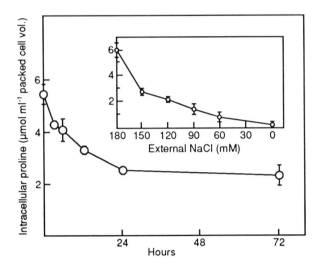

Figure 1-16. Kinetics of the reversibility of proline accumulation in NaCl-selected *Nicotiana* cells. Cells growing in 200 mM NaCl were transferred for 5 minutes to 150 mM NaCl to free them from 200 mM medium and then resuspended in fresh 150 mM NaCl medium for the periods indicated. Inset: Proline content of cells grown in 180 mM NaCl and then transferred at daily intervals through a series of NaCl concentrations, as indicated. (Watad et al., *Plant Physiol. 73*: 624-629, 1983; reproduced by permission of the ASPP.)

Figure 1-16 shows that the half-life for decreasing the proline concentration from the level characteristic of cells in 200 mM NaCl to that characteristic of 150 mM NaCl was about 6 hours. This process could be repeated at lower and lower external NaCl concentrations (*inset*, Fig. 1-16). The half-life for this process was far less than the generation time for this cell line (4 days). Thus, the disappearance of proline was not due to dilution caused by cell division. It can be said that, at least in this case, striking differences in proline content were observed between wild-type and NaCl-selected cells when exposed to increasing concentrations of NaCl.

There is a continuing controversy in the literature as to whether proline accumulation is related to resistance to salinity and drought. In these investigations (Fig. 1-16), cell lines that had the ability to grow and develop indefinitely in high NaCl media accumulated proline when grown in these media. When the two cell lines were grown at the highest external NaCl level at which the sensitive wild type could survive, far less proline accumulated in the sensitive cells than in the resistant ones. The conclusion, therefore, that there is a causal link between proline accumulation and resistance to salt stress is reinforced by the fact that, when cell lines are developed by gradually increasing stepwise the NaCl concentration in the medium, internal proline concentration rises at each step. There is, thus, a correlation between the ability of the cells to survive in a higher NaCl environment and proline content (Watad et al., 1983). Proline accumulation was also fully reversible. On sequential transfer to media of lower and lower NaCl concentration, the proline content of the selected cells fell progressively to the lower and lower steady-state levels corresponding to the external NaCl concentrations, another indication of the link between the NaCl level and the proline accumulation (Fig. 1-16).

Another point of significance and interest about these results (Watad et al., 1983) was that the stepwise development of the resistance to NaCl strongly pointed to the possibility of gene amplification as the mechanism of resistance. Such a stepwise response would not be expected in the case of mechanisms such as mutation or selection of a small number of resistant cells present from the start in the cell population.

It is wise to remember, however, that the interactions of plants with environmental salts are complex, involving uptake and translocation of the ions and their differential localization in various organs, tissues, cells, and subcellular compartments. It may not be surprising, therefore, to find that plants have developed a number of strategies with which to resist the deleterious effects of high soil salinity. The accumulation of high cytoplasmic levels of organic solutes, such as proline or glycine betaine, may be one such strategy in salt-stressed plants but there is not universal agreement among scientists that this is the only mechanism used by plants to combat high vacuolar ion concentrations. Two examples are given in which the hypothesis that proline accumulation is causally linked to salt stress was tested and found to be unsupported by the results obtained.

The availability of cell lines of *Nicotiana sylvestris* with enhanced resistance to growth inhibition by NaCl suggested the possibility of investigating the basis of salt resistance at the cellular level and against the relatively uniform genetic background of different cell lines within a single species (Dix and Pearce, 1981). A salt-resistant cell line, NCR 120, and a salt-sensitive line, NS, were used in experiments in which cells were exposed to 1.5 percent (wt/vol) of NaCl for a period of 10 days (Fig. 1-17). The highest level of proline attained was 11.8 µmol/g in the salt-sensitive strain, a greater than 20-fold increase. Prolonged culture of the tolerant strain led to an eventual 25 µmol/g accumulation. Dix and Pearce (1981) concluded that in their *N. sylvestris* cells, proline accumulation was not closely related to resistance to water stress or salinity but may have been a symptom of stress in general.

This conclusion was also supported by the results obtained with callus cultures initiated from hypocotyl sections of *Brassica napus* cv. Westar selected against sodium sulfate (Na_2SO_4). Figure 1-18 shows that unselected callus exposed to Na_2SO_4 produced proline, and that tolerant callus grown in the absence of Na_2SO_4 produced amounts comparable with unselected callus cultures in the absence of stress. Production of proline was, therefore, dependent on the presence of Na_2SO_4. Unselected callus produced more proline in the presence of

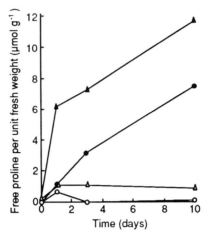

Figure 1-17. Free proline levels in NaCl-resistant (○,●) and NaCl-sensitive (△,▲) cell lines of *Nicotiana sylvestris* incubated in the presence (●,▲) or absence (○,△) of 1.5 percent NaCl. (Dix and Pearce, *Z. Pflanzenphysiol. 102*: 243-248, 1981; reproduced by permission of Gustav Fischer Verlag.)

Na_2SO_4 than tolerant callus grown on the same concentration (Fig. 1-18; Chandler and Thorpe, 1987).

Thus, once again it is noted that proline accumulation may be a general response to stress and not the sole mechanism of salt tolerance. Proline accumulation in salt tolerance in these last two cases, and one

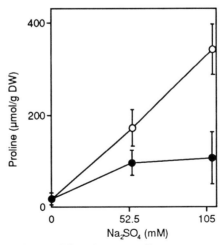

Figure 1-18. The effect of Na_2SO_4 on proline concentration in unselected (○) and tolerant (●) canola callus. (Chandler and Thorpe, *Plant Physiol. 84*: 106-111, 1987; reproduced by permission of the ASPP.)

can infer, therefore, in others, must be accompanied by other physiological/biochemical changes.

In summary, the role and importance of proline in salt stress in plants remains unresolved. In the case of microorganisms, the role of osmoprotectant molecules seems to be unequivocal. The *osm* genes of *E. coli*, for example, govern the production of molecules, such as proline and the betaines, that protect the cell and its constituents against dehydration (Rudulier et al. 1984). Thus, it seems likely that, because these compounds occur in plants, eventually it will be found that they also act as protectants, but perhaps not in all species. The cellular adaptation to osmotic stress is a fertile area for basic research with possible future applications in agriculture. The discovery of osmoprotective molecules and new classes of *osm* genes could lead to genetic enhancement of drought and salinity tolerance in crop plants.

1-2.3   Gene Expression in Response to Salt stress

At the present time, very little is known about the biochemical and genetic basis for the resistance exhibited by salt-selected cell lines of plants, although, as we have seen, it is evident that adaptation to salt entails some osmotic adjustment. To understand the possible mechanisms of resistance of lines of cells adapted to growth in high concentrations of NaCl, the protein patterns of a NaCl-adapted line of *Nicotiana tabacum* and a nonadapted line were compared. In the lines used, the resistance to salt was lost when the cells were returned to a medium without salt (Ericson and Alfinito, 1984).

The most significant difference between salt-adapted and control tobacco cells was found in three proteins that were either greatly enhanced (32-kD and 20-kD) or unique (26-kD) in salt-adapted cells (Fig. 1-19). The densitometer trace showed that the unique 26-kD protein band was 8 percent of the total sodium dodecyl sulfate (SDS)-extractable protein from salt-adapted cells and was not detectable in controls. The 32-kD band was enhanced from 9 percent in control cells to 17.5% and the 20-kD band from 0.2% to 13.3% in the salt-adapted strain. Furthermore, the 32-kD and 20-kD proteins decreased to control levels in 12 days after transfer of cells out of NaCl, whereas the 26-kD protein remained constant throughout and required two to three passages in unamended medium before disappearing (Ericson and Alfinito, 1984). However, the functions of the proteins are not known.

Recently, a number of other reports have given indications that salinity stress is associated with the expression of a number of genes at the messenger ribonucleic acid (mRNA) and protein levels. Thus, in barley and wheat, newly synthesized mRNAs and proteins have been detected that seem to be specifically induced by salt stress (Singh et al. 1987; Ramagopal, 1987a,b; Gulick and Dvorak, 1987). From this work,

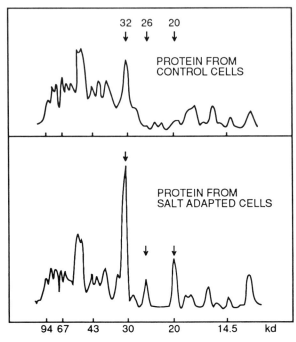

Figure 1-19. Densitometer tracing of gels showing the pattern of SDS-extractable protein in tobacco cells adapted to growth in medium containing 10 g/L NaCl and control cells. The arrows indicate salt adaptation proteins. (Ericson and Alfinito, *Plant Physiol.* 74: 506-509, 1984; reproduced by permission of the ASPP.)

however, it is also not apparent what the role of any of these gene products might be.

A promising approach to the problem of the role of these products is the work done by Singh et al. (1987) in experiments to demonstrate that ABA accelerates the rate of adaptation of tobacco cells to NaCl. They showed that the cells synthesize a predominant 26-kD protein upon exposure to ABA that is immunologically cross-reactive to the major 26-kD protein associated with NaCl adaptation. ABA also induces the synthesis of an immunologically cross-reactive 26-kD protein in cultured cells of several other plant species such as millet, soybean, carrot, cotton, potato and tomato. It appears that ABA is involved in the normal induction of the synthesis of the 26-kD protein and that the presence of NaCl is necessary for the protein to accumulate (Singh et al., 1987).

A close association between an increase in the endogenous ABA content, the synthesis of the 26-kD protein, and the subsequent adaptation of tobacco cells to NaCl, which involves considerable osmotic adjustment, suggests that these processes may be interdependent. Furthermore, the highest level of synthesis of the protein occurs in outer stem tissue of tobacco, which contains the water- and solute-transporting vascular tissues. Thus, the protein may possibly represent an *osm* gene product.

Perhaps the cloning of the gene for the 26-kD protein would be the best strategy in attempting to assign a role to the protein itself.

1-2.4   Conclusion

Thus, through the use of salt-tolerant and salt-sensitive tissues from natural and induced sources, much has been learned about the responses of plants to salt stress in their environment. Through the techniques of molecular biology, soon it may be possible to identify some of the genes involved in this important adaptation to environmental stress. The generation of mutant and variant genotypes is an important tool in this quest (Rhodes et al., 1989; Skriver and Mundy, 1990).

Case Study 1-3:   Alcohol Dehydrogenase and Anaerobiosis in Corn

Hageman and Flesher (1960) were the first to show that ADH activity increases as a result of flooding in maize seedlings. Comparable sets of corn seedlings were grown for short periods (24 hours) under aerobic and anaerobic conditions and the effect of environmental conditions on the levels of activity and the rates of development of several enzymes were examined. The single most noticeable effect was that the specific and total ADH activities from the embryonic axis and scutellum were increased between two and fivefold when the seedlings were grown under anaerobic conditions. The activities of several other enzymes were not significantly affected (Table 1-6). Freeling (1973) later showed that ADH inductions in maize under anaerobiosis proceeded at a near zero-order rate for at least 36 to 90 hours with an optimum rate of induction after 8 hours (Fig. 1-20).

There are three electrophoretically separable sets of ADH isozymes in maize. The isozymes are specified by two genes, *Adh1* and *Adh2*, that segregate independently of one another. The active enzyme is a dimer of 80-kD, therefore, the three isozymes are either homodimers or heterodimers of the two protein products of the two genes *Adh1* and *Adh2* (Fig. 1-21; Freeling and Schwartz, 1973). The homodimers form the set I and set III enzymes and the heterodimer, set II. Set I enzymes predominate in the mature embryo, but immersion of the entire seedling in an aqueous medium mediates a rapid accumulation of ADH specified by both genes (Fig. 1-22; Freeling, 1973).

Differential *Adh1:Adh2* gene expression appeared to be dependent on the mode of induction. The *Adh2* gene contributed far less product to the total ADH protein in aerobic than in anaerobic conditions (Fig. 1-22). And air halted ADH induction after about a 12-hour lag as did the inhibitor of protein synthesis, cycloheximide. Both treatments also

Table 1-6. Effect of environment (aerobic vs anaerobic) on the activity of several cytoplasmic enzymes extracted from 3-day-old corn seedlings (Millimicromoles substrate transformed/minute/mg.protein).

Enzyme Preparation	Embryonic axis		Scutellum	
	Air	N_2	Air	N_2
Alcohol dehydrogenase	92	270	207	323
Triosephosphate dehydrogenase	892	964	627	626
Glucose-6-phosphate dehydrogenase	107	72	125	101
Aldolase	318	466	172	172

(Hageman and Flesher, *Arch. Biochem. Biophys. 87*: 203-209, 1960.)

demonstrated that a nearly constant level of ADH activity was maintained for over 48 hours after the arrest of enzyme induction (Fig. 1-23; Freeling, 1973).

These observations created an interest in the significance of ADH and its induction to the survival of plants under anaerobic conditions.

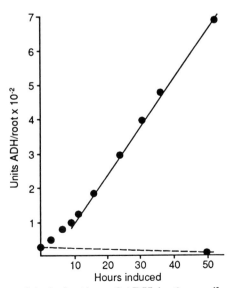

Figure 1-20. Anaerobic induction of ADH in 8-cm sib primary roots of intact maize seedlings. The 50-hour point locating the dashed line is from untreated sib roots. (Freeling, *Mol. Gen. Genet. 127*: 215-227, 1973.)

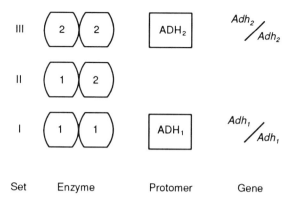

Figure 1-21. Schwartz's model for the genetic origins of maize ADH isozymes. (Freeling and Schwartz, *Biochem. Genet. 8*: 27-36, 1973; reproduced by permission of Plenum Publishing Corp.)

Some investigations arising from this curiosity about the enzyme and its role in anaerobiosis involving the use of mutants of maize are presented.

1-3.1   ADH and the Development of the Plant

It became possible to investigate the influence of ADH in maize plants after the isolation of mutants lacking the enzyme. Thus, Schwartz (1969) treated the seeds of maize with the mutagen ethylmethane sulfonate and induced a mutation at the *Adh1* locus that, in homozygous condition, failed to produce active enzyme. The mutant was designated *Adh1°*. At first, ADH did not appear to play an essential role in the development of the maize plant (Schwartz, 1969). Kernels that lacked an active *Adh1* gene developed normally, and in an F_2 segregating ear, mutant and nonmutant kernels could be distinguished only by enzyme assay. Furthermore, no difference could be detected in the germination and subsequent growth of mutant and nonmutant sibs (Schwartz, 1969).

The question arose, then, as to why, if the gene did not play an active role in plant development, it persisted in active form in the population? It was hypothesized that it must confer some selective advantage on the plant, otherwise it should be lost by accumulated mutations. A possible explanation was that, although the gene was not essential for normal development of the plant under the normal range of cultural conditions, it was essential for survival under adverse conditions. An educated guess was made about a possible role for the enzyme based on the function of ADH in anaerobic respiration (Schwartz, 1969).

If plant material is subjected to anaerobic conditions where aerobic metabolism is blocked, the liberation of free energy is completely dependent on anaerobic respiration. In nature, anaerobic conditions could result from flooding or from retention of an excessive amount of water in the soil after a heavy rainfall. The prediction was made that

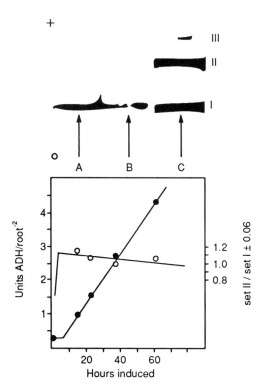

Figure 1-22. The simultaneous expression of two ADH genes during anaerobic induction in maize. The electrophoretograms reflect the ADH isozymes present in extracts prepared from sibling (A) dry seeds, (B) uninduced primary roots, and (C) primary roots induced anaerobically for 24 hours. The graph plots both the induction curve (● left ordinate) and the Set II : Set I ratio (○ right ordinate). + is the anode. (Freeling, *Mol. Gen, Genet. 127*: 215-227, 1973

seeds that contained active ADH and carried on anaerobic respiration would be capable of surviving flooding conditions much better than seeds that lacked the enzyme and this provided the selective pressure to keep the active gene in the population.

The prediction was tested by immersing F_2 maize seeds segregating for the mutant and nonmutant alleles (*Adh1°* and *Adh1ˢ*, respectively) under water for 3 days at 30°C. On the fourth day the seeds were germinated on moistened filter paper at the same temperature. The results showed a strong correlation between the *Adh1* genotype and the ability to germinate after exposure to the anaerobic environment (Table 1-7). All the kernels that germinated contained the active *Adh1ˢ* allele (Schwartz, 1969). None of the *Adh1°/Adh1°* mutants germinated (Table 1-7). When the F_2 seeds were germinated directly without prior soaking, 98 percent of them germinated, thus showing that differential viability

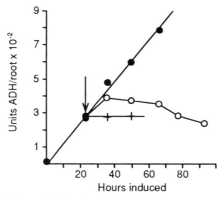

Figure 1-23. Air halts ADH induction in maize seedlings. The arrow marks the point of removal of the roots from anaerobiosis to air (○), or addition of cycloheximide to 10 g/ml (+). The normal course of anaerobic induction (●). (Freeling, *Mol. Gen. Genet. 127*: 215-227, 1973.)

between the wild-type and mutant seeds was not the reason for any differences in germination.

These results (Table 1-7) suggested that the role of the *Adh1* gene in maize, and perhaps in all plants, was to enable the seed to retain its viability when subjected to flooding conditions. This hypothesis was investigated further using *Adh1°* mutant plants under hypoxic conditions.

1-3.2   ADH and Anaerobiosis

Certain higher plant tissues, including maize roots, although requiring oxygen for normal functioning, can survive long periods (>18 hours) of anaerobiosis, with glycolysis continuing during most of this period. Most vertebrate tissues, on the other hand, can survive only short periods of anaerobiosis (<2 hours) after which glycolysis is greatly inhibited, and in most cases irreversible cell damage occurs (Roberts et al., 1984b). An explanation for the difference was thought to lie in the ability of higher

Table 1-7. Correlation between *Adh1* genotype and germination of soaked maize seeds kept under water for 3 days at 30°C.

	Germination	No Germination
Adh1ˢ/-	309	8
Adh1°/Adh1°	0	69

(Schwartz, *Amer. Nat. 103*: 479-481, 1969; copyright © 1969 by the University of Chicago. Used with permission.)

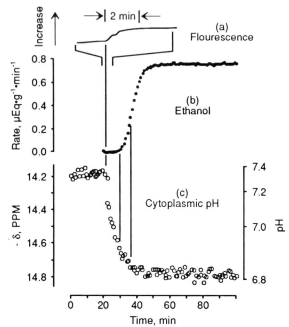

Figure 1-24. Time course of NADH fluorescence (A), rate of ethanol production (B), and cytoplasmic pH (C), in perfused maize root tip. Roots were initially perfused at 50 mL minute with O_2 saturated 50 mM glucose/ 0.1 mM CaSO4. After 20 minutes, they were perfused at 4 mL minute with the same solution now saturated with N_2. (Roberts et al., *Proc. Natl. Acad. Sci. USA 81*: 3379-3383, 1984b.)

plants to undergo a mainly ethanolic fermentation rather than the exclusively lactic fermentation seen in higher animals. This hypothesis was tested through the use of perfused maize root tips (hypoxic conditions).

In the cytoplasm of maize root tips, the pH fell to a stable value of 6.8, approximately 0.5 units below aerobic values, 20 minutes after transfer to anaerobic conditions, at which point it stabilized for at least 10 hours (Fig. 1-24). Reportedly, this contrasts with active, hypoxic vertebrate tissues where cytoplasmic pH falls throughout hypoxia because of continuous lactic acid accumulation until glycolysis ceases. After a 10-minute lag, ethanol appeared in the perfusion effluent and increased in amount for approximately 40 minutes, remaining constant thereafter (Fig. 1-24). No lactate was found in the effluent but other measurements taken showed that cytoplasmic acidification was apparently due to a transient lactic fermentation (Roberts et al., 1984b).

An explanation of the results summarized in Figures 1-24 was hypothesized on the basis of the difference in pH optima of lactic dehydrogenase (alkaline) and pyruvate decarboxylase (acidic), the first enzyme in the pathway to ethanol production (see Roberts et al., 1984b). It was suggested that in hypoxic plants it is the cytoplasmic pH that

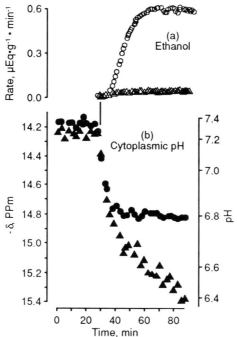

Figure 1-25. Time course of the rate of ethanol production (A) and cytoplasmic pH (B) in perfused 4-mm maize root tips from an *Adh1* line (△,▲) and a parent line (○,●). (Roberts et al., *Proc. Natl. Acad. Sci. USA 81*: 3379-3383, 1984b.)

controls which fermentation end-product is formed. The postulate was that initially lactic acid is formed in the alkaline cytoplasm. The resultant, lowered cytoplasmic pH then inhibits further lactic acid production while activating pyruvic decarboxylase, leading to ethanol production. Results in Figures 1-24 are certainly consistent with this hypothesis. Recent results with roots of rice and wheat also strongly indicate that pH of the cytoplasm may be the overriding factor that regulates activity of pyruvate decarboxylase in the cell (Morrell et al., 1990). But it is also the case, that although ethanol production is primarily regulated at the level of the decarboxylase, the regeneration of the oxidized form of nicotinamide adenine dinucleotide (NAD$^+$) is required for continued ethanol production, a process dependent on a functional ADH. Thus, the postulated control process hypothesized could be tested through the use of mutants that essentially lack a functional ADH under hypoxic conditions.

Maize root tips, homozygous for a mutation at the *Adh1* locus, contain less than 1 percent of the ADH found in wild type roots both under aerobic and hypoxic conditions. Moreover, when such root tips became hypoxic, the stabilization of the cytoplasmic pH at about 6.8 seen in normal root tips did not occur (Fig. 1-25; Roberts et al., 1984b).

Instead, cytoplasmic pH continued to fall throughout the hypoxic treatment, presumably due to continued lactate production. The implication in this explanation was that, not being able to generate NAD⁺ for glycolysis through ethanol fermentation, the plant continued to do so by the only route available to it, lactate production. But, because this continued acid production led to a rapid lowering of cytoplasmic pH to unacceptable levels, the process could go on for only short periods, therefore, roots lacking ADH cannot withstand hypoxia as well as comparable wild-type plants. Using F_3 progeny, it was found that ADH⁻ root tips died after about 12 hours of hypoxia, whereas ADH⁺ root tips survived about 24 hours (Roberts et al., 1984a).

In summary, the hypothesis suggested that when maize roots are subjected to hypoxia: oxidative phosphorylation ceases; lactic acid production begins; cytoplasmic acidification occurs; ethanol fermentation is triggered; lactic fermentation declines; energy production continues for a while; extreme cytoplasmic acidosis occurs because of leakage of acid from acidic vacuoles; death occurs as acidosis overcomes the maintenance processes by some means (Roberts et al., 1984a).

This hypothesis did not explain why roots of plant species differed in their ability to withstand hypoxia although apparently all undergoing a mainly ethanolic fermentation. In an attempt to address this question of the role of ADH in the metabolism and survival of hypoxic maize root

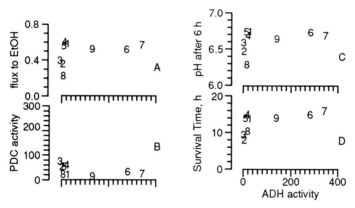

Figure 1-26. Dependence of metabolism and viability in hypoxic maize root tips on ADH activity in nine maize lines. ADH was assayed in extracts of root tips harvested before the start of hypoxia; units, nmol NADH produced/minute/mg protein. Data points in each graph are represented with numbers, which specify the line from which the *Adh1* alleles were derived. A. Maximum rate of ethanol production (about 1 hour after the onset of hypoxia); units, μmol/minute/g fresh weight. B. Pyruvate decarboxylase (PDC) activity in root tip extracts; units, nmol NADH consumed/minute/mg protein. C. Cytoplasmic pH after 6 hour of hypoxia. D. Time of root tip death under hypoxia. (Roberts et al., *Plant Physiol. 89:* 1275-1278, 1989; reproduced by permission of the ASPP.)

tips, Roberts et al. (1989) compared maize lines with ADH activities differing over a 200-fold range. It was found that root tips had similar capacities for fermentation to ethanol, except when ADH activity was reduced to about 10 milliunits per mg protein or lower (Fig. 1-26). Thus, although ADH activity does limit metabolism and survival in mutant maize root tips, which have the very lowest activities, in the "normal" maize root tips ADH was not limiting over wide concentration ranges.

Roberts et al. (1989) concluded from their results (Fig. 1-26) that the activity of ADH in normal maize root tips does not limit the capacity for energy production by fermentation and does not determine viability under extreme hypoxia. Only in *Adh1* nulls was there a clear link between ADH and survival under extreme hypoxia. The hypothesis as formulated was inadequate to explain tolerance to restricted oxygen availability in normal roots with a wide range of ethanolic fermentation capability.

The main conclusion from these investigations is that any hypothesis formulated to explain the tolerance of some plants to restricted oxygen availability cannot be based solely on the role played by ADH. The *Adh1* null mutants have yielded, and will continue to provide, useful information about the extent of ADH involvement in the process. However, others have suggested (Menegus et al., 1989) that the metabolic consumption of H^+, rather than just the arrest of acid production, seems to be important in the tolerance mechanism in several plant tissues. A combination of genetic and physiological investigations will have to be undertaken to arrive at a broad appreciation of the phenomenon of tolerance to hypoxia.

Case Study 1-4.   Correction of Iron Deficiency in the Tomato Mutant *chloronerva*

The primary function of Fe in plants is in oxidation-reduction (redox) reactions. Some of the important Fe-containing proteins that participate in such reactions are cytochrome oxidase, catalase, peroxidase, various cytochromes, leghemoglobin, ferredoxin, xanthine oxidase, aconitase, and the Fe-S proteins of the photosynthetic light reactions. Chlorophyll synthesis is also inhibited in Fe-deficient plants (Longnecker, 1988).

Iron status cannot always be assessed by simply measuring the amount in a tissue because Fe-sufficient and Fe-deficient tissues can contain similar quantities of the ion. This has caused a long-standing debate about "active" Fe, that is, Fe that is physiologically available compared with Fe in tissue that is metabolically nonfunctional.

Iron deficiency is the most commonly recognized micronutrient deficiency worldwide, although most soils contain adequate total amounts of Fe. The two most abundant forms of ionic iron, ferric (Fe^{3+}) and ferrous (Fe^{2+}), are only slightly soluble in a stable form in aerated soil solutions.

This low solubility results in free ferric ion concentrations that are an order of magnitude lower than those held to be necessary for growing plants (Bienfait, 1988a).

Higher plants have two strategies by which they can increase the availability of iron in the soil. In one of these, the grasses excrete compounds called siderophores that can solubilize Fe^{3+} ions, a process similar to that in microbes. In the latter cases, microbial siderophores are powerful chelators of Fe^{3+}, are produced and excreted under the influence of iron deficiency, and are reabsorbed as Fe^{3+}-chelate by specific sites on a cell membrane. Fe-deficient grasses have a family of similar interrelated compounds that they can excrete as do microbes (Bienfait, 1988a). These are all good chelators of ferric ions. Only the grasses have been shown to excrete the siderophores, which are all related to the compound nicotianamine (discussed later); the dicotyledons have either not developed or have lost this capacity. Nevertheless, the dicots contain nicotianamine, which is thought to be a ferrous ion carrier in these plants operating over short distances (i.e., within the cell and perhaps also from cell to cell).

Following in this Case Study is a discussion of the use of the *chloronerva* mutant of tomato to identify nicotianamine and to define its crucial role in the mineral nutrition of dicots.

1-4.1   Normalizing Factor in Tomato

More than 25 years ago experiments were conducted in the course of which tomato plants (*Lycopersicon esculentum* Mill., var. "Bonne Beste") were grafted onto tobacco rootstocks. Among the fruits obtained, one contained seeds that gave 67 normal and 22 mutant plants. The mutant was spontaneous, recessive, and monogenic, was characterized by severely retarded growth and distorted leaves of abnormal shape, and exhibited a pale yellow chlorosis of intercostal areas of young leaves (Prochazka and Scholz, 1984). Flower buds very rarely developed, did not unfold, and eventually died off. The mutant was given the name *chloronerva*.

Normal growth and development could be completely restored in the *chloronerva* mutant by grafts in which it was irrelevant whether the mutant was used as scion or as rootstock. This normalization of the phenotype also occurred in grafts between the mutant and other species. Grafting could be replaced by sprays or infiltrations with extracts from the wild type, as well as from other species. When the mutant was cultivated on a medium containing an excess of iron, chlorosis was relieved but the plants were still unable to produce flowers; organic iron chelate (FeEDTA) was more effective than inorganic iron. This suggested that the effect of iron was secondary and that iron alone was unable to compensate for the defect caused by the mutation (Prochazka and Scholz, 1984).

The "phenotypical normalization" of *chloronerva* by the application of extracts from different plant species containing an unknown "normalizing factor" was used as a basis for a biological test where sample solutions were "painted" onto young leaves of mutant seedlings. Normalization was recorded as regreening of the chlorotic leaflets after a few days. In this way, a semiquantitative test for the normalizing factor was established (Prochazka and Scholz, 1984).

1-4.2   Isolation and Characterization of the Normalizing Factor

With the development of a semiquantitative bioassay for the normalizing factor, it became possible to undertake the isolation of the substance in pure state, from alfalfa tissue at first (Scholz and Rudolph, 1968). The most likely structure of the compound was shown to be *N*-[*n*-(3-amino-3-carboxypropyl)-3-amino-3-carboxypropyl]-azetidine-2-carboxylic acid, identical to that of the imino acid nicotianamine (Fig. 1-27; Budesinsky et al., 1980), one of the important properties of which is to form complexes with iron and other heavy metal ions (Scholz, 1970; Scholz et al., 1987).

1-4.3   Occurrence and Physiological Role of Nicotianamine

Nicotianamine has been quantitatively determined or biologically assayed in a wide variety of plants, angiosperms and gymnosperms (shoot and root tissues), as well as liverworts, mosses, lycopods, and horsetails (Rudolph et al., 1985). In the tomato plant *chloronerva*, synthesis of the normalizing factor seems to be completely blocked (Table 1-8; Rudolph and Scholz, 1972).

The root system of the wild-type tomato in aerated nutrient solution developed to a long, very thin main root and many long lateral roots. Under the same conditions, the roots of *chloronerva* formed a brushlike system of stunted and thickened main and lateral roots (Scholz, 1983). Addition of nicotianamine to a concentration of 5×10^{-7} M led to the recovery of normal growth and appearance of the mutant roots. Because the synthesis of nicotianamine was completely blocked in the mutant, the sensitive response of the mutant with the addition of nicotianamine could be interpreted as a root growth regulation effect and, therefore,

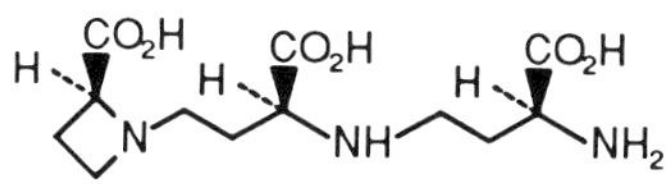

Figure 1-27. The structure of nicotianamine.

Table 1-8. Analysis of "normalizing factor" in *Lycopersicon esculentum* Mill. cv. "Bonne Beste" (*BB*) and in the mutant *chloronerva* (chln) organs.

Genotype	Organ	mg/100 g dry weight
BB/BB	Scion (Reis)	0.7
	Stock (Unterlage)	4.3
	Root (Wurzel)	5.7
chln/chln	Scion (Reis)	0
	Stock (Unterlage)	0
	Root (Wurzel)	0

(Rudolph and Scholz, *Biochem. Physiol. Pflanzen. 163*: 156-68, 1972; reproduced by permission of Gustav Fischer Verlag.)

nicotianamine as a morphogenetic agent somewhat like phytohormones. This view of its role as a morphogenetic agent was supported by the observed correlation between the concentration of nicotianamine applied and the root length of the mutant observed. Figure 1-28 shows the quantitative effect of the nicotianamine concentration in the nutrient medium on the main root length of *chloronerva* in the range between 10^{-7} and 10^{-5} M. Nicotianamine promoted root growth to an extent that is very close to that of the untreated wild type. At this concentration root elongation was accompanied by a regreening of chlorotic mutant leaves, an effect that was evident within 24 to 48 hours. The lowest concentration that gave a positive response in some cases was 10^{-8} M. The root growth of the wild type was unaffected by the addition of nicotianamine to the medium (Fig. 1-28a). A positive effect on mutant root growth was also observed after application of nicotianamine to the leaves (Fig. 1-28b). Various concentrations of nicotianamine were administered to the leaves of five plants. A concentration of 3.3 x 10^{-6} M gave a positive response and at 3.3 x 10^{-5} M a significant regreening of chlorotic leaves was seen (Scholz, 1983).

From this work (Fig. 1-28) it is reasonable to conclude that nicotianamine painted on the leaves is either transported to the roots and influences metabolism directly or causes some metabolic changes in the shoot, which indirectly enhance root growth (Scholz, 1983). The former assumption is more likely because mutant root growth is also enhanced by nicotianamine-containing plant extracts in sterile organ culture.

This is the first case where the biochemical analysis of a genetically well-defined auxotrophic mutant of a higher plant has led to the attribution of physiological significance to a hitherto rarely known metabolite. Because of this, some other details of its function are discussed.

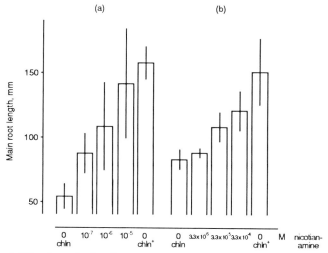

Figure 1-28. Correlation between nicotianamine concentration and main root length of the mutant *chloronerva* after 7 days' growth. (A) Application to the nutrient medium. (B) Application to the leaves. Application time was 5 days. Root length of the wild-type *chloronerva*⁺ is given for comparison. The bars represent SD. (Scholz, *Plant Science Letters 32*: 327-332, 1983; reproduced by permission of Elsevier Science Publishers.)

The leaves of young *chloronerva* mutant seedlings were grown in nutrient solution with 10^{-5} M FeEDTA and treated with different concentrations of nicotianamine. Five plants were used per treatment and five applications of nicotianamine were made per day for 7 days. The iron concentration as well as the total amount of iron in the untreated (control) mutant shoots exceeded that of the wild type by a factor of about two (Table 1-9; Scholz et al., 1985). The amount of iron in the mutant shoot was drastically reduced relative to that in the wild type by the application of nicotianamine to leaves. Both the decrease in iron concentration and the increase in root growth depended on the concentration of nicotianamine applied. The decrease in iron in the mutant shoot and the increase in the main root length seemed to be very closely linked (Table 1-9; Scholz et al., 1985).

The depressive effect of nicotianamine also occurred when both nicotianamine and iron were added to the nutrient solution (Table 1-10). The amount of iron, on a dry weight basis, was decreased by the presence of 10^{-5} M nicotianamine in the nutrient solution in both the shoot and the root (Table 1-10). This effect was especially evident with FeEDDHA (EDDHA, ethylenediamine N, N^1-bis-[o-hydroxyphenylacetic acid]) as the iron source where the iron content of the mutant dropped even below that of the wild type, mainly because of the sharp decrease in the amount of iron present in the roots.

Common to all results obtained in these experiments (Tables 1-9 and 1-10) was the depressive effect of nicotianamine on the high iron uptake

Table 1-9. Effect of nicotianamine (NA) applied to the leaves of seedlings of wild-type and of *chloronerva* tomato (*L. esculentum* L.) on the iron content in the shoot and on the main root length.
[a]Significant at the 0.1 percent level.
[b]Difference in mutant main root length between treatment and water control, significant at the 5 percent level.

Plant type	Treatment	μM NA	mg DW per shoot	nmol iron per shoot	nmol iron per mg DW	main root length mm+SD
Wild-type	Water	0	35	179	5.1	145± 16
Mutant	Water	0	46	419	9.1	116± 16
Mutant	NA	6	44	308	7.0	131± 5
Mutant	NA	60	56	190	3.4	166± 12
Mutant	NA	600	49	206	4.2	158± 27

(Scholz et al., *Physiol. Plant. 63*: 99-104, 1985.)

by the mutant at iron concentrations in the nutrient solution of 10^{-5} M and higher. Such results were obtained regardless of whether nicotianamine was supplied to the nutrient solution or to the leaves.

In addition to its excessive iron uptake the *chloronerva* mutant also accumulates higher concentrations of manganese (Mn^{2+}), zinc (Zn^{2+}) and copper (Cu^{2+}) than does wild-type tomato (Table 1-11). This extra uptake decreases after application of nicotianamine either to the nutrient solution or to the leaves of mutant seedlings. In contrast, the rate of rubidium (Rb^+) and phosphate (PO_4^{3-}) uptake is higher in the wild type than in the mutant (Scholz et al., 1987).

It might be concluded from these results (Table 1-11) that the role of nicotianamine is not confined to iron uptake but is aimed at the regulation of uptake of divalent heavy metal ions, in general. However, the way in which the chelating agent exerts its influence, if any, in the apparent feedback control mechanism is not yet understood.

1-4.4   Conclusions About the Role of Nicotianamine in Dicots

Some conclusions on the effect of nicotianamine on the uptake and distribution of iron in dicotyledonous plants may be drawn from experiments with the very useful *chloronerva* mutant (Scholz et al., 1985):

Table 1-10. Iron content of *chloronerva* mutant shoots and roots as affected by 10^{-5} M nicotianamine (NA) supply to the nutrient solution. The seedlings were grown in nutrient solution with 10^{-5} M Fe. EDDHA; ethylenediamine N, N'-bis-(o-hydroxyphenylacetic acid).

Plant type	Organ	FeCl$_2$.2 days		FeEDDHA.3 days	
		mg DW	nmolFe (mgDW)$^{-1}$	mg DW	nmolFe (mgDW)$^{-1}$
Mutant	shoot	25.5	1.1	35.5	4.3
		33.8	12.2	47.6	11.7
	root	8.3	46.4	12.1	33.6
Mutant +NA	shoot	23.2	0.6	35.5	3.1
		33.2	10.3	46.3	5.1
	root	10.0	32.5	11.0	11.8
Wild-type	shoot	35.8	0.2	64.2	1.5
		45.0	6.5	80.3	6.3
	root	9.2	31.3	16.1	25.5

(Scholz et al., *Physiol. Plant. 63*: 99-104, 1985.)

1. Nicotianamine supplied to external solutions interferes with the mechanisms of iron uptake by plant roots, perhaps by complex formation with Fe^{2+} after reduction of Fe^{3+}EDDHA at the plasmalemmae of cortical cells, thus inhibiting iron uptake.

2. Nicotianamine supplied to the leaves exerted the same decrease in iron uptake by the mutant as root application. The sites of iron reduction at the plasmalemmae of cortical root cells seem, therefore, to be accessible to nicotianamine transported from the shoot to the root.

3. Results do not support the view, held by some, that nicotianamine acts as a siderophore. Apart from the fact that bacterial siderophores are specific chelators for Fe^{3+}, whereas nicotianamine is not, siderophores are exclusively excreted under conditions of iron deficiency, conditions that did not occur in the experiments cited. Moreover, the role of nicotianamine seems to be more complex than that of just a chelator in that it seems to be responsible for the availability of Fe^{2+} by regulating its uptake, simultaneously increasing the elongation growth of the root and overcoming the chlorosis of intercostal areas in the leaf.

4. Iron absorbed by the root as Fe^{2+} is believed to be reoxidized and subsequently transported to the shoot as a Fe^{3+}-citrate complex. This process seems to be unimpaired by the *chloronerva* mutation and is, therefore, not sensitive to the supply of nicotianamine. In the vessels of the leaf, Fe^{3+} is again reduced to Fe^{2+}, perhaps by means of photochemical events. In young mutant leaves iron is immobilized in the veins.

Table 1-11. Heavy metal concentrations in tomato seedlings after 1 week of cultivation in nutrient solution with and without the addition of nicotianamine (NA) (10 μM).

Wild type	Number of plants	mg dry weight per plant	heavy metal concentration nmol per mg dry weight/%			
			Cu	Fe	Mn	Zn
wild-type	6	126	0.41/100	7.43/100	3.84/100	3.22/100
mutant	10	69	0.99/241	9.30/125	10.45/273	5.72/184
mutant + NA	9	68	0.84/205	7.29/ 98	7.83/204	4.79/154

(Scholz et al., *Biochem. Physiol. Pflanzen. 182*: 189-194, 1987; reproduced by permission of Gustav Fischer Verlag.)

Nicotianamine could play a role in remobilizing this iron by interacting with Fe^{2+}, thus eliminating the chlorosis of intercostal areas. Very recent evidence suggests that it acts as a symplastic iron transporter (Stephan et al., 1990).

The results reported in this Case Study represent a first approach to understanding the functions of nicotianamine in plants. The *chloronerva* mutant has already provided, and continues to provide, a way to investigate the physiological role of iron in plants (Becker et al., 1989; Stephan and Grun, 1989; Stephan and Prochazka, 1989). More experiments are necessary to resolve whether the regulatory effect of nicotianamine on uptake and short-distance transport of iron is the primary function of this substance or an indirect consequence of other metabolic interactions. What is quite clear is that *chloronerva* is an auxotrophic mutant the survival of which is closely linked to the provision of the compound, nicotianamine.

1-4.5   Nicotianamine Derivatives in Monocots

The marked susceptibility of rice to iron chlorosis led to the discovery that root washings of water-cultured oat and rice contained several amphoteric, iron-chelating compounds, and that the release of the chelators into the washings increased greatly under iron stress conditions (see Prochazka and Scholz, 1984). These chelators were shown to be new amino acids: mugineic acid, isomugineic acid, and 2´-dehydroxymugineic acid, all of which are derivatives of nicotianamine (Shojima et al., 1990). Therefore, it might be hypothesized that nicotianamine is the parent substance of a group of chelators containing an azetidine ring. Other substances of similar type have been isolated from oats (avenic acids A and B). Thus, these substances would appear to belong to a new class of important physiologically active chelators in plants, derived from 2-azetidinecarboxylic acid. In some cases, they act as true siderophores (Romheld and Marschner, 1986) (i.e., they are excreted from roots under iron stress, they are able to chelate Fe^{3+} strongly and specifically, and they increase the iron and chlorophyll content of chlorotic leaves). Recently, these studies have been extended to include other aspects of the Fe-efficient response in Fe-deficient plants (Bienfait, 1988b). There is no evidence, however, that nicotianamine acts as a phytosiderophore.

In plants the assimilation systems of other minerals are not as well documented and studied as are those of iron. However, one very promising investigation reported recently by Goldstein et al. (1989) suggests that the assimilation system for phosphate might be open to the same kind of study as that of iron. In conditions of phosphate starvation, higher plants excrete acid phosphatases to solubilize organic phosphates in the rhizosphere of the soil. These enzymes have been termed *excreted, phosphate-starvation-inducible acid phosphatases* (epsi-APases) (Goldstein et al., 1989). Recently, it has been found possible to select for mutants in *Arabidopsis* which are apparently deficient in the epsi-APases. A dye, 5-bromo-4-chloro-3-indolyl-phosphate-*p*-toluidine, when cleaved by epsi-APases can then enter living cells, but appears to be toxic. Thus, plants having the epsi-APases are killed by the toxic component of the dye. It will be interesting to discover whether this selection system is sufficiently stringent to produce mutants that are useful for investigating the phosphate-scavenging mechanism in higher plants.

References

Becker, R., Pich, A., Scholz, G., and Seifert, K. 1989. Influence of nicotianamine and iron supply on formation and elongation of adventitious roots in hypocotyl cuttings of the tomato mutant chloronerva (*Lycopersicon esculentum*). Physiol. Plant. 76: 47-52.

Ben-Hayyim, G. and Kochba, J. 1983. Aspects of salt tolerance in a NaCl-selected stable cell line of *Citrus sinensis*. *Plant Physiol.* 72: 685-690.

Bienfait, H.F. 1988a. Mechanisms in Fe-efficiency reactions of higher plants. *J. Plant Nutr.* 11: 605-629.

Bienfait, H.F. 1988b. Proteins under the control of the gene for Fe efficiency in tomato. *Plant Physiol.* 88: 785-787.

Bradford, K.J. 1983. Water relations and growth of the *flacca* tomato mutant in relation to abscisic acid. *Plant Physiol.* 72: 251-255.

Bradford, K.J., Sharkey, T.D., and Farquhar, G.D. 1983. Gas exchange, stomatal behavior and *delta* ^{13}C values of the *flacca* tomato mutant in relation to abscisic acid. *Plant Physiol.* 72: 245-250.

Budesinsky, M., Budzikiewicz, H., Prochazka, Z., Ripperger, H., Romer, A., Scholz, G., and Schreiber, K. 1980. Nicotianamine, a possible phytosiderophore of general occurrence. *Phytochemistry 19*: 2295-2297.

Chandler, S.F. and Thorpe, T.A. 1987. Characterization of growth, water relations, and proline accumulation in sodium sulfate tolerant callus of *Brassica napus* L. cv. Westar (canola). *Plant Physiol. 84*: 106-111.

Daines, R.J. and Gould, A.R. 1985. The cellular basis of salt tolerance studied with tissue cultures of the halophytic grass *Distichlis spicata*. *J. Plant Physiol. 119*: 269-280.

Dix, P.J. and Pearce, R.S. 1981. Proline accumulation in NaCl-resistant and sensitive cell lines of *Nicotiana sylvestris*. *Z. Pflanzenphysiol. 102*: 243-248.

Epstein, E., Norlyn, J.D., Rush, D.W., Kingsbury, R.W., Kelley, D.B., Cunningham, G.A., and Wrona, A.F. 1980. Saline culture of crops: A genetic approach. *Science 210*: 399-404.

Ericson, M.C. and Alfinito, S.H. 1984. Proteins produced during salt stress in tobacco cell culture. *Plant Physiol. 74*: 506-509.

Freeling, M. 1973. Simultaneous induction by anaerobiosis or 2,4-*D* of multiple enzymes specified by two unlinked genes: Differential *Adh1-Adh2* expression in maize. *Mol. Gen. Genet. 127*: 215-227.

Freeling, M. and Schwartz, D. 1973. Genetic relationships between the multiple alcohol dehydrogenase of maize. *Biochem. Genet. 8*: 27-36.

Freeling, M. and Bennett, D.C. 1985. Maize *Adh1*. *Ann. Rev. Genet. 19*: 297-323.

Goldstein, A.H., Baertlein, D.A., and Danon, A. 1989. Phosphate starvation stress as an experimental system for molecular analysis. *Plant Molecular Biology Reporter 7*: 7-16.

Gulick, P. and Dvorak, J. 1987. Gene induction and repression by salt treatment in roots of the salinity-sensitive Chinese Spring wheat and the salinity-tolerant Chinese Spring x *Elytrigia elongata* amphiploid. *Proc. Natl. Acad. Sci. USA. 84*: 99-103.

Hageman, R.H. and Flesher, D. 1960. Effect of an anaerobic environment on the activity of alcohol dehydrogenase and other enzymes of corn seedlings. *Arch. Biochem. Biophys. 87*: 203-209.

Imber, D. and Tal, M. 1970. Phenotypic reversion of *flacca*, a wilty mutant of tomato, by abscisic acid. *Science 169*: 592-593.

Jones, H.G., Sharp, C.S., and Higgs, K.H. 1987. Growth and water relations of wilty mutants of tomato (*Lycopersicon esculentum* Mill.). *J. Exp. Bot. 38*: 1848-1856.

Longnecker, N. 1988. Iron nutrition of plants. *ISI Atlas of Science: Animal and Plant Sciences*, Institute for Scientific Information, Philadelphia, PA, pp. 143-150.

Matters, G.L. and Scandalios, J.G. 1986. Changes in plant gene expression during stress. *Dev. Genet. 7*: 167-175.

Menegus, F., Cattaruzza, L., Chersi, A., and Fronza, G. 1989. Differences in the anaerobic lactate-succinate production and in the changes of cell sap pH for plants with high and low resistance to anoxia. *Plant Physiol. 90*: 29-32.

Morrell, S., Greenway, H., and Davies, D.D. 1990. Regulation of pyruvate decarboxylase in vitro and in vivo. *J. Exp. Bot. 41*: 131-139.

Neill, S.J. and Horgan, R. 1985. Abscisic acid production and water relations in wilty tomato mutants subjected to water deficiency. *J. Exp. Bot. 36*: 1222-1231

Neill, S.J., Mcgaw, B.A., and Horgan, R. 1986. Ethylene and 1-aminocyclopropane-1-carboxylic acid production in *flacca*, a wilty mutant of tomato, subjected to water deficiency and pretreatment with abscisic acid. *J. Exp. Bot. 37*: 535-541.

Prochazka, Z. and Scholz, G. 1984. Nicotianamine, the 'normalizing factor' for the auxotroph tomato mutant *Chloronerva*; A representative of a new class of plant effectors. *Experientia 40*: 794-801.

Ramagopal, S. 1987a. Differential mRNA transcription during salinity stress in barley. *Proc. Natl. Acad. Sci. USA. 84*: 94-98.

Ramagopal, S. 1987b. Salinity stress induced tissue-specific proteins in barley seedlings. *Plant Physiol. 84*: 324-331.

Rhodes, D., Rich, P.J., Brunk, D.G., Ju, G.C., Rhodes, J.C., Pauly, M.H., and Hansen, L.A. 1989. Development of two isogenic sweet corn hybrids differing for glycinebetaine content. *Plant Physiol. 91*: 1112-1121.

Roberts, J.K.M., Callis, J., Jardetzky, O., Walbot, V., and Freeling, M. 1984a. Cytoplasmic acidosis as a determinant of flooding intolerance in plants. *Proc. Natl. Acad. Sci. USA. 81*: 6029-6033.

Roberts, J.K.M., Callis, J., Wemmer, D., Walbot, V., and Jardetzky, O. 1984b. Mechanism of cytoplasmic pH regulation in hypoxic maize root tips and its role in survival under hypoxia. *Proc. Natl. Acad. Sci. USA. 81*: 3379-3383.

Roberts, J.K.M., Chang, K., Webster, C., Callis, J., and Walbot, V. 1989. Dependence of ethanolic fermentation, cytoplasmic pH regulation, and viability on the activity of alcohol dehydrogenase in hypoxic maize root tips. *Plant Physiol. 89*: 1275-1278.

Romheld, V. and Marschner, H. 1986. Evidence for a specific uptake system for iron phytosiderophores in roots of grasses. *Plant Physiol. 80*: 175-180.

Rudolph, A. and Scholz, G. 1972. Physiologische Untersuchungen an der Mutante *chloronerva* von *Lycopersicon esculentum* Mill. IV. Über eine Methode zur quantitativen Bestimmung des "Normalisierungsfactors" sowie über dessen Vorkommen im Pflanzenreich. *Biochem. Physiol. Pflanzen:163* 156-168.

Rudolph, A., Becker, R., Scholz, G., Prochazka, Z., Toman, J., Macek, T., and Herout, V. 1985. The occurrence of the amino acid nicotianamine in plants and microorganisms. A reinvestigation. *Biochem. Physiol. Pflanzen 180*: 557-563.

Rudulier, D. Le, Strom, A.R., Dandekar, A.M. Smith, L.T., and Valentine, R.C. 1984. Molecular biology of osmoregulation. *Science 224*: 1064-1068.

Rush, D.W. and Epstein, E. 1976. Genotypic responses to salinity. Differences between salt-sensitive and salt-tolerant genotypes of the tomato. *Plant Physiol. 57*: 162-166.

Salisbury, F.B. and Ross, C.W. 1985. *Plant Physiology*, 3rd edition, Wadsworth, Publishing Co., Belmont, CA., chapter 3.

Scholz, G. 1970. Physiologische Untersuchungen an der Mutante *chloronerva* von *Lycopersicon esculentum* Mill. III. Über die Komplexbildung des 'Normalisierungsfaktors' mit Eisen und Kupfer. *Biochem. Physiol. Pflanzen 161*: 358-367.

Scholz, G. 1983. The amino acid nicotianamine, an effector of root elongation for the tomato mutant *chloronerva (Lycopersicon esculentum* Mill.). *Plant Sci. Lett. 32*: 327-332.

Scholz, G., and Rudolph, A. 1968. A biochemical mutant of *Lycopersicon esculentum* Mill. Isolation and properties of the ninhydrin-positive normalizing factor. *Phytochemistry 7*: 1759-1764.

Scholz, G., Schlesier, G., and Seifert, K. 1985. Effect of nicotianamine on iron uptake by the tomato mutant *chloronerva*. *Physiol. Plant. 63*: 99-104.

Scholz, G., Seifert, K., and Grun, M. 1987. The effect of nicotianamine on the uptake of Mn^{2+}, Zn^{2+}, Cu^{2+}, Rb^{2+} and PO_4^{3-} by the tomato mutant *chloronerva*. *Biochem. Physiol. Pflanzen 182*: 189-194.

Schwartz, D. 1969. An example of gene fixation resulting from selective advantage in suboptimal conditions. *Am. Nat. 103*: 479-481.

Shojima, S., Nishizawa, N-K., Fushiya, S., Nozoe, S., Irifune, T., and Mori, S. 1990. Biosynthesis of phytosiderophores. In vitro biosynthesis of 2'-deoxymugineic acid from L-methionine and nicotianamine. *Plant Physiol. 93*: 1497-1503.

Singh, N.K., LaRosa, P.C., Handa, A.K., Hasegawa, P.M., and Bressan, R.A. 1987. Hormonal regulation of protein synthesis associated with salt tolerance in plant cells. *Proc. Natl. Acad. Sci. USA. 84*: 739-743.

Skriver, K. and Mundy, J. 1990. Gene expression in response to abscisic acid and osmotic stress. *The Plant Cell 2*: 503-512.

Stephan, U.W. and Grun, M. 1989. Physiological disorders of the nicotianamine auxotroph tomato mutant *chloronerva* at different levels of iron nutrition. II. Iron deficiency response and heavy metal metabolism. *Biochem. Physiol. Pflanzen 185*: 189-200.

Stephan, U.W. and Prochazka, Z. 1989. Physiological disorders of the nicotianamine-auxotroph tomato mutant *chloronerva* at different levels of iron nutrition. I. Growth characteristics and physiological abnormalities related to iron and nicotianamine supply. *Acta Bot. Neerl. 38*: 147-153.

Stephan, U.W., Scholz, G., and Rudolph, A. 1990. Distribution of nicotianamine, a presumed symplast iron transporter, in different organs of sunflower and of tomato wild type and its mutant *chloronerva*. *Biochem. Physiol. Pflanzen. 186*: 81-88.

Taylor, I.B. and Tarr, A.R. 1984. Phenotypic interactions between abscisic acid deficient tomato mutants. *Theor. Appl. Genet. 68*: 115-119.

Watad, A-E.A., Reinhold, L., and Lerner, H.R. 1983. Comparison between a stable NaCl-selected *Nicotiana* cell line and the wild type. K⁺, Na⁺, and proline pools as a function of salinity. *Plant Physiol. 73*: 624-629.

2   Mutants in Metabolism

The physiological geneticist working with plants faces a number of major problems when approaching the isolation, characterization, and use of metabolic mutants. First, there is inadequate knowledge of metabolic pathways and their regulation. Without information of this kind, the devising of selection methods and of analyzing mutants is that much more difficult. Second, because of the prevalence of polyploidy throughout the plant kingdom, the production of true haploids containing only one basic chromosome set (monohaploids) is difficult. In the case of an autotetraploid, such as potato or alfalfa, reduction to the gametic chromosome number creates a dihaploid that still possesses two similar sets of chromosomes. Thus, recessive mutations occurring in one gene copy will be concealed by the remaining normal allele. For example, in the case of auxotrophy, a mutation leading to a defective enzyme in a dihaploid will be masked by the presence of an unmutated allele that will continue to give rise to normal enzyme copies. Auxotrophy is manifested only under monohaploid conditions (discussed later).

Faced with these difficulties in investigating the role of genes, the physiological geneticist in the past usually attempted to determine the physiological and biochemical bases of already known dominant hereditary traits. The most serious limitation of this approach is that the investigator must, in general, confine study to nonlethal heritable characters. Such characters are likely to involve nonessential, terminal reactions. A classic example of this approach is the subject of Case Study 2-1, the study of anthocyanin pigments. Because of their visually striking phenotype, mutants altered in pigment biosynthesis can be identified readily at the level of the intact plant and, therefore, were among the earliest plant mutant types to be isolated and characterized. Studies of such mutants inaugurated the biochemical genetic investigation of higher plants. Certain of the photosynthetic (chlorophyll-deficient) mutants are another case in which a highly visible phenotype was used to isolate material of use to the physiological geneticist (see Chapter 3).

Limitations such as those in the case of anthocyanins led Beadle and Tatum (1941) to look for other ways to investigate the problem of the genetic control of developmental and metabolic reactions by reversing the recognized procedure for mutant isolation. Instead of trying to work out the chemical bases of known genetic characters, they set out to determine if and how genes control known biochemical reactions (Beadle and Tatum, 1941). Their procedure was based on the assumption that treatment with a mutagenic agent (in their case, x-rays) would induce mutations in genes concerned with the control of known specific chemical reactions. If the organism must be able to carry out a certain reaction to survive, a mutant unable to do this would obviously be lethal. Such a mutant could be maintained and studied, however, if it would grow on a medium to which had been added the essential product of the genetically blocked reactions.

The method devised by Beadle and Tatum is important to describe in some detail as it has more recently been applied successfully with higher plant material. X-rayed single-spore cultures of *Neurospora sitophila* were established on a *complete* medium (i.e., one containing as many of the normally synthesized constituents of the organism as was practicable). Then, growing colonies of the fungus were tested by transferring them to a *minimal* medium (i.e., one requiring the organism to carry on all the essential syntheses of which the wild type is capable). The complete medium included agar, inorganic salts, malt extract, yeast extract, and glucose. The minimal medium contained agar, inorganic salts, and biotin, the one growth factor that wild-type *N. sitophila* could not synthesize, and sucrose or some other carbon source. Any loss of the ability to synthesize an essential substance was indicated by a strain growing on the complete but not the minimal medium. These *auxotrophic strains* were then tested in a systematic way to determine what substance or substances they were unable to synthesize. Using this method it was possible, for example, by finding a number of mutants unable to carry out a particular step in a given synthesis, to determine whether only one gene was associated with a given chemical reaction. The power of the technique, called the *total isolation method*, was amply demonstrated by its use by Beadle and Tatum in formulating the "one gene, one enzyme" hypothesis, and in the elucidation of many metabolic pathways in microoganisms.

With the advent of tissue culture and, specifically, the availability of technique to culture and grow cells and protoplasts from monoploid plants, it became possible to apply the methods of Beadle and Tatum to higher plant species. By far the greatest experimental advantage of cell culture for the geneticist is that it makes available large numbers of plant cells for screening for variants. The methods of cell culture can be applied to isolate biochemically defined mutants of plants but, because plant regeneration from cell or tissue cultures often cannot be accomplished, and as it is a rather slow process even when it can be done, of the many variant cell lines that have been isolated so far, only small numbers have been shown to be mutants (King, 1986).

Auxotrophs and other conditional mutants have been isolated frequently from among the lower green plants, notably the algae, mosses, and liverworts. Greater difficulty has been encountered in generating such variants among the flowering plants (King, 1986). There is a desire to provide auxotrophs for a number of purposes, such as the selection of somatic hybrids, the determination of genetic linkage relationships, genetic transformation, and the elucidation of metabolic pathways. In the last dozen years, sterile cultures of true haploid (monohaploid) cells from a number of plants have become available. These have allowed the isolation of conditional lethal variants (among which are the auxotrophs) using the techniques devised by Beadle and Tatum for microoganisms (King, 1986). These techniques and the mutants isolated using them are the subject of discussion in Case Study 2-2.

A negative selection method, such as that represented by *total isolation* normally must be used to isolate recessive mutants like auxotrophs. Another method, however, has been used to identify individuals within populations of mutagenized protoplasts, cells, or whole plants that specifically lack the enzyme nitrate reductase (the NR- auxotrophs). This second method involves the use of potassium chlorate as a selection agent. Chlorate is toxic to plants. It is commonly thought to be toxic because of the ability of a plant cell to reduce chlorate to chlorite, which is the actual toxic substance. The reduction of chlorate is brought about by the enzyme nitrate reductase (NR). Thus, any cell not having NR activity or having a substantially reduced NR activity will be resistant to chlorate. Growing mutagenized plant material in the presence of sufficiently high concentrations of chlorate then allows the selection of NR- deficient lines. Case Study 2-3 presents an account of the identification, properties and use of some NR- variants isolated using the chlorate method.

The nature of the desired mutants will dictate the ploidy of the material to be used in a mutant isolation experiment. As described previously, for the purposes of producing recessive mutants, such as auxotrophs, true haploid cells are preferred. When it is anticipated that the desired mutant is dominant or semidominant, however, mutants may be recovered equally efficiently from diploid or haploid cells. In Case Study 2-4, examples are described of resistance to amino acids and their analogues in diploid cell cultures where enzymes are produced that are insensitive to the toxic agents or where some other modification of metabolism protects the cell against the toxic effect of the selection agent. Variants of this class are useful in the study of pathways of amino acid biosynthesis and their control.

Many novel plant mutants have been isolated by the application of selective growth conditions to cultured cells, as in the case of amino acids. However, few of the variant phenotypes that have been reported thus far are of potential agronomic value. This limited practical success of mutant selection in vitro results to a large degree from the genetic and developmental complexity of agronomically important characteristics. Many such traits are exclusively whole plant functions that are not expressed and, therefore, cannot be selected at the cellular level.

One trait of agronomic interest that may be expressed by cultured cells is sensitivity to certain herbicides. Herbicides that interfere with basic metabolic activities can be expected to inhibit growth of cultured cells as well as the whole plant. In such cases, herbicide-tolerant mutants can be selected by culturing cells in the presence of a herbicide concentration that is toxic to normal cells. Examples of mutants of this class are described in the first part of Case Study 2-5. In the second part, another group of mutants is discussed that could only have been isolated from populations of tissues where photosynthesis was in operation as the herbicides in question (the s-triazines) act through inhibition of photosynthetic processes.

Case Study 2-1: Pigment (anthocyanin)-deficient Mutants

The anthocyanins are the colored pigments that commonly occur in red, purple, and blue flowers. One of their useful functions in flowers appears to be the attraction of insects for pollination. Certainly their abundance suggests some function favorable to their evolution. In addition to the innumerable investigations of these pigments in flowers, the genetic control of anthocyanin pigmentation in the aleurone layer of the maize kernel has long been regarded as a textbook example of genic interaction. The conspicuous and nonvital nature of anthocyanin pigments and the suitability of the endosperm as an experimental system are the main factors that have caused the genetics of anthocyanin formation in the aleurone to be more fully studied than that of any other constituent of maize.

Anthocyanins belong to the more general class of phenolic compounds, known as flavonoids, that have a common basic structural unit, the C-15 skeleton of a flavone (Fig. 2-1). The primary metabolites from which flavonoids are derived are the phenylpropanoid aromatic amino acids, phenylalanine and tyrosine. The first three steps in the conversion of phenylalanine to cinnamic acid derivatives are common to all phenylpropanoid pathways (Fig. 2-2). The first enzyme in this pathway, phenylalanine ammonia lyase (PAL), catalyzes the deamination of phenylalanine to cinnamic acid and is considered to be a key enzyme because it channels aromatic amino acid metabolism in the direction of lignin and flavonoid biosynthesis. The next two enzymes are cinnamic acid hydroxylase and p-coumaroyl coenzyme A (CoA) ligase (Fig. 2-2). Chalcone synthase (CHS) is the first enzyme unique to flavonoid

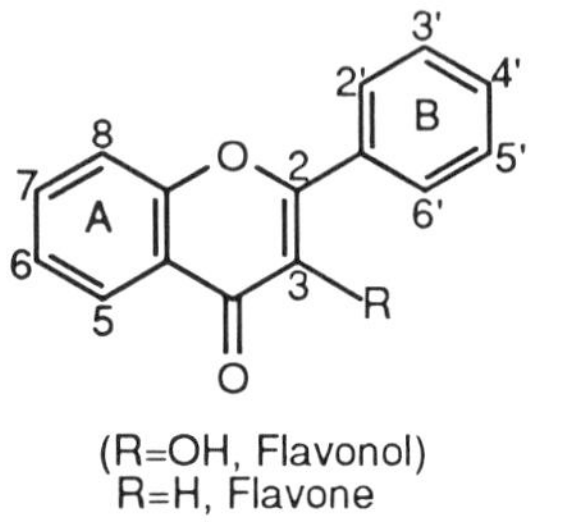

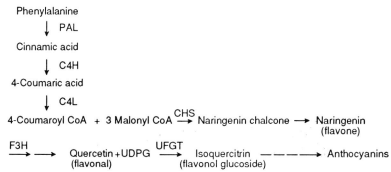

(R=OH, Flavonol)
R=H, Flavone

Figure 2-1. General structure of a flavone. R=OH, flavonol; R=H, flavone. Hydroxyl groups may be found at almost any position but more commonly at the 3, 5, 7, 3' and 4' carbons.

biosynthesis. It catalyzes the synthesis of the C-15 flavonoid skeleton from the activated precursors *p*-coumaroyl CoA and malonyl CoA. The anthocyanidins are, then, one of the groups of flavonoid pigments derived from this basic C-15 unit (Martin et al., 1987; Mol et al., 1989)

At least 17 different loci across the maize genome are known to be involved in the synthesis of the anthocyanins in various tissues. Variations exist as to the kinds of compounds, their concentrations, tissue specificities, and patterns within a tissue. Some are structural genes such as the *c2, a1, a2, bz1*, and *pr* loci; others are involved in the regulation of color development in various parts of the plant such as the *c, r* and *vp* loci. Clearly, it is not possible to discuss all of these factors or the vast amount of literature that has been gathered about them. Included here are a few examples of the use of maize mutants to work out what are the products of some of the structural genes responsible for anthocyanin biosynthesis. After is a discussion of some of the ways in which the occurrence of these brightly colored pigments has been

Phenylalanine

↓ PAL

Cinnamic acid

↓ C4H

4-Coumaric acid

↓ C4L

4-Coumaroyl CoA + 3 Malonyl CoA —CHS→ Naringenin chalcone → Naringenin
                                                                    (flavone)

—F3H→ —→ Quercetin+UDPG —UFGT→ Isoquercitrin — — — → Anthocyanins
            (flavonal)   (flavonol glucoside)

Figure 2-2. Some early steps in the biosynthesis of flavonoids and anthocyanins in maize. PAL, phenylalanine ammonia lyase; C4H, cinnamate 4-hydroxylase; 4CL, 4-coumarate: CoA ligase; CHS, chalcone synthetase; F3H, (NADPH): flavanone 3'-*O*-hydroxylase; UFGT, UDPG: 3-*O*-flavonoid glucosyl transferase.

exploited to uncover important genetic and physiological principles, such as the presence and role of transposons in plants.

2-1.1. Characteristics of Maize Mutants Defective in Anthocyanin Biosynthesis

2-1.1.1 *Chalcone Synthase*

Chalcone synthase is the first enzyme unique to flavonoid biosynthesis; it catalyzes the synthesis of the C-15 skeleton (Fig. 2-2). The activity of the enzyme could be demonstrated in extracts of corn freed from contamination by hydroxylapatite chromatography (Fig. 2-3; Dooner, 1983). Several genes were found to influence CHS activity (see a later discussion in this Case Study) but the only gene that did so without affecting the expression of any other appeared to be *c2* (Table 2-1). The data suggested that *C2* is the structural gene for CHS in part because of the very clear dose-dependent effect of the gene illustrated in Table 2-1. As we shall see later in this Case Study, other evidence gathered by Dooner (1983) indicated that the expression of this and other structural genes, which specify the synthesis of the anthocyanins, are modified by a number of regulatory genes.

2-1.1.2 *Flavonoid 3'-Hydroxylation*

After the action of CHS in the synthesis of the basic C-15 skeleton of the flavonoids, the hydroxylation at the three position on the B ring

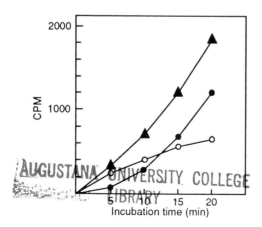

Figure 2-3. Time-course of chalcone (○) and flavonone (●) formation, and the sum of both (▲), in the assay of chalcone synthase purified through hydroxylapatite chromatography. (Dooner, *Mol. Gen. Genet. 189*: 136-141, 1983.)

Table 2-1. Dosage effect of *c2* on chalcone synthase activity in developing corn endosperms.

Genotype	%+
++ / +	100
++ / c2	63
c2c2 / +	10
c2c2 / c2	0

(Dooner, *Mol. Gen. Genet. 189*: 136-141, 1983.)

of flavonoid compounds is a key reaction in the biosynthesis of anthocyanins in maize (Fig. 2-2). The activity has long been attributed to the action of the *Pr* gene in maize (see Larson et al., 1986). The isolation and characterization of a microsomal enzyme, flavanone 3'-hydroxylase, made it possible to investigate the activity of the *Pr* gene at the enzymatic reaction level. And indeed, gene dosage to enzyme ratios (Table 2-2) and other data suggested that *Pr* is the structural gene for reduced nicotinamide adenine dinucleotide phosphate (NADPH):flavanone 3'-*O*-hydroxylase (Larson et al., 1986).

Thus, aleurone tissue homozygous for *pr* had no identifiable hydroxylase activity, apparently confirming a structural gene role for *Pr*. What could not be explained by these data was the occurrence of cyanidin, a later product of the pathway, in this same aleurone tissue, and hydroxylase activity in leaf sheaths and seedlings of plants homozygous for *pr* (Larson et al., 1986). Although there seems no doubt that *Pr* is the structural gene for the hydroxylase, the data indicate that other factors (genes?) influence the activity of *Pr*. These matters, however, can only be resolved through the further use of specific genotypes at all stages of growth. Clearly, there are controls of the

Table 2-2. Flavanone 3'-hydroxylase activity vs. genotypes in maize seedlings and mature maize aleurone. Terms in parentheses are for triploid aleurone only.

Genotype	Specific Activity (units/mg protein)	
	Seedlings	Aleurone
Pr Pr (Pr)	205.5±16.94	13.45±1.27
Pr (Pr) Pr	68.4±4.77	5.55±0.67
(Pr) Pr Pr	10.5±2.65	3.95±0.57
pr pr pr		0

(Larson et al., *Biochem. Genet. 24*: 615-624, 1986; reproduced by permission of Plenum Publishing Corp.)

expression of *Pr* and other structural genes in the pathway of anthocyanin biosynthesis about which nothing is known at present.

2-1.1.3 Uridine Diphosphoglucose (UDPG): Glucosyltransferase (UFGT)

The effect of the *bronze (bz)* mutation in maize is to cause the husk and sheath tissues homozygous for it to undergo a browning reaction and become necrotic. The data obtained by assaying the transferase in different tissues support a direct gene-transferase relationship in the case of *Bz* (Table 2-3; Larson and Coe, 1977). The direct influence of *Bz* on the transferase activity is evident in the easily identified gene dosage effect in these data (Table 2-3). For example, in pollen the ratio of *Bz* from dominant to recessive is 2:1:0, and the ratio of specific activities for

Table 2-3. UFGT activity: Genotype and maize tissue relationships.
[a] Expressed as μg isoquercitrin formed/hr/mg protein.
[b] Heterozygotes derived from reciprocal crosses between homozygous strains; female parent listed first.
[c] Endosperm and embryo samples derived from the same seed.
[d] Heterozygotes derived from reciprocal crosses between homozygous strains; female origin listed in first two symbols.

Tissue and genotype	Specific activity [a]
Pollen source	
BzBz	210
Bzbz	102
bzbz	0
Seedling [b]	
BzBz	308
Bzbz	157
bzBz	151
bzbz	0
Embryo [bc]	
BzBz	117
Bzbz	57
bzBz	50
bzbz	0
Endosperm [cd]	
BzBzBz	2850
BzBzbz	921
bzbzBz	484
bzbzbz	0

(Larson and Coe, *Biochem. Genet. 15*: 153-156, 1977; reproduced by permission of Plenum Publishing Corp.)

the transferase is also 2:1:0. Thus, the suggestion is a very strong one that *Bz* codes for the transferase (Fig. 2-2) in maize.

2-1.1.4   *Dihydroflavonol-4-reductase*

The *A1* locus of maize has also been investigated with the use of anthocyanin mutants. This locus is of particular interest because its expression is required for pigment production in almost all plant tissues provided all other anthocyanin genes are in a dominant condition. Data indicate that the *A1* locus is the structural gene for the enzyme dihydroflavonol-4-reductase (DFR) (Reddy et al., 1987).

As can be seen from these few examples, substantial progress has been made in recent years in the gene-enzyme correlation of anthocyanin (flavonoid) biosynthesis. However, the steps in the pathway leading to anthocyanin formation had already largely been worked out by purely biochemical methods (Hahlbrock and Scheel, 1989). Thus, genetic analyses have tended merely to confirm what had already been learned in other ways. Where maize pigment mutants have been of unique value is in the use of precisely characterized genetic lesions in the study of genetic regulation as illustrated in the following examples.

2-1.2.   The Use of Anthocyanin Mutants in Physiological Studies

2-1.2.1   *The Control of Gene Expression in Maize Tissue*

It is known that dominant alleles, such as *A, A2, C, C2, R, Bz* and *Bz2*, are required for anthocyanin formation in the aleurone layer of the maize seed (see Dooner and Nelson, 1977). The phenotypes conditioned by recessive mutations at each one of these loci are given in Table 2-4. As can be seen from these data normal activity levels of the enzyme UFGT were found in endosperms containing the recessive mutations *c2, a, a2*, and *bz2* suggesting that none of these loci has an influence on the formation of UFGT. No activity of the enzyme was detected in *bz*, however, confirming that this locus is the structural gene for UFGT (see Section 2-1.1.3). In addition, only very low levels of activity of the enzyme were found in *c* and *r* mutants. Nothing is known about the direct products of *C* and *R*, but they are probably regulatory genes, a conclusion reinforced by later work involving still another gene, *Vp*, that reduced UFGT activity in mature endosperms to the same extent as did mutations in *C* and *R*. Some of the data illustrating the effect of these three loci (*C, R*, and *Vp*) on the formation of UFGT during endosperm development are shown in Figure 2-4 (Dooner and Nelson, 1979a).

The data in Figure 2-4 show, first, that all genotypes tested, except

Table 2-4. UFGT activity in mature seed endosperms of normal (all dominant pigment genes present) and various pigment mutants in a particular maize (W22) inbred background.

[a.] Except for the recessive mutant listed, all dominant anthocyanin factors are present.

[b.] μmoles isoquercitrin/hour/mg protein.

Genotype[a]	Aleurone phenotype	Specific activity[b]	Percent normal
Normal	purple	0.633	100.0
c	colorless	0.018	2.8
c2	colorless	0.630	100.0
a	colorless	0.727	115.0
a2	colorless	0.698	112.0
r	colorless	0.015	2.4
bz	bronze	0.00	0.0
bz2	bronze	0.635	100.0

(Dooner and Nelson, *Biochem. Genet. 15*: 509-519, 1977; reproduced by permission of Plenum Publishing Corp.)

bz, had roughly comparable levels of UFGT activity very early in development (21 days after pollination); second, that enzyme activity rose continuously during development; third, that enzyme activity in *c*, *r*, and *vp* endosperms remained at a basically low level through the fifth week of development and then declined slightly; fourth, that the *bz* endosperms lacked detectable UFGT activity at all stages of development.

One interpretation of these data (Fig. 2-4) is that early in development, normal (+), *vp*, *c*, and *r* genotypes have roughly the same residual enzyme level, whereas *bz* endosperms lack activity. Later, *c*, *r*, and *vp* endosperms lack a developmental signal that, in normal endosperms, triggers the *Bz* locus to produce increasingly higher levels of UFGT beginning at the fourth week of development and continuing to maturity. Thus, early in development, a low level of UFGT, independent of *C*, *R*, and *Vp* control and solely dependent on the presence of *Bz*, represents an uninduced condition of the *Bz* locus. Later in development, in response to signals generated by the action of *C*, *R*, and *Vp*, *Bz* is induced to produce higher levels of UFGT. The enzyme then accumulates in the developing normal endosperms, reaching a maximum at maturity (Dooner and Nelson, 1979a).

The nature of the products of these regulatory genes is not known but through such biochemical genetic analyses, progress can be made in determining which regulatory genes modify the expression of which structural genes. This task is entirely dependent on the provision of suitable, defined mutants and is important to an understanding of the

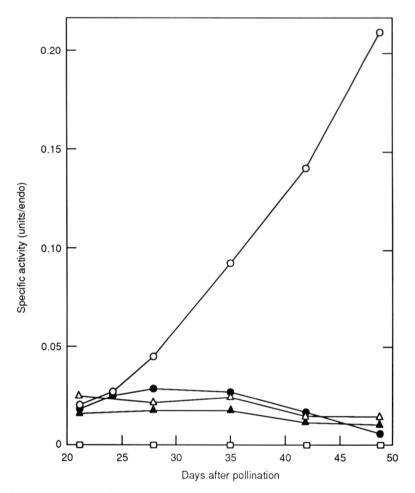

Figure 2-4. UFGT activity during endosperm development in the + (○) and *vp* (●) types of maize, and in *c* (△), *r* (▲) and *bz* (□) homozygotes. (Dooner and Nelson, *Genetics 91*: 309-315, 1979.)

control of the expression of the plant genome at the physiological level during development.

2-1.2.2   *The Coordinate Regulation of Flavonoid Biosynthetic Enzymes*

UFGT and several other enzymes responsible for the formation of flavonoids are known to be coordinately induced in parsley cell suspension cultures (see Hahlbrock and Scheel, 1989). The argument has been made that the genetic information for this group of enzymes is transcribed simultaneously as well. In view of the evidence for the coordinate regulation of flavonoid biosynthesis enzymes in parsley, it became of interest to determine whether in corn, the genes involved in the regulation of UFGT also affected other enzymes in the flavonoid pathway; that is, whether regulatory genes could be identified that coordinately controlled a set of functionally related structural genes (Dooner, 1983).

As we have seen earlier, CHS is the first enzyme unique to flavonoid biosynthesis (Fig. 2-2; Section 2-1.1.1). Mutations at the same loci as those discussed earlier in regard to UFGT also affect CHS simultaneously. Thus, a functional allele at each of the loci *C, R*, and *Vp* (as well as at the *Clf* locus, not discussed here) is required for the coordinate expression of CHS, UFGT and, in all likelihood, other flavonoid biosynthetic enzymes (Table 2-5). These results lend credence to the hypothesis that the structural genes in question are coordinately controlled by products of several regulatory genes including *C, R*, and *Vp* (Dooner, 1983).

2-1.2.3 The Physiological Role of Transposable Elements

Knowledge of the structural gene products in anthocyanin biosynthesis, such as those illustrated so far in this Case Study, can be used to

Table 2-5. Dosage effect of C, C2, R, and Vp on chalcone synthase activity in developing endosperms of maize.
[a.] 1 unit = 1 pmol malonyl CoA converted to products per minute.

	Units[a]/endo	%+
+ +/+	2.64	100
+ +/c2	1.67	63
c2 c2 /+	0.26	10
c2 c2/c2	0	0
+ +/-	2.70	102
r r/+	1.58	60
r r/r	0	0
+ +/+	3.47	100
vp vp /+	3.04	88
vp vp/vp	0	0
+ +/+	6.92	100
+ +/c	4.90	78
c c/+	2.63	38
c c/c	0	0

(Dooner, *Mol. Gen. Genet. 189*: 136-141, 1983.)

investigate the nature and function of transposable elements that are thought to play a modifying role in many physiological processes (e.g., McClintock, 1984). The relatively detailed information available from many years of investigation of the genetics of these very visible pigments has provided a background against which the complexities of the interactions between transposable elements and other genes can be probed. What has been learned so far should make physiological geneticists pause when they attempt to interpret data about gene activity in cases where little or nothing is known about the involvement of these elements.

It may be useful, however, to digress and give a brief account of some elementary facts about transposable elements (see Fedoroff, 1984, for an excellent, simple explanation of what these elements are and how they are thought to act): A gene can be subjected to either a stable or an unstable mutation. For example, the *Bz* gene function, already discussed in this Case Study, allows the production of purple anthocyanins in the kernels of maize (of course, assuming that all other structural genes in the flavonoid pathway are present). As we have seen, however, if a mutation occurs in the gene (yielding *bz*), then the enzyme UFGT is not functional and the kernel containing the mutation is colorless. When genes come under the control of transposable elements, on the other hand, variegated or mosaic phenotypes consisting of normal-appearing sectors in an otherwise mutant (colorless) kernel are produced. Thus, for instance, in the case of anthocyanin pigmentation, the colorless *bz* mutant kernels can become unstable in the presence of transposable elements and display coarse or fine pigmented spots in an otherwise colorless background (see Fedoroff, 1984, for examples).

What are these transposable elements? In structure, they can be quite simple, consisting of a piece of deoxyribonucleic acid (DNA) that can move from place to place in the genome of an organism, promoting transposition. The movement of these elements can generate mutations or chromosomal rearrangements and thus affect the expression of other genes. Apparently, chromosomes are littered with these mobile elements, or so-called jumping genes, and it is suspected that their ability to modify the expression and structure of an entire genome is an important mechanism in the long-term genetic changes central to development and to evolution (McClintock, 1984; Fedoroff, 1989).

To illustrate how transposable elements work, one can take an example investigated by McClintock, who was the first to understand that a genetic element can transpose. The locus manifested itself as a specific site of chromosome breakage, or dissociation, therefore, McClintock called it the dissociation (*Ds*) locus. Although *Ds* is the site of breakage, it is not itself responsible for breakage, which occurs only if another locus is present. This second locus became known as activator (*Ac*) for its ability to activate breakage at the *Ds* locus. *Ac* acted like a conventional locus most of the time, but occasionally it disappeared or moved to a new position on either the same or a different chromosome.

The *Ds* locus could also transpose, McClintock discovered, but just as it was incapable by itself of breaking the chromosome on which it resides, so, too, it was unable to transpose in the absence of *Ac* (Fedoroff, 1984). In short, *Ac* can move by itself but *Ds* can move only when it is activated by *Ac*. The relationships between *Ds* and *Ac* are illustrated in Figure 2-5.

The activity of *Ac-Ds* can be demonstrated by reference to its action in the case of the locus *Bz* in the anthocyanin biosynthetic pathway (Fedoroff, 1984). A mutation occurs when the *Bz* locus (Fig. 2-6a) is invaded by *Ds* (Fig. 2-6b). The mutation modifies the gene, the pigment is not made, and the aleurone tissue is colorless, not purple. If *Ac* is present in the genome, however, it promotes the transposition of *Ds* away from the locus in some cells during kernel development (Fig. 2-6c). The mutation reverts when *Ds* leaves giving rise to cells in which the *Bz* locus is functional. Each such cell gives rise, in turn, to a pigmented sector in the aleurone (Fig. 2-6c).

The types of *Ds*-induced modifications are also varied. Often, gene action is totally inhibited, and the resulting phenotype is that of the null,

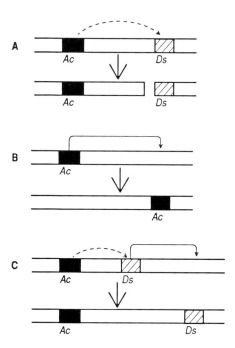

Figure 2-5. The *Ac-Ds* transposable element system. Activator (*Ac*) can activate chromosome breakage at the locus, dissociation (*Ds*) (**A**). The two loci can be on the same or on different chromosomes. *Ac* can promote its own transposition (**B**), or that of *Ds* (**C**) to another position either on the same chromosome or a different one. *Ds* cannot move unless *Ac* is in the same cell. (From *Transposable genetic elements in maize*, N.V. Fedoroff. Copyright © 1984 by Scientific American, Inc. All rights reserved.)

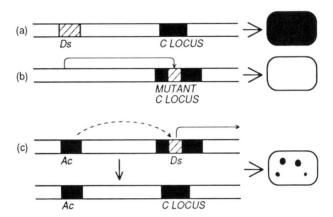

Figure 2-6. A maize kernel *Bz* locus showing the effect of mutation and transposition in the presence of the *Ds-Ac* System. (See the text for explanation.) (From *Transposable genetic elements in maize*, N.V. Fedoroff. Copyright © 1984 by Scientific American, Inc. All rights reserved.)

recessive, mutant allele (the colorless *bz* kernels in Fig. 2-6b). However, the association of *Ds* with a gene need not result in total suppression of gene action; genes may retain partial function after coming under the control of *Ds*, that is, a gene may be variously "leaky" as we see in the case of the *bz-m4* mutant of maize, and derivatives from it.

The mutable allele *bz-m4* arose by transposition of *Ds* to the *Bz* locus. An examination of UFGT in *bz-m4* and *Bz* endosperms throughout development revealed that in early developmental stages enzyme levels were much higher in *bz-m4* than in *Bz* (Fig. 2-7) (Dooner, 1980). For example, at 18 days after pollination, UFGT activity in *bz-m4* was six times higher than normal. Activity in the mutant peaked at about 22 days after pollination and then declined steadily until at maturity (the time of maximum UFGT activity in normal endosperms) it amounted to less than 1 percent of normal. In short, an apparently normal enzyme was made in *bz-m4* endosperms, but too early in development.

A particular puzzle in these studies was that the amount of anthocyanin present in *bz-m4* aleurones at early developmental stages was only a trace of that found in *Bz* aleurones at corresponding stages, although UFGT levels were much higher in the mutant. A possible explanation was that the tissue distribution of the enzyme was altered in the mutant endosperms. Therefore, the localization of UFGT in *Bz* and *bz-m4* immature endosperms was compared (Table 2-6) (Dooner, 1980).

Most UFGT activity in *bz-m4* was localized in the subaleurone endosperm. In contrast, the activity in *Bz* was equally distributed in aleurone and subaleurone tissues, whereas, at a later stage, the aleurone was deeply pigmented in *Bz* and contained high UFGT activity (Table 2-6). Thus, in this case, the association of *Ds* with the *Bz* locus appeared

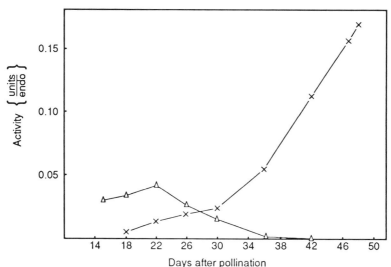

Figure 2-7. UFGT activity during endosperm development in *Bz* (x) and in *bz-m4* (△). (Dooner, *Cold Spring Harbor Symp. Quant. Biol. 45*: 457-462, 1980.)

to have caused an alteration in the mechanism that normally controlled the tissue-specific expression of the *Bz* locus in the endosperm.

Further information about the role and interaction of genes with transposable elements was gained from analyses of the reversion products in the *Bz* locus. Mutable *bz* alleles produced, on reversion, more or less purple kernels. These derivatives were designated *Bz'* to identify their

Table 2-6. Localization of UFGT in *Bz* and *bz-m4* maize endosperms. [a] Milliunits of enzyme per subaleurone endosperm or per aleurone.

Genotype	Tissue			
	subaleurone activity[a]	aleurone activity[a]	Subaleurone / aleurone activity	*bz-m4*/*Bz* total activity
Bz (21 DAP)	12	14	0.86	3.0
Bz-m4 (21 DAP)	72	5	14.4	
Bz (42 DAP)	37	237	0.16	0.07
bz-m4 (42 DAP)	15	3	5.0	

(Dooner, *Cold Spring Harbor Symp. Quant. Biol. 45*: 1980.)

origin as selections from a mutable gene system (Dooner, 1980). Seven derivatives of *bz-m4* with higher levels of expression were established in this study (Table 2-7). Five of the derivatives (*Bz'-1* to *Bz'-5*) gave fully purple kernels; the other two (*Bz'-6* and *Bz'-7*) were pale purple. In the presence of *Ac*, these last two gave a pale-to-dark purple variegated phenotype, indicating that *Ds* was still at the *Bz* locus. Of the five fully purple derivatives, two were found to change infrequently to unstable types in response to *Ac*, thus disclosing the existence of *Ds* in the vicinity of *Bz*.

Bz' derivatives 1 to 5 could be interpreted as arising by transposition of *Ds* away from the *Bz* locus, which restored the gene to its wild-type condition. Derivatives 6 and 7, it was surmised, arose by rearrangements of the genetic material in *bz-m4* that led to the expression of the new phenotypes. All derivatives, however, were regarded as similar in one respect, that is, they made an enzyme with properties similar to those of the normal UFGT found in *Bz* endosperms (Dooner, 1980).

Other observations made with *Bz'* derivatives from another mutable *bz* allele, *bz-m2/(D1)*, provided evidence of further modifications of gene activity caused by the interaction of the *Ac-Ds* system with the gene. The allele in question also had *Ds* at the *Bz* locus; in the absence of *Ac*, it lacked UFGT activity throughout endosperm development. Although phenotypically similar to one another, these *Bz'* derivatives fell into two groups with respect to UFGT properties in mature endosperms; those with high levels of an essentially normal enzyme (group 1) and those with low to intermediate levels of an altered, thermolabile, and electrophoretically unstable enzyme (group 2) (Table 2-8) (Dooner and Nelson, 1979b).

Table 2-7. Characterization of mature endosperm UFGT in seven *Bz'* derivatives from *bz-m4 + Ac*.
[a.] Enzyme milliunits per endosperm.

		Activity	
Allele	Kernel phenotype	specific [a]	*Bz* (%)
Bz-McC	purple	216	100
Bz'- 1	purple	199	92
Bz'- 2	purple	225	104
Bz'- 3	purple	196	91
Bz'- 4	purple	197	91
Bz'- 5	purple	250	116
Bz'- 6	pale	9	4
Bz'- 7	pale	0	5

(Dooner, *Cold Spring Harbor Symp. Quant. Biol. 45:* 457-462, 1980.)

Table 2-8. Characterization of mature endosperm UFGT in *Bz* derivatives from *bz / m2(D1)* + *Ac*.

Allele	% *Bz* activity	Group
Bz:McC	100	
Bz' I-1	2	II
I-2	99	I
I-3	2	II
I-4	0.4	II
I-5	0	II
I-6	111	I
I-7	92	I
I-8	0.6	II
I-9	113	I
I-10	106	I
I-11	2	II
I-12	3	II
I-13	5	II
I-14	3	II
I-15	21	II

(Dooner and Nelson, *Proc. Natl. Acad. Sci. USA 76*: 2369-2371, 1979b.)

The interpretation of these results by Dooner and Nelson (1979b) was that the organization of the gene before its association with *Ds* was restored only in some cases and that group 2 *Bz* derivatives might have arisen by the imprecise excision of *Ds* from the *Bz* locus. According to this hypothesis, therefore, controlling elements like *Ds* appeared to have the potential to generate new gene forms and may have been implicated in the origination of present-day genetic diversity. The more recent identification of two alleles of the single-copy CHS gene in parsley, which differ by a transposon like element (Herrmann et al., 1988), provides a specific example that can be used to investigate whether more or less subtle differences in the regulation of the two alleles occurs during development and, in turn, what might be the potential importance of the presence of such elements or portions of such elements in general.

In summary, the mobility of these activated elements allows them to enter different gene loci and to take over control of action of a gene wherever one may enter. In addition to modifying gene action, as seen in the examples discussed earlier, these elements can restructure the genome at various levels, from small changes involving a few nucleotides, to gross modifications involving large segments of chromosomes, such as duplications, deficiencies, inversions, and other more complex reorganizations. Such a broad spectrum of responses might give rise to dramatic physiological events and provide an explanation for as yet unexplained plant responses (McClintock, 1984). Therefore, it is of considerable importance to physiologists that more, and more detailed, investigations be performed with systems known to involve transposable elements. To this end, the recent increasing use of transposons to clone specific genes is of particular interest (see Wienand et al., 1986). Such an advance will allow the finer study of the tissue-specific expression of genes, of gene interactions. and of the regulation of structural genes by other genes (Kuhn and Klein, 1987). Physiologists would also be wise to consider that the variations they see, for example, in enzyme expression at different stages of plant development or under different environmental conditions, may be influenced by these elements in ways that, at present, are being attributed to other causes. Interactions at the genetic and physiological levels require further careful study so that the full extent of the control of cellular events can be understood.

Case Study 2-2:   The Auxotrophic Mutants

As already described in the introduction to this Chapter, auxotrophs are considered to be genetically altered organisms or cells having at least one more nutritional requirement than those from which they were derived. They often differ from parental wild types by lacking, or having a defect in, a single step of the de novo pathway leading to an essential metabolite. Usually, the defect occurs because of a point mutation in the genes associated with the formation of an enzyme.

The isolation of plant auxotrophs has proved to be difficult and still cannot be achieved routinely. In the case of microorganisms, replica plating (see Beadle and Tatum, 1941) was a method that could be used to screen large numbers of colonies for mutants. With plants, the rapid screening of large numbers of individual plants or their cells is far from routine. An early technique was to use the visual phenomenon of chlorophyll deficiency as a screen for mutation and then the further testing of these variants for auxotrophy. Large numbers of whole-plant auxotrophs were isolated using this approach (see King, 1986), but nearly all of them had a requirement for the vitamin thiamine (discussed later). Other similar visual techniques have been used to uncover a very few auxotrophic variants at the whole plant level, one of which, the proline-requiring mutants of corn recognized by their collapsed endosperm in the seed, is described. One auxotroph in tomato, the *chloronerva* mutant, which arose spontaneously, is the subject of Case Study 1-4.

The greatest degree of success in isolating conditional lethal variants in general, and auxotrophs in particular, from green plants has come with the application of the techniques pioneered by Beadle and Tatum with microorganisms (see introduction to Chapter 2) to true haploid cells or protoplasts derived from tissues of the plant and grown in sterile culture conditions. An example of this approach is described here in which variants with recessive biochemical lesions in a number of essential metabolic pathways were isolated. An additional specific technique involving the use of chlorate resistance to isolate variants lacking the enzyme nitrate reductase, is described separately in Case Study 2-3.

2-2.1 Isolation and Characterization of Some Whole Plant Auxotrophs

2-2.1.1 *The Thiamine Mutants of Tomato and Arabidopsis*

In 1961, Langridge and Brock described the first mutant in flowering plants shown to be lethal in the absence of a specific organic compound. The mutant was spontaneous, occurring in two tomato lines that were being bred for disease resistance. The segregation of the mutation in selfed progeny was completely recessive to the wild type allele and resulted from a single gene change. Plants homozygous for the mutation survived for several months under good growing conditions but they grew extremely slowly. Cotyledons were normal presumably because of the diffusion of metabolites into the seed from the heterozygous maternal parent. The few leaves formed were small and, at first, pale yellow in color. Later, the leaves often became light green but soon lost their chlorophyll, turned white, withered, and died. Plants were only 3 to 6 inches in height after growing for 4 months, were very chlorotic but were still alive. The only remedy found to repair this condition was to spray repeatedly onto the leaves, a mixture containing the vitamin thiamine (2 mg/100 mL of water, three times per week) (Langridge and Brock, 1961).

Thiamine is derived by the condensation of pyrimidine and thiazole precursors, specifically the pyrophosphate ester of 2-methyl-4-amino-5-hydroxymethylpyrimidine (OMP-PPi) and the monophosphate ester of 4-methyl-5-hydroxyethylthiazole (thiazole-P) to yield thiamine monophosphate (TMP) that, in turn, gives rise to thiamine and thiamine pyrophosphate (Fig. 2-8). The mutant showed no response to spraying with thiazole, thus demonstrating that the genetic lesion appeared to be

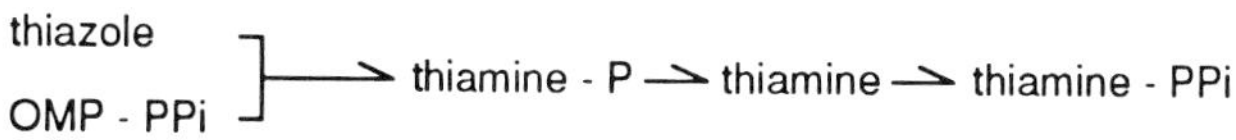

Figure 2-8. Biosynthesis of thiamine pyrophosphate.

in the pyrimidine arm of the pathway, but the precise lesion was not identified.

Very recently, interest in thiamine auxotrophs of the crucifer *Arabidopsis thaliana* has increased along with the realization that this species is an ideal experimental material in plant molecular biology. Thiamine-requiring mutants may be useful, for example, not only for the traditional study of the genetic control of metabolic pathways in plants, as other auxotrophs among the microoganisms have been, but also as markers in gene transfer experiments, as the starting gene in "gene walking," and as a model for the breeding of artificial plants that produce excess amounts of thiamine (Komeda et al., 1988).

Among the thiamine auxotrophs of *Arabidopsis* isolated so far is *th-1*, which requires thiamine itself to grow (Komeda et al., 1988). The formation of thiamine requires, first, the production of TMP that is manufactured from pyrimidine and thiazole precursors by the enzyme TMP pyrophosphorylase and then, the dephosphorylation of TMP to thiamine, a reaction catalyzed by the enzyme TMP phosphatase (Fig. 2-8). The *th-1* mutant would appear to lack the first of these two enzymes. Enzymic activities were examined after separation of proteins

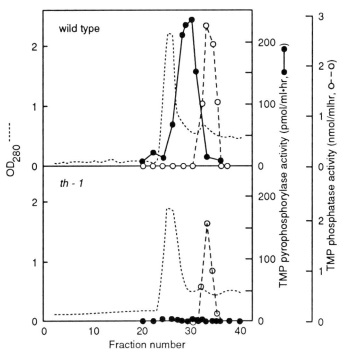

Figure 2-9. Gel filtration profiles of extracts from wild type and a *th-1* mutant of *A. thaliana*. Alternate fractions were examined for the activities of thiamine phosphate pyrophosphorylase (●) and of thiamine phosphate monophosphatase (○). (Komeda et al., *Plant Physiol.* 88: 248-250, 1988; reproduced by permission of the ASPP.)

by gel filtration chromatography. The pattern of activity clearly showed that the two enzymes could be separated from one another when extracts from wild-type tissue were analyzed (Fig. 2-9). The *th-1* mutant, on the other hand, was deficient in the peak corresponding to the activity required for the production of TMP, whereas TMP phosphatase activity was found in the mutant as well (Komeda et al., 1988). These results conformed to the suggested pathway of thiamine pyrophosphate production in many organisms and provided one clear example to illustrate the value of auxotrophic mutants in defining metabolic pathways.

2-2.1.2   *Proline-requiring Mutants of Corn*

In contrast to the clear results obtained in the case of thiamine mutants of *A. thaliana* are the proline-requiring mutants of corn that remain a puzzle even though much time has been devoted to defining the lesion causing their auxotrophic requirement.

The proline-requiring (*pro-1*) mutants in *Zea mays* have been recognized by their collapsed endosperm morphology and lethality arising at the seedling stage of growth (Gavazzi et al., 1975). The mutants arose spontaneously in a genetic stock with the W22 maize inbred background. The growth requirements of the mutants were determined from in vitro cultures of detached primary roots or excised embryos (Fig. 2-10).

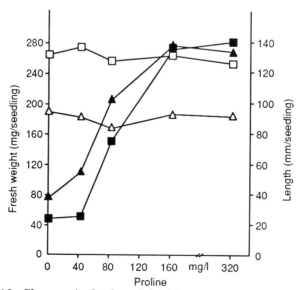

Figure 2-10. Changes in fresh weight (▲ mutant, △ normal) and length (■ mutant, □ normal) of seedlings with increasing concentrations of proline in the medium. (Gavazzi et al., *Theor. Appl. Genet. 46*: 339-345, 1975.)

The length of normal seedlings was not affected by the addition of proline, whereas that of the mutants showed an initial lag followed by a linear increase, reaching a maximum at a dose of 160 mg/L of proline (Fig. 2-10). The weight of normal seedlings also remained constant with increasing doses of proline, while that of the mutants showed a linear increase in growth up to the dose of 160 mg/L of proline, followed by a stationary phase (Fig. 2-10). Thus, this became the first case of the absolute genetic requirement of an amino acid in a flowering plant (Gavazzi et al., 1975). The seven mutants isolated were all found to be recessive and to be characterized by four phenotypic characters, namely, the dull appearance of the kernel, the collapsed endosperm morphology, the striation of the first two leaves, and the stunted seedling growth followed by lethality at the one-two-leaf stage. Complementation did not

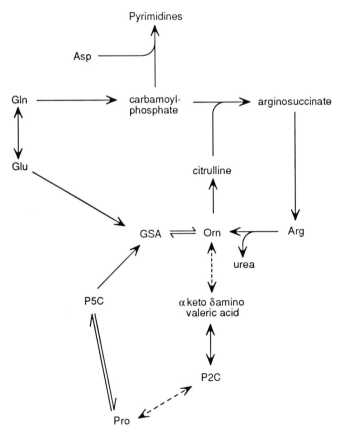

Figure 2-11. Abbreviated scheme of pathways involved in proline biosythesis. --> denotes enzyme activity not yet demonstrated in higher plants. Abbreviations: P5C, Δ'-pyrroline-5-carboxylic acid; GSA, L-glutamic semialdehyde; P2C, Δ'-pyrroline-2-carboxylic acid. (Dierks-Ventling and Tonelli, *Plant Physiol. 69*: 130-134, 1982; reproduced by permission of the ASPP.)

occur between any two mutants tested with regard to any of these phenotypic characters suggesting that all mutants were allelic.

The biosynthetic pathways leading to proline, as they are understood from results obtained with bacteria, fungi, and animal cells, are given in Figure 2-11. In plants, it is probable that proline is synthesized by glutamic semialdehyde, which is in equilibrium with Δ^1-pyrroline-5-carboxylic acid (P5C). Because P5C is an intermediate in both arginine and proline biosynthesis, it appears to be a link between the two metabolic pathways. To show whether it would allow normal growth resumption of mutant maize embryos, it was added, either alone or with proline, to the medium on which embryos had been placed (Table 2-9; Racchi et al., 1978). However, P5C did not support mutant embryo growth (Table 2-9), which, in turn, suggested that the *pro* mutant was the result of a genetic block in the metabolic pathway between P5C and proline.

In spite of these clear indications of the site of the auxotrophic lesion, further analysis has served only to confuse rather than clarify the issue. For example, in most cases, the endosperm of the mutant seed was shown to contain levels of free proline in excess of the wild-type level (Table 2-10; Racchi et al., 1981). There appears to be no shortage of proline in the tissues of any of the mutant plants and, indeed, there is a higher total content of this and other amino acids in mutants than in the wild type. Furthermore, the enzyme that brings about the final step in proline synthesis from P5C in the glutamate pathway has been shown in mutant seedlings (see Racchi et al., 1981).

In agreement with the collapsed seed character of the mutant, the endosperm was found to contain much less protein than the normal seed. The loss of endosperm protein was mainly due to the reduced accumulation of zein, which normally makes up 50 percent of the endosperm protein

Table 2-9. Growth of normal and mutant (*pro*) embryos of maize in the presence of P5C and proline. [a] Concentrations are: 2 and 6 mM DL-P5C, 1 and 3 mM L-pro in experiments A and B, respectively.

Growth Medium	Average Length (mm)			
	A [a]		B [a]	
	normal	mutant	normal	mutant
Basic (D)	221.0(8.0)	34.6(3.0)	156.7(8.9)	21.6(1.7)
D + P5C	176.6(5.8)	30.0(2.9)	79.1(5.3)	26.3(2.5)
D + P5C + pro	165.8(8.1)	85.0(7.5)	94.5(5.3)	55.3(8.0)
D + pro	201.6(10.2)	121.0(12.5)	170.3(10.7)	162.4(8.0)

(Racchi et al., *Z. Pflanzenphysiol. 101*: 303-311, 1981; reproduced by permission of Gustav Fischer Verlag.)

Table 2-10. Analysis of free amino acids in the endosperm of normal and *pro-1* seeds.

	Amino acids (μmol/g endosperm) Homoallelic combinations					
	pro 1-1		*pro 1-2*		*pro 1-3*	
	normal	mutant	normal	mutant	normal	mutant
Glu	1.55	5.85	1.12	1.40	0.58	1.61
Pro	4.31	1.13	1.16	1.79	0.61	1.14

(Racchi et al., *Z. Pflanzenphysiol. 101*: 303-311, 1981; reproduced by permission of Gustav Fischer Verlag.)

(Racchi et al., 1981). In general, the zein content of all the mutant endosperms was decreased by between 60 and 80 percent (Table 2-11). One possible explanation for the zein deficiency is that proline is confined in the cell compartment in which the amino acid synthesis occurs. Mobilization of limited storage proteins and possible leakiness of the compartments is sufficient to support germination and early growth of the seedlings in the case of the mutant, but the leakiness is not sufficient to support normal seedling growth of the mutant in the autotrophic state and the seedling dies. This and other hypotheses have been put forward to explain this auxotrophic phenomenon (for example, see Dierks-Ventling, 1982), but the nature of the genetic lesion remains obscure.

Table 2-11. Analysis of proteins in *pro-1* and normal kernels of maize.

pro 1 Constitution	Endosperm proteins (mg per 10 endosperms)			Embryo proteins (mg per 10 embryos)	
	Zeins	Albumins	Globulins	Albumins	Globulins
pro 1-1	4.4	35.2	23.2	43.2	13.2
norm	22.3	28.2	19.6	56.8	10.4
pro 1-2	7.3	10.5	12.6	37.6	34.2
norm	19.1	15.9	21.5	46.9	35.3
pro 1-3	4.0	17.2	20.7	35.0	16.9
norm	25.7	13.8	20.3	43.2	18.0

(Racchi et al., *Z. Pflanzenphysiol 101*:   303-311, 1981; reproduced by permission of Gustav Fischer Verlag.)

These mutants provide an example, even when they remain puzzling, of how auxotrophs can still be used in metabolic investigations. Here, the embryos and endosperms of *pro* mutants show alterations in protein composition that appear to be a direct consequence of a defective proline metabolism and provide a valuable tool to explore the production, distribution, and use of this amino acid (Dierks-Ventling, 1982). The mutants also serve to illustrate that auxotrophy can result from more than just a mutation leading to the formation of a defective enzyme in a metabolic pathway. The lesion in the *pro* mutants may be, for example, a defective membrane that will not allow the transport of L-proline to sites where it is required for continued growth of the plant.

2-2.2 The Use of Cell Culture Techniques to Isolate Auxotrophic Variants of Plants

The technology of plant tissue culture offers many possible methods for auxotroph recovery. Attempts have been made to devise enrichment schemes for the isolation of auxotrophs (see, for example, Suter et al., 1988), but the greatest success so far has been achieved using the "total isolation" method (see introduction to this Chapter) in which individual cloned colonies of cells are tested nonselectively for nutritional defects. The application of this method to higher plants involved the production of methods to cultivate and manipulate suspensions of true haploid (monohaploid) cells and protoplasts. Several such haploid systems have been employed (see King, 1986), one of which was to culture the cells of the true haploid (n=x=12), *Datura innoxia* P. Mill.

The method to isolate auxotrophs from *Datura* cultures required provision of suspensions of mainly single cells (King, 1986). After mutagenesis with ethylmethane sulfonate, these cell suspensions were rescued by growing them for 8 days in liquid medium supplemented with essential nutrients. Any colonies that grew were transferred to supplemented solid medium and allowed to grow into small calli. Each callus was then tested for auxotrophy by cutting it in two and transferring one half to fresh supplemented solid medium and the other half to minimal medium lacking nutritional supplements. Any colony that grew on supplemented but not unsupplemented medium was tested further to determine which of the many added nutrients in the supplemented medium was required by the putative auxotroph for growth.

The first auxotrophic cell line isolated using this method required the metabolite pantothenate for continued growth (Fig. 2-12; Savage et al., 1979). The cell line did not grow in a medium supplemented with pantothenate at 0.25 mg/L or less but grew nearly as well as the wild type in the presence of 2 mg/L of the supplement (Fig. 2-12). In further tests, the line was shown to grow in minimal medium with either ketopantoate, pantoate, or pantothenate, but not 2-oxo-*iso*-valerate, as the sole

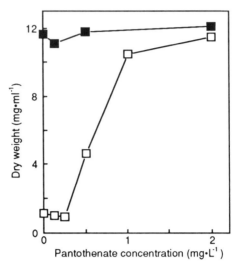

Figure 2-12. Pantothenate requirement of a *Datura* auxotroph in suspension culture: wild type (■), auxotroph (□). (Savage et al., *Plant Science Letters* *16*: 367-376, 1979; reproduced by permission of Elsevier Science Publishers.)

supplement (Fig. 2-13; Sahi et al., 1988). These data (Figs. 2-12 and 2-13), together with the knowledge that β-alanine cannot support the growth of these cells (King, 1986) suggested that the auxotroph lacks the ability to convert 2-oxo-*iso*-valerate to ketopantoate and, hence, to form the essential metabolite coenzyme A. It can be hypothesized that the cells have a defective ketopantoate hydroxymethyltransferase that catalyzes the conversion of 2-oxo-*iso*-valerate to ketopantoate (Fig. 2-14), a conclusion that has not yet been confirmed by enzyme assay.

Datura cell suspensions also yielded another auxotroph in which the biochemical lesion has been more precisely defined than in the case of the pantothenate requirer. Growth tests indicated that the cell line would not grow continuously unless the medium was supplemented with either casein hydrolysate or both of the branched-chain amino acids isoleucine and valine (Horsch and King, 1983). The branched-chain amino acids are synthesized in a common pathway from pyruvate (leucine, valine) or pyruvate and threonine (isoleucine) (Fig. 2-15; Wallsgrove et al., 1986b). When a number of enzymes in this pathway were assayed, there was no detectable dihydroxyacid dehydratase activity in the variant cells (Table 2-12; Wallsgrove et al., 1986a).

If it is presumed that a single mutational event led to the complete loss of enzyme activity, then it might also be hypothesized that these *Datura* cells have only one isoenzyme that is responsible for the synthesis of both 2-oxomethylvalerate and 2-oxo-*iso*-valerate. Such a conclusion would reinforce the understanding from biochemical studies that the reactions in the two metabolic pathways, illustrated in Figure 2-15, are catalyzed by a common set of enzymes.

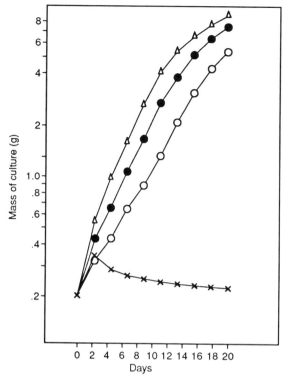

Figure 2-13. Growth analysis of a pantothenate-requiring variant of *Datura* (Pn1) on medium containing equimolar concentrations (10 μM) of ketopantoate (△), pantoate (●), pantothenate (○), and on basal medium (control) (×). (Sahi et al., *J. Plant Physiol. 133*: 277-280, 1988; reproduced by permission of Gustav Fischer Verlag.)

A number of other defined mutants in the isoleucine-valine pathway have been isolated using cell systems from other plant species. Of the eight known auxotrophs in the branched-chain amino acid pathway, in two different plant species, four lack the enzyme threonine dehydratase

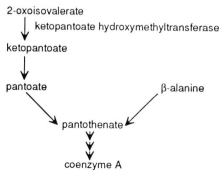

Figure 2-14. The pathway of coenzyme A biosynthesis.

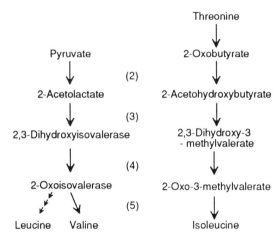

Figure 2-15. Biosynthesis of the branched-chain amino acids. (1). Threonine dehydratase; (2). Acetohydroxyacid synthase; (3). Acetohydroxyacid reductoisomerase; (4). Dihydroxyacid dehydratase; (5). Branched-chain amino acid aminotransferase.

and four others, dihydroxyacid dehydratase (King, 1986; Wallsgrove et al., 1986b). No other auxotrophic lesions affecting any other enzymes of this pathway have yet been discovered. Of course, auxotrophs in other species and in other metabolic pathways are known in higher plants (see, for example, King, 1986), but most of them have not yet been precisely characterized biochemically. A notable recent exception is the determination that a tryptophan auxotroph of *Hyoscyamus muticus* lacks the enzyme tryptophan synthase (Fankhauser et al., 1990).

2-2.3   The Use of Auxotrophs in Physiological and Molecular Studies

2-2.3.1   The Elucidation of Metabolic Pathways

Traditionally, auxotrophs have proved to be ideal to determine the sequences of steps in metabolic pathways. By supplying metabolic intermediates to cells blocked at some point in a pathway leading to an essential metabolite, it is not only possible to determine which step in the pathway has been affected by the mutational event but in so doing, the participation of these intermediates in the pathway can be verified, at least indirectly.

An example of the participation of metabolites in a pathway by implication is provided in the case of the pantothenate-requiring *Datura* cell line, described in Section 2-2.2. The pathway leading to the formation of coenzyme A has not been determined in higher plants, but

Table 2-12. Enzyme activities in wild type and mutant *Datura* cell lines (expressed as pkat/g fresh weight).

	Threonine dehydratase	Aceto-hydroxyacid synthase	Dihydroxyacid dehydratase	Branched chain amino acid aminotransferase
Wild-type	24	23	207	234
IV-1	16	34	0	75

(Wallsgrove et al., *Plant Science 43*: 109-114, 1986a; reproduced by permission of Elsevier Science Publishers.)

it is presumed to be the same as that found in microorganisms. The fact that ketopantoate, pantoate, and pantothenate satisfied the growth requirement of this variant (Fig. 2-13) implicates all of these compounds, directly or indirectly, in the pathway and provides some preliminary evidence of the metabolic sequence leading to coenzyme A synthesis in plants.

In another example, although histidine biosynthesis in higher plants has not been studied extensively, again, the pathway is considered to be similar to that in microorganisms. Histidinol phosphate, histidinol, and histidine were equally effective in supporting the growth of a histidine-requirer of *Hyoscyamus muticus* (Fig. 2-16; Shimamoto and King, 1983), results that provide preliminary support for the hypothesis that the pathway of histidine biosynthesis in plants is the same as in microorganisms.

It may not seem important to ascertain facts of this kind directly when many metabolic pathways have already been worked out in microorganisms and when it seems likely that the pathways are similar

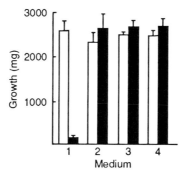

Figure 2-16. Growth of histidine-requiring variant (PO3) and wild-type cells of *Hyoscyamus muticus* on media supplemented with histidine or presumptive histidine precursors. Media: 1-minimal; 2-minimal + 0.5 mM histidine; 3-minimal + 0.5 mM histidinol; 4-minimal + 0.5 mM histidinol phosphate. PO3 (■); wt (□). (Shimamoto and King, *Mol. Gen. Genet. 189*: 69-72, 1983.)

in plants. Significant differences may be uncovered in the case of higher plants, however, if not in the individual steps of a pathway, at least in the control of a metabolic sequence. The molecular biologist who may be attempting, for example, to transfer a gene for a certain metabolic reaction from one species to another will need to have precise knowledge not only of the reaction itself but also of its control. Therefore, auxotrophs will continue to be important tools in the dissection of plant metabolism.

2-2.3.2   *The Physiological Effects of D-proline in* Zea mays

Several D-amino acids and their derivatives can be found in living organisms. Inhibition of growth, morphological abnormalities, and chlorosis of the leaves are common consequences of feeding D-isomers of several amino acids, but their precise role remains unknown. The *pro* mutants of corn provide a possible way to investigate the profound effects of D-amino acids on organisms.

The *pro-1* mutant of corn is lethal at the seedling level, a trait that makes it easy to determine whether a particular compound is able to overcome the effect of the mutation. For instance, *pro-1* seedlings recover only if L-proline is added to the culture medium; other L-amino acids and L-proline precursors do not restore mutant growth (Tonelli, 1985).

When a number of D-amino acids were added to media on which *pro-1* or wild type maize embryos were growing, only D-proline among those tested stimulated growth whereas other D-isomers were inhibitory. Clearly, D-proline was taken up by the mutant embryos and used for phenotypic repair, although a higher concentration of the D-isomer was required to duplicate the influence of L-proline (Table 2-13; Tonelli, 1985).

Although the biochemical basis of the *pro-1* mutation is not known (see Section 2-2.2) two hypotheses may be proposed to explain the effect of D-proline on mutant growth. First, D-proline could be converted into L-proline by direct racemization, the latter being the only isomer utilized by the mutant (Fig. 2-17). The observation that the concentration of D-proline needed to restore mutant growth is greater than that of L-proline (Tonelli, 1985) supports this hypothesis as does the fact that the compound a-keto-d-amino valerate cannot be used to substitute for the D-isomer, that is, we can exclude the conversion of D- into L-isomer through oxidative deamination followed by transamination of the resulting a-keto-d-amino valerate (Fig. 2-17). The other hypothesis is that *pro* mutants can recognize both D- and L-isomers as activators of a metabolic step not yet identified (Tonelli, 1985). To clarify this point it would be necessary to determine in more detail how D-proline is utilized and metabolized. For example, these results should encourage a search for a racemase enzyme capable of interconverting D- and L-proline.

Table 2-13. Effects of D-amino acids on growth of *pro 1-1* mutant maize embryos. Growth of *pro 1-1* mutant and normal embryos on media supplemented with 4 mM of D- and L-isomers of proline, glutamate and alanine after 28 days of growth at 26°C and 16-hour photoperiod. Figures in parentheses represent the standard errors.

Amino acid added	Shoot length (cm)	
	Wild type	*pro1-1*
None	23.7 (1.3)	1.5 (0.1)
L-Proline	25.5 (1.4)	28.4 (1.1)
D-Proline	25.9 (1.4)	15.3 (0.3)
L-Glutamate	30.8 (1.9)	1.9 (0.1)
D-Glutamate	15.7 (1.8)	1.4 (0.1)
L-Alanine	35.4 (1.7)	1.6 (0.2)
D-Alanine	7.4 (0.7)	1.5 (0.1)

(Tonelli, *Plant Cell Physiol. 26*: 1205-1210, 1985.)

These results do not explain why, in other cases, the D-amino acids inhibit or distort the growth of plants. D-proline does not affect maize in this way. What is clear, however, is that amino acid-requiring mutants, such as *pro-1*, can be used effectively to produce information about the influence of D-isomers without the complications resulting from the

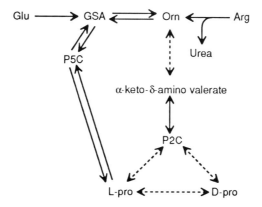

Figure 2-17. Simplified scheme of pathways involved in proline biosynthesis. --> enzyme activity not yet demonstrated in higher plants. Abbreviations: Arg, arginine; Glu, glutamic acid; GSA, L-glutamic semialdehyde; Orn, ornithine; P5C, Δ'-pyrroline-5-carboxylic acid; P2C, Δ'-pyrroline-2-carboxylic acid; Pro, proline. (Tonelli, *Plant Cell Physiol. 26*: 1205-1210, 1985.)

formation and utilization of L-isomers in situ. In the particular case discussed here, the use of the *pro-1* mutant provided guidance in the search for an explanation of the influence of D-proline on plant growth. Other auxotrophs, such as the isoleucine-valine requirer of *Datura*, could also be used to investigate the role of such stereoisomers.

2-2.3.3   *Somatic Hybridization Using Auxotrophs*

Somatic hybridization is brought about by first forming protoplasts from cells of parents that differ from one another genetically. Suspensions of the protoplasts of two different parents are then mixed in the presence of an agent (such as polyethylene glycol) that will cause the protoplasts to fuse and then merge. In some instances, the fusion products are made up of two protoplasts, one from each of the two parents in the mixture. In that event, a hybrid has been formed that is different from either parent. Such hybrids may be useful for a variety of purposes. If regeneration can be brought about it may be possible to produce plants that are not just different from either parent but also not possible to produce by normal sexual crosses. Thus, protoplasts of sexually incompatible species can be fused and caused to grow, whereas such species cannot be crossed sexually. By this means, it may be possible to introduce into a particular species traits from another species that cannot be transmitted in any other way. Such interactions could be of considerable physiological as well as practical interest.

A major difficulty in the application of the hybridization technique is the ability to recognize genuine hybrids in a mixture of many thousands of fusion products involving protoplasts of the same species as well as protoplasts that have not fused at all. One protoplast looks much like another and to distinguish fusion products when there are often no suitable, visible markers to indicate when a hybrid has been formed can be a tedious matter. The provision of selectable genetic markers is, therefore, of great importance in somatic hybridization.

Auxotrophs provide one means to select hybrids with great efficiency from mixtures of protoplasts. Thus, if protoplasts of two different auxotrophs are fused, one would expect genetic complementation to occur. In this event, a hybrid should be capable of growth in or on a medium lacking the nutrients required by *both* parents. Any cell colonies, therefore, found growing in a medium lacking the compounds required by the parents would be putative hybrids (or, alternatively, revertants of one of the parents). One of many examples (see Kishinami and Widholm, 1987) is given. Nitrate reductase-deficient lines have been described in *Nicotiana plumbaginifolia* and can be classified into two major groups (see Case Study 2-3 for a full discussion of NR-deficient auxotrophs). One type is defective in the functional apoenzyme of NR (*nia* mutants, NA) but retains xanthine dehydrogenase (XDH) activity; the other is defective in XDH as well as NR (*cnx* mutants, NX) and can be further subdivided. NR activity could be restored by fusing protoplasts of the different NR⁻ cell lines. Genetic complementation would be expected only if the lines were defective in different genetic factors (Table 2-14; Marton et al., 1982).

Results in Table 2-14 show that the four NA (*nia*) lines did not complement one another and can be presumed to be allelic, as expected. The complementation that occurred between the NA2 and the NX (*cnx*) lines was also to be expected and it can be presumed that the other NA lines would do the same thing. Among the NX lines, two complementation groups could be distinguished. NX1 and NX9 did not complement one another and are likely to be mutants in the same gene. NX21 and NX24 complemented each other as well as NX1 and NX9, which indicates that these lines represent different genes from one another and from NX1 and NX9 (Marton et al., 1982). Thus, through the use of hybridization it was possible, in these cases, to distinguish between different lines of mutants and to put them into complementation groups for further biochemical and physiological analyses, a task that would have been much more tedious and complicated without this in vivo technique.

Table 2-14. Complementation of nitrate reductase (NR) activity by fusion of NR⁻ cell lines of *Nicotiana plumbaginifolia*. NA = *nia* mutants; NX = *cnx* mutants.

Fusion combination	Comple- mentation	NR activity	Plant regeneration
NA1 + NA2	-	-	-
NA1 + NA9	-	-	-
NA1 + NA18	-	-	-
NA1 + NA36	-	-	-
NA2 + NX1	+	+	+
NA2 + NX9	+	+	+
NA2 + NX21	+	+	+
NA2 + NX24	+	+	+
NX1 + NX9	-	-	-
NX1 + NX21	+	+	+
NX1 + NX24	+	+	+
NX9 + NX21	+	+	+
NX9 + NX24	+	+	+
NX21 + NX24	+	+	+

(Marton et al., *Mol. Gen. Genet. 187*:1982.)

A further advance in the use of auxotrophic markers for somatic hybridization has been made recently with the selection of *universal hybridizers*, double mutant cell lines that are both auxotrophic and have a second dominant mutation. Thus, for example, Toriyama et al. (1987) have isolated a double mutant of *Sinapis turgida* that is both NR⁻ and resistant to 5-methyltryptophan (5MT). Toriyama et al. (1987) were able to demonstrate that when protoplasts from this cell line were fused with protoplasts from wild-type *Brassica* cells (i.e., having no usable genetic markers), somatic hybrids could be selected on a medium containing NO_3^- as sole nitrogen source and 5MT. The parent double mutant *Sinapis* cells would not grow on the selection medium because it had in it only NO_3^- as a nitrogen source; *Brassica* cells would not grow because of the lethal level of 5MT in the medium; hybrid cells did grow because genetic complementation during fusion had restored the ability to utilize NO_3^- as a nitrogen source and the dominant trait of resistance to 5MT had been inherited from the *Sinapis* parent. Theoretically, therefore, it should be possible to fuse such double mutants with any other wild-type cell line having high susceptibility to compounds like 5MT and select hybrids. The technique provides an efficient way to answer the question as to whether and with what efficiency crosses can be made between widely separated species. The most recent universal hybridizer was created in the pantothenate-requiring (Pn1) auxotroph of *Datura* (see earlier discussion in this Case Study) by direct transfer of the gene for kanamycin resistance (Saxena et al., 1990).

2-2.3.4   Genetic Transformation Using an Auxotrophic Receptor

Cells are usually auxotrophic because they have a defect in a structural gene for an essential enzyme (but not always, as we have seen in the case of the *pro* mutants of maize). The lesion could be repaired by introducing a foreign gene into the affected genome to substitute for the defective one. In the case of a complete auxotroph recognition of repair through such a transformation event would be made easier as only transformed cells would grow on a minimal medium; cells which remained auxotrophic would continue to grow only if supplied with the essential metabolite they could no longer produce for themselves. Of course, genetic revertants would also grow on minimal medium and could be confused with transformants. If transformation was performed with an auxotrophic cell line that did not revert under any growth conditions tested, however, then the confusion between transformation and reversion would be virtually eliminated. Such an auxotroph does exist; the pantothenate-requiring cell line of *Datura innoxia* (Pn1) has an absolute requirement for pantothenate to grow (see Section 2-2.2) and has not given rise to revertants since it was isolated in 1978. The transformation of Pn1 was carried out by introducing nuclei isolated from cells of the legume *Vicia hajastana* into protoplasts produced from the auxotroph (Saxena et al., 1986). After a period of several days to allow the introduced *Vicia* DNA to integrate with the genome of Pn1, treated cells were plated on medium lacking pantothenate; only putative transformants would be expected to grow under these circumstances. An example of the production of growing prototrophic clones is shown in Figure 2-18 (Saxena et al., 1986). Only those clones that underwent some genetic change rendering them no longer auxotrophic continued to grow on minimal medium; all other clones died. Upon further testing a number of the growing clones were shown not only to be prototrophic but also to have incorporated into their genomes DNA originating from *Vicia*. Dot-blot hybridization using *Datura* and *Vicia* DNA as probes showed the presence of *Vicia* DNA in two of the three putative transformants tested (Fig. 2-19; Saxena et al., 1986).

These experiments in somatic hybridization and genetic transformation demonstrate the power of auxotrophs as a tool in bringing about genetic modifications. The fact that auxotrophic cells will not grow unless they are genetically changed results in a stringent selection system. With such a system to hand it should be possible to study, for example, cytoplasmic-cytoplasmic and nucleo-cytoplasmic interactions in a way not before possible. For instance, if nuclei are used for genetic transformation that have marker genes, it ought to be possible to answer

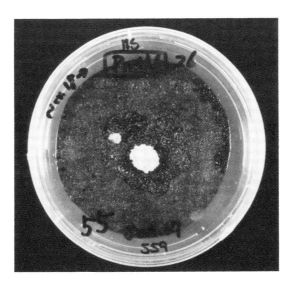

Figure 2-18. Selection of prototrophic clones after the uptake of isolated nuclei from *Vicia hajastana* by protoplasts of a pantothenate-requiring auxotroph (Pn1) of *Datura innoxia*. Two surviving prototrophic clones can be seen surrounded by many dead auxotrophic colonies on a black filter paper background. (Saxena et al., *Planta 168*: 29-35, 1986.)

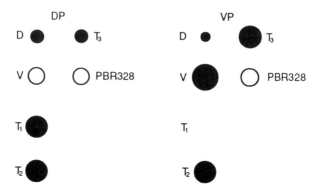

Figure 2-19. Dot-blot hybridization of DNA from three clones with nick-translated *Datura* (Pn 1) and *Vicia* probes. *DP, Datura innoxia* (Pn 1) probe; *VP, Vicia hajastana* probe; T1, T2, T3, the prototrophic clones (PnV) of Pn1, tested for transfer of *Vicia* DNA. (Saxena et al., *Planta 168*: 29-35, 1986.)

such questions as: Can particular genes be transferred between genetically, widely separated species of plants and if so, with what efficiency? How much foreign DNA can be transferred through the use of nuclei (or isolated chromosomes) as a vehicle and can the amount be in any way controlled? With what frequency can normal plants be regenerated from material that has been transformed by variable amounts of foreign DNA? Can polygenic traits be transferred using nuclei (or chromosomes) as the vehicle of transfer? Through the use of universal hybridizers with both recessive and dominant markers it is already possible to attempt crosses between widely separated species to test the degree to which such crosses can be expected to produce viable useful plants. The addition of transformation using organelles should add another significant tool to the array of plant molecular techniques.

Case Study 2-3:   The Nitrate Reductase-less Mutants

The nitrate assimilation pathway is the major point of entry of inorganic nitrogen into organic combination in most plants. This pathway converts nitrate to ammonium by a nitrate uptake mechanism at the cell surface and two enzymes, NR and nitrite reductase (NiR). Nitrate and hormone availability, light and end-products have all been shown to play a role in the regulation of this pathway and much has been learned about the biochemistry and physiology of the entire process through the use of natural genetic variants. Increasingly, however, plant physiologists have turned to induced mutations to study this important process in detail (Warner et al., 1985; Wray, 1986, 1988). During the last decade there has been progress made in understanding the genetics of nitrate assimilation; it is now one of the best characterized pathways in higher plants. These genetic studies are an important first step in developing a molecular understanding of the pathway (Wray, 1988). Central in these studies has been the isolation of a large number and variety of mutants of NR.

Higher plant NR's are flavohemomolybdoproteins that catalyze the two electron reduction of nitrate to nitrite. Almost all higher plants so far examined possess NADH-NR (E.C.1.6.6.1), which has a pH optimum of about 7.4 and a Michaelis constant (K_m) for NO_3^- and NADH of about 200μM and 2μM, respectively. Some plants also have a bispecific NAD(P)H-NR (E.C.1.6.6.2) that is a distinct species. The NADH-NR is presumed to be a homodimer made up of subunits, each with a molecular weight of about 115-kD (Fig. 2-20; Wray, 1986; Caboche and Rouze, 1990). The suggestion is that there are one flavin adenine dinucleotide (FAD), one heme, and one molybdenum (Mo) per 115-kD subunit. The Mo is carried on a dissociable, dialyzable, and oxygen-labile structure, the molybdenum cofactor (Mo-Co).

In addition to the overall physiological reaction of NADH-dependent nitrate reduction, the higher plant enzymes carry a dehydrogenase (diaphorase) function that allows the in vitro transfer of electrons from NADH to a variety of electron acceptors such as nitroblue tetrazolium, dichlorophenolindophenol, and cytochrome c. It also has reduced flavin mononucleotide ($FMNH_2$) and reduced viologen (MV) dye NR activity. The dehydrogenase function is usually measured as cytochrome c reductase. The $FMNH_2$ reduction activities involve the Mo-Co component but not the diaphorase (Fig. 2-20).

Two approaches have been used to identify NR-defective individuals within populations of mutagenized cells and protoplasts or whole plants. The first is a nonselective, total isolation procedure in which each plant is tested for NR activity. The other is through the use of chlorate as a selective agent (see introduction to this Chapter). Using these techniques, two categories of NR mutants have been identified.

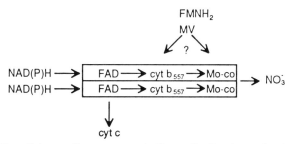

Figure 2-20. Schematic representation of nitrate reductase showing probable sites of interaction of substrates and electron donors. FAD = flavin adenine dinucleotide; $FMNH_2$ = reduced flavin mononucleotide; MV = methyl viologen dye; Mo-Co = molybdenum cofactor.

Apoprotein Gene Mutants

A very wide range of NR mutants having defects in an apoprotein gene locus have been isolated in higher plants. It seems evident that the apoprotein of the NADH-NR is coded by a single gene, indicating that the enzyme is a homodimer. These are known generally as *nia* mutants (or *nar1* mutants in barley) (Table 2-15; Warner et al., 1985).

Molybdenum Cofactor Mutants

Molybdenum cofactor mutants are designated as *cnx* mutants (cofactor for *n*itrate reductase and *x*anthine dehydrogenase) (Table 2-15). The *cnx* mutants may fall into as many as seven complementation groups (Wray, 1986). At least six or seven genes can mutate to give the *cnx* phenotype and the reason for this lies in the structure of Mo-Co (Fig. 2-21). Two metabolic pathways are probably required: one, leading from mainstream pterin metabolism and involving modification of the pterin nucleus to generate the phosphorylated pterin derivative, molybdopterin; the other, including molybdenum uptake into cells, generation of a form of molybdenum that is in the correct redox and ligandable state, and the liganding of molybdenum to molybdopterin to generate the functional Mo-Co. Many Mo-Co mutants have been described in a wide variety of higher plants, an identification based largely on the pleiotropic loss of NR and XDH activity with retention of a flavohemoprotein subunit that continues to have NADH-cytochrome c activity (Wray, 1986).

Table 2-15. Genes controlling NR activity in fungi and higher plants. [a]. Duplicate genes (*N. tabacum* is an amphidiploid).

Species	NR Apoenzyme	Molybdenum Cofactor
Aspergillus nidulans	1	5
Neurospora crassa	1	4
Hordeum vulgare	1	3
Nicotiana plumbaginifolia	1	3
Nicotiana tabacum	2[a]	2
Pisum sativum	1	2

(Warner et al., in *Exploitation of Physiological and Genetic Variability To Enhance Crop Productivity*, J.E. Harper, L.E. Schrader, R.W. Howell (eds), pp. 23-30, 1985; reproduced by permission of the ASPP.)

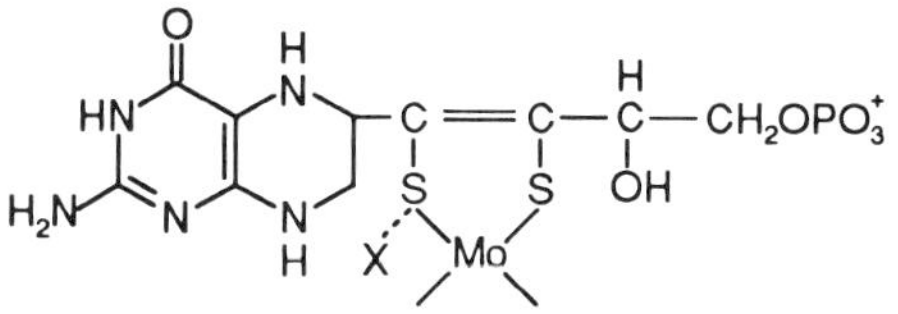

Figure 2-21. Proposed structure of the molybdenum cofactor of nitrate reductase.

2-3.1   Isolation and Characterization of Some NR⁻ Mutants

One of the earliest detailed analyses of chlorate-resistant mutants in higher plants was that of Oostindier-Braaksma and Feenstra (1973) using seeds of *Arabidopsis thaliana*. Wet seeds were submerged in a solution of 1 mM *N*-methyl-*N*-nitro-*N*-nitrosoguanidine for about 4 hours and then sown in petri dishes on a mineral medium. The resulting M_1 seeds were grown in the same way until they were 6 days old and then transferred to medium supplemented with a solution containing chlorate. Progeny that survived this treatment were tested for chlorate resistance and then for NR activity (Table 2-16). The chlorate-resistant mutants in no case had reduced NO_3^- uptake and in two of the three cases had an elevated NR activity. Those mutants with elevated NR activity had an impaired ability to take up chlorate. Thus, it was concluded that chlorate resistance in higher plants can be caused by at least two different mechanisms: (1) a lower rate of chlorate reduction because of a low rate of NR or (2) a lower uptake of chlorate.

Warner et al. (1985) carried out a comprehensive analysis of NR-deficient mutants in barley, which were derived from sodium azide-treated seed. M_2 seedlings were screened for NR activity and about 35 mutants, thus, partially characterized. All mutants analyzed had low NADH-NR and high NiR activities (Table 2-17; Warner et al., 1985). The

Table 2-16. Nitrate content and nitrate reductase activity of 12-day-old plants of *Arabidopsis thaliana*.

Line	mμg equiv. nitrate/ μg protein	Specific nitrate reductase activity
Wild type	2.6	53.2
BI	2.3	84.2
B2-I	4.2	9.6
B3	2.5	82.3

(Oostindier-Braaksma and Feenstra, *Mutat. Res. 19*: 175-185, 1973; reproduced by permission of Elsevier Science Publishers.)

Table 2-17. Representative mutants of genes controlling NR in barley.
[a.] Presence (+) or absence (-) of activity.
[b.] Not determined.

Gene Symbol	NR		Cyt. c Reductase	NiR	Xanthine Dehydrogenase [a]
	NADH	FMNH$_2$			
			% of Steptoe		
nar1a	1	0	7	174	+
nar1b	1	0	61	174	+
nar1d	2	0	227	193	+
nar1h	2	167	10	156	+
nar2a	6	0	64	170	-
nar3b	3	0	42	128	+
nar4y	1	0	nd[b]	135	-

(Warner et al., in *Exploitation of Physiological and Genetic Variability To Enhance Crop Productivity*, J.E. Harper, L.E. Schrader, R.W. Howell (eds), pp. 23-30, 1985; reproduced by permission of the ASPP.)

mutants could be distinguished from one another by their NR associated activities. For example, *nar1h* had a high FMNH$_2$ but a low NADH-NR activity. NR associated cyt c reductase activities in the mutants varied from low (*nar1a, nar1h*) to intermediate (*nar1b, nar2a, nar3b*), to high (*nar1d*) (Table 2-17). The mutants were representative of four loci; the *nar1* alleles were codominant with the wild-type allele whereas *nar2* and *nar3* alleles were recessive. All three exhibited segregation patterns characteristic of single Mendelian genes.

What can be drawn from the analysis of Warner et al. (1985) is that *nar2, nar3*, and *nar4* probably control Mo-Co functions, a conclusion based on the pleiotropic loss of XDH activity (Table 2-17). The *nar1* locus is the NR structural gene given the considerable variations among *nar1* alleles for NR-associated activities such as cytochrome c reductase and FMNH$_2$-NR.

Some of the most detailed genetic analyses of NR have been carried out in *Nicotiana tabacum* and *N. plumbaginifolia*. Apart from intensive investigations of the *nia* mutations, leading to defects in the structural gene for NR, the so-called *cnx* loci, involved in the synthesis and processing of Mo-Co have also been studied in depth. Table 2-18 summarizes a great deal of the knowledge presently available for *N. plumbaginifolia* and *N. tabacum* mutants of the *cnx* type (Mendel et al., 1986).

The complementation analysis in this study (Table 2-18) led to descriptions of three *Nicotiana* loci, *cnxA, cnxB*, and *cnxC*. The numerous biochemical data available for these mutants allowed the drawing of

Table 2-18. Comparison of biochemical data for *N. tabacum* (*N.t.*) and *N. plumbaginifolia* (*N.p.*) *cnxA*, *cnxB* and *cnxC* mutant cell lines. [a] Xanthine dehydrogenase is another molybdoenzyme containing the Mo-Co. [b] Restoration of NADH-NR activity by common homogenization of *cnx* and *nia* cells. [c] Restoration of NADH-NR activity by addition of active Mo-Co from bovine milk xanthine oxidase to *cnx* extracts. [d] Restoration of in vivo-NR activity by growth on medium containing 1 mM Na$_2$MoO$_4$. [e] Restoration of NADH-NR activity by extraction of *cnx* cells in buffer containing 20 mM Na$_2$MoO$_4$, thiol reagents and EDTA. [f] Restoration of NADH-NR activity in *cnx* cell extracts by addition of own Mo-Co which was reactivated separately in vitro. [g] Donation of Mo-Co from *cnx* extracts to restore NADPH-NR activity in extracts of *N. crassa nit-1*. [h] NR monomers occur dimerized (Dim.) or exclusively monomeric (Mon.).

	cnx A		*cnxB*	*cnx C*
	N.t .68/2;101	N.p. Nx9	N.p. Nx24	N.p. Nx21
NADH-NR	-	-	-	-
Xanthine dehydrogenase [a]	-	-	-	-
Molybdenum content in cells	+	+	+	+
Complementation in vitro of *nia*[b]	+	+	+	+
Repair in vitro by Mo-Co from milk xanthine oxidase[c]	+	+	+	+
Repair in vivo by molybdate[d]	+	+	-	-
Repair in vitro by molybdate[e]	+	+	-	-
Repair in vitro by own Mo-Co[f]	+	+	-	-
Complementation in vitro of *N. crassa nit -1*[g]	+	+	-	-
Assembly state of NR apoprotein[h]	Dim.	Dim.	Mon.	Mon.
Mo-Co fluorescence	High	High	Low	Low

(Mendel et al., *Plant Science 43*: 125-129, 1986; reproduced by permission of Elsevier Science Publishers.)

some preliminary conclusions about the function of the gene products. Thus, the gene product of *cnxA* is probably involved in the incorporation of Mo into the cofactor because high Mo, among other methods, can be used to restore Mo-Co activity. The gene products of *cnxB* and *cnxC* seem to participate in the synthesis of the molybdopterin moiety of Mo-Co. This moiety has a pterin ring with a Mo-Co-specific C-4 side chain having a terminal phosphate group and sulfur atoms at C-1 and C-2 (Fig. 2-21).

The sulfur atoms are thought to be important for dimerization as well as for liganding Mo. Because the pterin ring also forms part of other compounds, some of which are essential for cellular metabolism (e.g., folic acid flavins), a block in the synthesis of pterin would be lethal. Hence, the products of *cnxB* and *cnxC* are more likely to be involved in the synthesis of the Mo-Co-specific side chain or in some other aspect of Mo-Co processing in the cell. Since the report of these analyses (Mendel et al., 1986) preliminary involvement of three other genes, *cnxD, cnxE* and *cnxF*, in Mo-Co assembly has been reported (see Gabard et al., 1988). None of these studies, however, has so far provided knowledge of either the enzymes involved in the synthesis of Mo-Co or the existence of regulatory genes both of which presumably occur.

2-3.2   The Use of NR-deficient Mutants in Physiological and Biochemical Studies

2-3.2.1   *The Appearance of NAD(P)H-NR in* nar1 *Barley Mutants*

Many of the barley mutants isolated by Warner et al. (1985) were capable of continued growth with NO_3^- as the sole source of nitrogen, contrary to what would be expected of mutants apparently lacking NR. When grown in the field or in the greenhouse, for example, mutants *nar1a* and *nar1b* produced nearly as much total dry weight and reduced nitrogen as the wild-type cultivar. Similar results were obtained when excised embryos were grown under sterile culture conditions with NO_3^- as the nitrogen source. In an attempt to resolve this apparent anomaly, the NR present in *nar1* mutants was characterized (Table 2-19). All *nar1* mutants were found to have greater activity with NADPH than with NADH (Table 2-19B). The NR from *nar1a* had an optimum pH of 7.7 for both the NADH and NADPH activities, whereas the NADH enzyme from the wild type had an optimum pH of 7.5 for the NADH activity and 6.0 with NADPH (Table 2-19A).

The hypothesis arising from these results (Table 2-19) was that the NADH-NR and NADPH-NR activities of *nar1* mutants are properties of a single NAD(P)H, bispecific NR. Further evidence suggested that this was the case. Thus, during several purification steps the NADH and NADPH activities of the *nar1a* enzyme remained together in about the same proportion. The K_m for NO_3^- was the same with NADH or NADPH as the electron donor and was about five times greater than that of the wild-type K_m (Table 2-19A). The wild-type NADH-NR was insensitive to low concentrations of dithionite and was inactivated by less than half the antiserum required to inactivate the NAD(P)H bispecific NR (Table 2-19A).

Table 2-19. Characteristics of NADH- and NADPH-nitrate reductases (A) and nitrate uptake and assimilation and total NR activity (B) in wild type (Steptoe) and *nar1* mutants from barley.
[a] μmol/g dw root/24 hours.
[b] μmol/g fw seedling/24 hours.

A.

	Wild Type (EC 1.6.6.1)		*nar1a* (EC 1.6.6.2)	
	NADH	NADPH	NADH	NADPH
pH optimum	7.6	6.0	7.7	7.7
K_m NO$_3$ mM	0.13	-	0.62	0.61
K_m NAD(P)H(μM)	12	270	68	10
Inhibition by:				
1 mM dithionite (%)	15	-	100	100
2 μl NADH NR antiserum %	70	-	30	25
Ratio NADPH/NADH at pH 7.5	<0.01	1.8		

B.

	Nitrate		NR[b]	
Genotype	Uptake[a]	Assimilation[b]	NADH	NADPH
nar1a	2160	41.6	4.1	15.4
nar1b	2045	47.9	4.6	10.7
nar1h	1920	45.1	3.8	8.5
Steptoe	2230	58.4	97.5	-

(Warner et al., in *Exploitation of Physiological and Genetic Variability to Enhance Crop Productivity*, J.E. Harper, L.E. Schrader, R.W. Howell (eds), pp. 23-30, 1985; reproduced by permission of ASPP.)

These results support the conclusion that the NADH-NR and NADPH-NR activities of *nar1* mutants of barley are properties of a single enzyme and that this enzyme is the major or only pyridine nucleotide-dependent NR in most *nar1* mutants. The enzyme is not present in the wild type and its mode of regulation is not known (Warner et al., 1985). Thus, when placed in the position of not being able to assimilate NO$_3^-$ in the normal way, these barley plants, nevertheless, were capable of producing NR. The physiological (and evolutionary) origin of this mechanism is not apparent and remained undetected until mutants lacking normal NR were isolated. Recently, this bispecific NR has been shown to be encoded by the *nar7* gene, which is not expressed in wild-type leaves; in *nar1* mutants, however, the *nar7* NR is expressed in leaves as if to compensate for the loss of the *nar1* (see Crawford and Campbell, 1990).

2-3.2.2   *Improving NO₃⁻ Assimilation in Crop Plants*

The ability of a plant to acquire nitrate from the environment influences both the nitrate flux and also NR activity as the NO_3^- assimilatory system is inducible by nitrate (Wray, 1986). Thus, nitrate uptake is a major point of control in the process. Furthermore, estimates show that more than two-thirds of the edible dry matter and half of the protein produced in the world are contributed by the cereals. The increased use of fertilizer nitrogen has been a significant factor in increasing productivity of these crops. Concerns about the high cost of energy required to produce fertilizer nitrogen and the effects of nitrate pollution on the quality of the environment, however, suggests that more efficient management and utilization of fertilizer nitrogen is a worthwhile goal. These objectives can be approached at the plant level by improving the efficiency of assimilation of nitrate nitrogen and the factors influencing these processes in the plant. Some areas in which such modifications might be explored are discussed.

Although at present there seems to be no way of analyzing nitrate uptake at the molecular level, it may be possible to isolate nitrate transport mutants using NR⁻ mutants as the starting material. Many NR⁻ mutants cannot use NO_3^- as a source of nitrogen. If NO_3^- is fed to such mutants it accumulates but is not assimilated. Such systems provide models with which to study the uptake process (see King et al., 1980). It may be possible, for example, to screen large numbers of clones of mutagenized NR⁻ mutants for variations in NO_3^- uptake capacity with a view to significantly increasing the rate at which nitrate is taken up. Certainly, mutants of higher plants that are changed in NO_3^- uptake characteristics can be obtained (Doddema and Telkamp, 1979).

Grain yield, protein yield, grain reduced nitrogen, or plant-reduced nitrogen are all related to NR levels in a wide variety of plants. Curiously, however, grain yield and plant-reduced nitrogen were little affected in NR-deficient mutants of barley possessing less than 10 percent of wild-type NR activity, results that indicate that these mutants are capable of reducing nitrogen at a rate sufficient to support growth rates approaching that of the control (Warner and Kleinhofs, 1981). Thus, contradictory evidence exists to suggest that, on the one hand, NR levels control primary nitrogen metabolism in plants and, on the other, are greatly in excess of what is required to maintain normal nitrogen levels.

The ability to introduce multiple copies of cloned NR apoprotein genes into crop plants would allow the effect of NR gene dosage to be analyzed in the same genetic background and perhaps resolve the question as to the degree of control NR exerts on nitrogen assimilation. Relevant to such an aim is the observation by Muller (1983) that in tobacco, the level of NR activity is independent of the number of nia^+ genes, indicating a complete compensation for gene dosage effects, presumably by regulatory mechanisms (Table 2-20). Stepwise substitution of nia^+ genes for the null alleles in a nia mutant of *Nicotiana tabacum* indicated that NR activity was independent of gene dosage and, thus, determined exclusively by regulatory mechanisms (Table 2-20). One nia^+ gene seems sufficient to ensure maximum NR activity in tobacco. However, this may not be true in all species.

These results (Table 2-20) suggest that gene dosage cannot be used reliably to improve the assimilation of NO_3^- and that an alternative approach would be required to improve nitrogen assimilation in higher plants. One possibility might be to have more efficient nia gene promoters or NRs that are kinetically more efficient either as a result of natural variation or due to site-directed mutagenesis (Wray, 1986; Caboche and Rouze, 1990; Crawford and Campbell, 1990). In this respect, the NR of *Erythrina senegalensis* is interesting. NR levels in this tropical leguminous tree are about 100-fold those of most crop plants (Table 2-21; Stewart and Orebamjo, 1979). It was possible to measure extremely high levels of NR in this species. Activities of $20\mu mol/hour/g$ fresh weight are more normal than those seen in this legume. Characterization of the cloned gene would allow the reasons for the greater enzyme activity to be determined and, then, the possibility of transferring the higher NR capability to other species to be explored.

Table 2-20. Genotype and phenotype of selected F_2 *Nicotiana tabacum* plants. Shoot height and in vivo NR activity (units/100 mg fr.wt.) were determined at 82 days after germination.
[a.]The genotype of each F_2 plant was inferred from the F_3 segregation. Homozygous single mutants were identified by intercrossing them.

Genotype[a] (*nia1* ; *nia2*)	No. of plants	NR activity	Shoot height (cm)
WT/*Nia28r5* selfed:			
-/-; -/-	9	0-4	(grafts)
+/-; -/- or -/-; +/-	14	744	54
+/-; +/-	2	710	48
+/+; -/-	5	736	51
-/-; +/+	1	696	62

(Muller, *Mol. Gen. Genet. 192*: 275-281, 1983.)

Table 2-21. In vivo nitrate reductase levels in *Erythrina senegalensis* leaf tissue.

Sampling site	Nitrate reductase activity (μmol NO_3/h / g fresh wt)	Total nitrogen (mg /g dry wt)
1	129	27
2	159	29
3	200	-
4	278	42
5	50	-

(Stewart and Orebamjo, *New Phytol. 83*: 311-319, 1979; reproduced by permission of Cambridge University Press.)

2-3.2.3   Nitrate and Nodulation in Peas

It is generally known that supply of NO_3^- to actively nitrogen-fixing leguminous plants considerably decreases fixation (i.e., acetylene reduction, by which fixation is measured experimentally). The nitrogen requirements of the plant are then met by reduction of nitrate. There are different explanations possible for the mechanism of inhibition: It can be the regulatory effect of NO_3^- itself or the result of the reduction of NO_3^-, either by the bacteroid in the nodule or by the plant, or by both. The rapid use of NO_3^- supplied to the plant obscures the ability to distinguish between these various possibilities. One way to solve this problem would be to use mutants blocked in the reduction of nitrate so that the effect of NO_3^- could be separated from those of reduction products of nitrate, such as NH_4^+ (Feenstra et al., 1982).

Through the use of a mutant of pea that had only about 20 percent of the NR activity in its leaves of the wild type, the effect of nitrate, ammonia, and chloride on acetylene reduction was determined (Table 2-22). In the wild-type plants (cv. Rondo), acetylene reduction in the presence of nitrate was 47 percent less than in the presence of chloride. In the mutant (E1) the inhibition by nitrate was only 19 percent (Table 2-22). On the other hand, both in the wild type and the mutant, the presence of ammonia lead to a considerable inhibition of acetylene reduction, which was the same in both cases. These results seem to indicate clearly that it is not the presence of nitrate itself that brings about the decrease in acetylene reduction, but that reduction of nitrate by the plant must take place to evoke the inhibiting effect. The observation that nitrate also had a slight inhibiting effect in the mutant can be explained by the fact that the in vivo NR activity of the mutant is not zero, but rather, 20 percent of that of the wild type (Feenstra et al., 1982).

Incidentally, but interestingly, nitrate has been found not only to

Table 2-22. Acetylene reduction characteristics of pea mutant E_1 and its parent, cv. Rondo, in the presence of nitrate or ammonium and chloride. Number of sets of two plants are in parentheses.
Different letters after data sets indicate significant differences ($P=0.05$). All data are given ± SD.

	cv. Rondo	E_1
μmol C_2H_2 reduced per h per g fresh weight of nodules in the presence of Cl^-	10.1±3.0(5) a	9.0±3.6(6) a
in the presence of NO_3^-	5.4±1.3(8) b	7.3±1.0(7) a
inhibition by NO_3^-	47%	19%
μmol C_2H_2 reduced per h per g fresh weight of nodules in the presence of Cl^-	117.4±4.6(7)	17.4±3.6(6)
in the presence of NH_4^+	1.3±4.5(6)	9.7±0.8(5)
inhibition by NH_4^+	35%	44%

(Feenstra et al., *Z. Pflanzenphysiol.* *105*: 471-474, 1982; reproduced by permission of Gustav Fischer Verlag.)

influence acetylene reduction in legumes but nodulation itself, which is reduced if excess nitrate is supplied to seedlings at the stage of nodule formation. For the selection of persistently nodulating mutants, mutagenized pea seeds (cv. Rondo) were selected at the M_2 stage by growing them in medium containing 15 mM KNO_3. A mutant was obtained that had a high nodulation character and that segregated as a single, recessive gene (Table 2-23; Jacobsen and Feenstra, 1984).

The results in Table 2-23 show that under standard conditions of growth the nodulation of the mutant was more rapid and gave larger numbers of nodules and a higher nodule mass than the wild type. After growth on 15 mM KNO_3 the difference between the genotypes was even more striking, results that emphasize the efficient nodulation of the mutant even in the presence of high nitrate concentrations (Jacobsen and Feenstra, 1984). Separately, it was shown that acetylene reduction in this mutant was the same as that in the wild type and was affected to the same degree by NO_3^- when placed on a per nodule basis (Table 2-24).

Thus, the difference in the nodulation is not attributable to a change in the resistance of the acetylene reduction mechanism to nitrate but to some other change in the physiological processes governing nodulation. Investigation of the reasons for the observed differences between mutants such as this, so-called, *nod3* mutant in pea, and others in other species (e.g., see Gremaud and Harper, 1989), and the wild type could throw light on the processes that control nodulation in legumes, an understanding of which may have economic value.

Table 2-23. Mean nodulation of *Pisum sativum* cv. Rondo and the mutant cultured in medium with or without nitrate. SMS = Standard Mineral Solution.

Genotype	Total		Conditions of	
	No. of nodules	Total fresh nodule wt. (mg)	Germination	Culture
cv. Rondo	59.3	125	Water	SMS
Mutant	>300	589	Water	SMS
cv. Rondo	16.6	26	Water	SMS+15 mM KNO$_3$
Mutant	>250	682	Water	SMS+15 mM KNO$_3$
cv. Rondo	12.7	21	SMS+15 mM KNO$_3$	SMS+15 mM KNO$_3$
Mutant	>250	505	SMS+15 mM KNO$_3$	SMS+15 mM KNO$_3$

(Jacobsen and Feenstra, *Plant Science Letters 33*: 337-344, 1984; reproduced by permission of Elsevier Science Publishers.)

In addition to these major studies (Sections 2-3.2.1 and 2-3.2.2), two other smaller investigations illustrate how mutants with disturbed NR can be valuable in plant physiological investigations.

Table 2-24. Acetylene reduction in *Pisum sativum* cv. Rondo and a nodulation mutant after growth in medium with and without added nitrate. [a] μmoles of C$_2$H$_4$ produced per hour. SMS = Standard mineral solution.

Genotype	Acetylene reduction[a]		Culture conditions
	per plant	per g of fresh wt. of nodules	
Rondo	2.0±0.5	16.1±	SMS
Rondo	0.2±0.04	8.5±0.6	SMS+15 mM KNO$_3$
Mutant	6.8±0.4	11.6±2.5	SMS
Mutant	4.1±0.04	5.6±0.3	SMS+15 mM KNO$_3$

(Jacobsen and Feenstra, *Plant Science Letters 33*: 337-344, 1984; reproduced by permission of Elsevier Science Publishers.)

2-3.2.4A   *Chlorate-hypersensitive, High Nitrate Reductase Mutant*

The collection of chlorate-resistant mutants of *Arabidopsis thaliana* is made up mainly of those that are deficient in NR or have an impaired uptake of NO_3^- or of chlorate. In one study, however, a population of *A. thaliana* was produced by self-fertilization of mutagenized plants exposed to chlorate in the watering solution. Plants showing early susceptibility symptoms were rescued. Among the progeny lines of these plants was one (designated as C-4) that possessed elevated activity of NR throughout its life cycle (Fig. 2-22; Wang et al., 1986).

Inheritance studies of NR indicated that the elevated activity in C-4 was responsible for the hypersusceptibility to chlorate and was controlled by a single recessive gene. One of the important uses of such a high NR mutant would be in comparative molecular biology. The NR of *A. thaliana* represents one enzyme for which low, normal, and high activity variants now exist in similar genetic backgrounds. Comparative biochemical and molecular studies of these could advance knowledge of the control of nitrogen metabolism in higher plants (Wang et al., 1986).

2-3.2.5   *The Consequence of the Absence of Nitrate Reductase Activity on Photosynthesis*

Chlorate-resistant *Nicotiana plumbaginifolia* cv. Viviani mutants were found to be NR⁻ and were cultivated by grafting onto wild-type *N.*

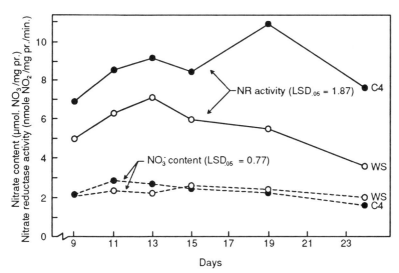

Figure 2-22. Nitrate reductase activity and tissue nitrate content of mutant (C-4) and normal genotypes of *Arabidopsis thaliana* as affected by plant age. (Wang et al., *Theor. Appl. Genet.* 72: 328-336, 1986.)

tabacum plants. The grafts of mutant plants were chlorotic compared with the grafts of wild type. Mutant leaves did not accumulate more nitrogen and nitrate, but contained less malate and more glutamine than wild-type leaves (Table 2-25; Saux et al., 1987).

Thus, as a consequence of NR deficiency, grafted mutants displayed a chlorotic leaf phenotype. Clearly this was not as a result of nitrate poisoning or nitrogen starvation. The mutants did not accumulate nitrate but did contain as much total nitrogen as, and more reduced nitrogen than, the wild type. One of the most striking manifestations that preceded the deterioration of the photosynthetic apparatus was the accumulation of excessive numbers of starch grains in the chloroplasts. Thus, the partitioning of carbon in photosynthesis of mutant leaves between starch (storage) and sucrose (utilization) seems to be a consequence of the NR⁻ genotype. Such mutants may, therefore, provide a means to investigate the control at this important point in the photosynthetic metabolism (Saux et al., 1987).

These few examples illustrate the wide range of physiological processes that can be investigated using NR-deficient variants. They also represent the best case to date to illustrate the point made in the introduction to this Chapter: that auxotrophs, in general, provide unique material for exploring connections between the genetics and physiology of plants.

Case Study 2-4:   Resistance to Amino Acids or Their Analogues

Variants resistant to amino acids or their analogues have been sought among plants or in plant cell cultures for a number of reasons. Most often

Table 2-25. Nitrogen, nitrate, malate, glutamine, and starch content of leaves of *Nicotiana plumbaginifolia* NR⁺ and NR⁻ after two weeks of growth as graftings on *N. tabacum* scion.

	Total Nitrogen	NO_3^-	Malate	Glutamine	Starch
	mg/g dry wt	*µmol/g dry wt*			*mg/g dry wt*
Control NR⁺	72±5	2000±180	68±5	40±4	18±3
Mutant NR⁻	69±	1900±170	15±3	52±4	79±9

(Saux et al., *Plant Physiol. 84*: 67-72, 1987; reproduced by permission of the ASPP.)

the search for variants has been motivated by a desire either to improve the nutritional value of food crops or to provide more knowledge about metabolic regulation.

Resistance to analogues of amino acids is the simplest and most widely employed procedure for selecting mutations affecting the control of amino acid biosynthesis or catabolism. The toxicity of an amino acid analogue to cultured plant cells is due either to its incorporation into proteins in place of the natural amino acid or to its inhibition of an allosterically regulated enzyme in the pathway leading to the biosynthesis of an amino acid. Resistance can be conferred by mutations that lead either to the degradation of the analogue, to its exclusion from the cell, or, if it interferes with cellular metabolism by binding to an allosteric enzyme, by reducing the affinity of the enzyme for the analogue. This last type of mutation will also allow the formation of more of the natural end-product amino acid that, in turn, will lead to the dilution of the effect of the analogue in protein synthesis by reducing its effective intracellular concentration (Chaleff, 1981).

In this Case Study, two brief examples are given of amino acid variants in plants and their value in physiological investigations. The first example is the use of an amino acid analogue to isolate variants in the pathway of tryptophan biosynthesis; the second, with the isolation of mutants resistant to high levels of certain natural amino acid products in the aspartate pathway.

2-4.1   Mutants in the Pathway of Tryptophan Biosynthesis

2-4.1.1   *The Isolation and Characterization of Mutants*

The tryptophan biosynthetic pathway that operates in higher plants appears to be the same as that in microorganisms (Fig. 2-23; Widholm, 1972a). Mutants in this pathway were isolated from tobacco (*Nicotiana tabacum* var. Xanthi) by growing pith cells in liquid culture medium in the presence of varying concentrations of DL-5-methyltryptophan (5MT). The growth of the cells was inhibited completely by the presence of 2 mg/L of the 5MT (Fig. 2-24; Widholm, 1972b). If tobacco cells were inoculated into medium containing 10 mg/L of 5MT and incubated for periods of 30 to 60 days, clumps of cells began to appear in the culture flasks. When these cells were further tested at various concentrations of 5MT they were shown to be highly resistant to the analogue (Fig. 2-24). The growth of these cells was only about 50 percent inhibited by 250 mg/L of 5MT and the resistance persisted over many generations (cell doublings) (Fig. 2-24).

After isolating these resistant variants the task was then to determine the origin of resistance that required some biochemical knowledge of the pathway in plants leading to *trp* synthesis. In plants, *trp* biosynthesis

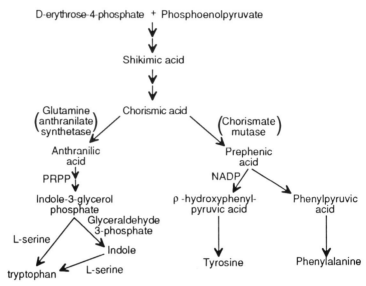

Figure 2-23. The pathway for aromatic amino acid biosynthesis in plants. (Widholm, *Biochim. Biophys. Acta 261*: 44-51, 1972a; reproduced by permission of Elsevier Science Publishers.)

is controlled by feedback inhibition of the branch-point enzyme, anthranilate synthetase (AS) (see Fig. 2-23). The 5MT inhibits the synthetase, thus stopping cell growth because of *trp* starvation. Cells resistant to 5MT had a synthetase less sensitive to inhibition by the analogue and, incidentally, *trp* (Fig. 2-25; Widholm, 1972b). The data

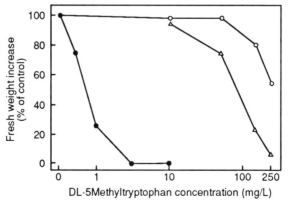

Figure 2-24. The effect of 5-methyltryptophan (5MT) on tobacco cell growth (increase in fresh weight in 10 days). Normal cells (●); resistant cells grown in a medium containing 10 mg/L of 5MT (○); resistant cells grown for 60 cell mass doublings in a medium lacking 5MT (△). (Widholm, *Biochim. Biophys. Acta 261*: 44-51, 1972a; reproduced by permission of Elsevier Science Publishers.)

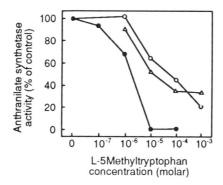

Figure 2-25. Inhibition of anthranilate synthetase by DL-5-methyltryptophan. Enzyme from normal cells (●); (○) and △, as in Fig. 2-24. (Widholm, *Biochim. Biophys. Acta 261*: 52-58, 1972b; reproduced by permission of Elsevier Science Publishers.)

show that 2×10^{-5} M of 5MT completely inhibited normal tobacco cell AS. The same concentration of the analogue, however, inhibited the enzyme from the resistant cells by only about 50 percent. The occurrence of the resistant enzyme was accompanied by a *trp* pool size in resistant cells of 80 to 150 μM as opposed to 3 to 10 μM in normal cells (Widholm, 1972b). This indicated that the normal feedback control of *trp* biosynthesis was indeed relaxed, which, in turn, indicated that feedback control of AS is important in the control of the size of the *trp* pool.

2-4.1.2   The Use of 5MT-Resistant Mutants in Physiological Studies

2-4.1.2.1   Investigations of the Expression of Anthranilate Synthetase in Whole Plants.

When plants were regenerated from the 5MT-resistant tobacco cell lines isolated by Widholm (1972a,b), the altered AS enzyme was not expressed in the plants (see Ranch et al., 1983). Cultures reinitiated from the leaves of these plants, however, did express the altered enzyme, showed increased free *trp* and expressed 5MT resistance. Therefore, the question arose as to whether potentially useful genetic variation established in tissue culture could be carried over into the whole plant. If it could not, then tissue culture as a technique to generate rapidly genetic variants of agronomic value, for example, would be much diminished.

To address this problem, Ranch et al. (1983) used a suspension culture of haploid cells of another solanaceous species, *Datura innoxia*, to generate 5MT-resistant variants and, then, whole plants. As shown in Table 2-26, the resistant AS was expressed in the leaves of the plants

Table 2-26. Inhibition of anthranilate synthetase activity from leaves of wild type *Datura innoxia* and three 5MT-resistant variants, MT-2/1, MT-2/2, and MT-2/3.
[a.] nmole anthranilate formed mg protein hour.

Trp	Anthranilate Synthase Activity				
	Wild type leaf	MT-2/1	MT-2/2	MT-2/3	MT-2 suspension
μM	*% of control*				
0	100	100	100	100	100
1	53.3	67	89	91.6	
6.1	25.4	58.9	77.8	72.7	76.7
11	13.7	48.2	57.8	72	56.7
61	4.28	33.3	22.2	38.8	7.3
110	0	10.2	11.8	16.7	0
610		0	0	0	0
μM Trp for 50% inhibition of control activity	1.2	12	18	30	15
Specific activity [a]	1.04	1.19	0.87	0.59	5.6

(Ranch et al., *Plant Physiol. 71*: 136-140, 1983; reproduced by permission of theASPP.)

regenerated from cultures resistant to 5MT thus allaying the fear, to an extent, that tissue culture could not be used as a source of genetic variation at the whole-plant level. The information available from these experiments, however, did not allow a determination of why regenerated *D. innoxia* and not *N. tabacum* plants express the same AS as the cultured cells. Speculation arose as to whether the difference lay in the fact that *D. innoxia* is a true diploid (i.e., monohaploid) species, whereas *N. tabacum* is amphidiploid. Thus, it was suggested that tobacco could possess two separate gene loci for AS that could be expressed differentially in cultures and in plants.

Some important results were obtained through the use of cell cultures of another solanaceous species, the potato (*Solanum tuberosum* L. cv. Merrimac), to explain the different expression of an enzyme in various growing conditions (Carlson and Widholm, 1978). Potato cells were isolated with resistance to 5MT and the resistance was shown to be accompanied by a resistant AS. Crude extracts of proteins from wild-type (PO1) cells were made and separated on gels. The analysis, illustrated in Figure 2-26, clearly showed a peak of AS activity with a R_m

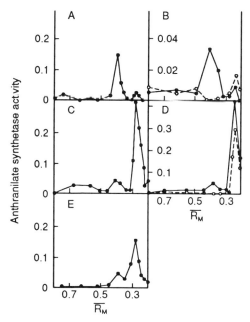

Figure 2-26. Preparative polyacrylamide gel electrophoresis of potato cell crude extracts. **A.** PO1 extract (1 mL); **B.** PO1 extract (2 mL); **C.** PO2 (1 mL); **D.** PO2 (2 mL); **E.** equal activities in PO1 and PO2 extracts mixed. (●) 0.5 mL each fraction assayed for 0 and 60 minutes at 37°C without inhibitor; (○) with 100 µM tryptophan. Activity is in anthranilate produced by 0.5 mL/hour. (Carlson and Widholm, *Physiol. Plant. 44*: 251-255, 1978.)

value of 0.40 and a much smaller one at about 0.28. The main peak of PO1 AS activity was inhibited by 100 µM *trp* whereas the small peak was unaffected. In the case of the resistant cell line, PO2, the bulk of the AS activity was in a peak at R_m 0.27 and much less at R_m 0.40. The 0.40 peak was completely inhibited by 100 µM *trp* whereas the 0.27 peak was not much affected (Fig. 2-26).

Thus, in the case of these potato cells, two isozymes of AS, one sensitive and one resistant to inhibition by 5MT and *trp*, were detected. Normal cell lines contained more of the feedback-sensitive form, whereas 5MT-resistant cell lines contained more of the feedback-insensitive form. Presumably, resistance to 5MT was achieved by increasing the rate of synthesis of an AS isozyme that is not inhibited by the analogue and correspondingly decreasing the rate of synthesis of the sensitive isozyme. In the case of potato, whether these alterations were due to mutations or to adaptations that altered the AS isozyme levels is unclear but it may be that in other cases, where resistance is manifested in cell culture conditions but not in the whole plant, the 5MT-sensitive isozyme is more strongly expressed at the plant level even in the resistant lines. Of course, this possibility exists in any plant species in which more than one form of an enzyme is normally found.

2-4.1.2.2   Studies of Embryogenesis in Carrot.

Growth and development in plants are mediated by growth hormones. The action of the hormones varies depending on their concentration and their location in the plant. The regulation of the hormone levels within and between different cells of the plant, however, is still not well understood. Plant cell cultures consisting of large, relatively homogeneous cell populations provide a convenient system to study hormone regulation in cells. Two examples are provided to illustrate the use of cultures of normal and variant carrot cells to investigate the influence of hormones on aspects of development (Sung, 1979).

Plant cells in culture may exhibit two types of growth: (1) undifferentiated growth as callus or (2) differentiation of organs or embryos. To maintain callus growth, cultured plant cells, as a rule, require exogenously supplied hormones. In the case of carrot, only one type of hormone, auxin, is required. In the absence of auxin, carrot cultures usually differentiate into embryos (see Sung, 1979).

In 5MT-resistant cell lines of carrot, as in the case of tobacco already described, higher levels of endogenous *trp* accumulated than in normal cells and some, but not all, also were capable of continuing to grow as callus in the absence of exogenous IAA (Sung, 1979). Because *trp* is known to be the natural precursor of IAA, a simple explanation for the IAA-independent growth in the 5MT-resistant cell lines would be the conversion of the "excess" *trp* into IAA, resulting in elevated IAA levels that were sufficient to support callus growth in the absence of exogenous IAA.

To test this hypothesis, endogenous levels of *trp* were determined in wild-type and in 5MT-resistant cultures. Generally, the resistant cell lines contained more *trp* than did the wild type (W001C) (Table 2-27). In the medium without 2,4-dichlorophenoxyacetic acid (2,4-D), the wild-type cells regenerated 100 percent to plantlets, whereas the 5MT-resistant (W001) cells remained as callus tissue and had a rate of plantlet formation of less than 10^{-5} per cell input. The question then arose as to whether the level of IAA in the cells was related to the *trp* level

Table 2-27. Endogenous concentrations of tryptophan in cell lines of wild carrot grown on medium with and without the auxin 2,4-D. *Trp* was determined in 17-day-old wild type (W001C) and 5MT-resistant (W001) cultures grown on agar.

Cell Line	Tryptophan (μg/g fresh wt.)	
	+ 2.4-D	- 2.4-D
W001C	70	52
W001	21.3	148

(Sung, *Planta 145*: 339-345, 1979.)

(Table 2-28). In further experiments, it was observed that, in the absence of 2,4-D, 5MT-resistant (WOO1) cells contained about 40 times more of IAA than did wild-type (WOO1C) cells. It was also observed that the level of IAA in W001 cells in the presence of 2,4-D was reduced by 15 times more than in the absence of the artificial hormone (Table 2-28). Thus, either 2,4-D interferes with IAA synthesis, or removal, or else there is a phenomenon of auxin self-regulation in these cells.

The main conclusion to be drawn from these observations is that in carrot, auxin alone can have a crucial influence on the degree of differentiation of cells. In contrast, in tobacco, in the classic work of Skoog and Miller (see Sung et al., 1979 for reference), and in many other species, organogenesis was found to be dependent on the ratio of auxin to kinetin (ie., cytokinins). In view of these contradictory findings it seemed useful to examine the interaction of auxin and cytokinin in embryogenesis using cell lines with different endogenous levels of IAA, namely, normal (W001C) and 5MT-resistant (W002 not W001, as above) (Sung et al., 1979).

When normal cells were cultured in a medium lacking 2,4-D for 20 days, as expected, no calli were observed, only embryos in various stages of development (globular, heart, torpedo) and plantlets, and some greatly enlarged cells. If the frequency of embryogenesis in wild-type cells was arbitrarily set at 1.0, the 5MT-resistant cell line, W002, had a frequency of 0.0005 under the same conditions. To study the effect of the cytokinin 2-isopentenyladenine (2-ip) on embryogenesis in wild-type cells, 2,4-D was added to the culture to reduce the frequency of regeneration and to simulate the situation in the 5MT-resistant cultures where regeneration is impaired by a high endogenous concentration of auxin. Table 2-29 illustrates the effects of varying concentrations of 2-ip on the regeneration frequency of W001C grown in 0.04 mg/L of 2,4-D.

Results showed that a small amount of regeneration occurred in the control, that is wild-type cells grown in 0.04 mg/L 2,4-D and zero 2-ip. Although the total number of calli remained relatively constant, the frequency of embryo formation increased to 0.4 in 0.1 mg/L 2-ip and then decreased at higher 2-ip concentrations (Table 2-29). Most embryos

Table 2-28. Endogenous concentration of indole-3-acetic acid in wild-type (W001C) and 5MT-resistant (W001) cell lines of wild carrot grown on medium with and without 2,4-dichlorophenoxyacetic acid (2,4-D).

Cell Line	IAA (µg/g dry wt.)	
	+ 2.4-D	- 2.4-D
W001C	0.28	1.15
W001	3.17	46.31

(Sung, Planta 145: *339-345, 1979.)*

Table 2-29. Effect of 2-isopentenyladenine (2-ip) on embryo formation in W002 (5MT-resistant) and W001C (wild type) wild carrot cells, the latter grown in a medium containing 0.04 mg/L of 2,4-D.

Conc. of 2-ip (mg/l)	Ratio embryos/(calli - embryos)	
	W001C in 2.4-D	2002
0	0.0^7	0.0005
0.01	0.17	0.005
0.05	0.23	0.010
0.10	0.40	0.013
0.50	0.08	0.020
1.00	-	0.012

(Sung et al., *Planta 147*: 236-240, 1979.)

scored in these cultures at 20 days were at the globular stage; at 35 days, a few had advanced to the heart stage but no torpedo stage or plantlets were observed. Similar results were found in W002 cultures (Table 2-29), except that the globular embryos that appeared in these cells failed to differentiate to more advanced stages although they continued to increase in diameter. These results (Table 2-29) give a clear indication that cytokinin can promote embryogenesis but only in a narrow range of concentrations. The effect of 2,4-D on carrot culture is twofold. First, the auxin inhibits the induction of adventive embryo formation. Second, once embryogenesis is initiated the auxin can stop subsequent organogenesis. The 2-ip can partly overcome the former effect of 2,4-D but not the latter, indicating that these two effects are based on different mechanisms. By what mechanisms 2,4-D influences embryogenesis adversely is not understood and was not addressed in these experiments (Sung et al., 1979). The comparison of cell lines differing in hormone concentration, however, allows an investigation of the interaction of plant growth substances in development and provides material useful in separating the effects of one hormone from another in differentiation.

2-4.2.   Mutants in the Pathway of Aspartate-derived Amino Acids

The series of reactions by which lysine, threonine, isoleucine, and methionine are synthesized from aspartate is of particular interest as all four of these amino acids are essential to human nutrition. Moreover, lysine is generally the nutritionally limiting amino acid in cereals, whereas small amounts of methionine limit the quality of legume seed proteins, for example. The hope, then, is to select regulatory mutations that will cause increased production of these amino acids. The biosynthesis of lysine and methionine is regulated by complex mechanisms that differ from one species to another. This complexity is compounded by the production during plant development of several enzyme forms that exhibit different sensitivities to inhibition by the end products of the pathway (Fig. 2-27; Chaleff, 1981).

The three enzymes of lysine and methionine synthesis that apparently function as key control points are aspartokinase, dihydropicolinate synthetase, and homoserine dehydrogenase. These enzymes are allosterically controlled by different combinations of lysine and threonine in different species. Because of these allosteric control mechanisms, the flow of carbon through the aspartate pathway can be interrupted by

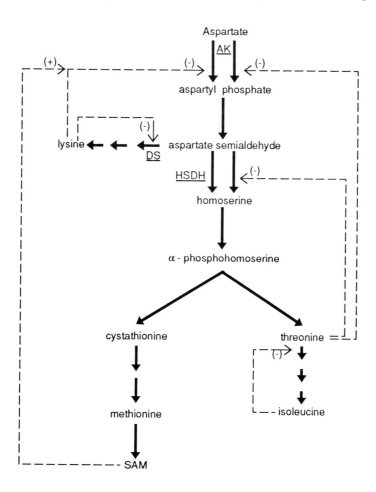

Figure 2-27. Biosynthesis and regulatory control of the aspartate family of amino acids. AK, aspartate kinase; DS, dihydrodipicolinate synthase; HSDH, homoserine dehydrogenase. (Gonzales et al., *Plant Physiol.* 74: 640-644, 1984; reproduced by permission of the ASPP.)

exposing the cell to either a lysine analogue (which, presumably, would inhibit aspartokinase and dihydropicolinate synthetase) or an excess of lysine plus threonine (which leads to the cessation of methionine synthesis). Selection of mutations affecting regulation of the aspartate pathway, then, could lead to the isolation of variants exhibiting an overproduction of one or other of the end products of the pathway (Chaleff, 1981). A case in point is the isolation of a series of barley mutants.

2-4.2.1   The Isolation and Characterization of Aspartate Pathway Mutants in Barley

From an M_2 (M_1 treated with sodium azide) population of 10^4 mature barley embryos three mutants resistant to lysine plus threonine (2-3 mM) were selected for further study and named R (Rothamsted) 2501, R3004 and R3202 (Bright et al., 1982). Both R3202 and R3004 plants grew better than controls in a range of concentrations of lysine and threonine and also lost the synergistic inhibition of growth by lysine plus threonine (Fig. 2-28). Results of genetic crosses in the three mutants suggested that resistance was due to dominant genes designated as *Lt1a, Lt1b,* and *Lt2* which were present, in R2501, R3202, and R3004, respectively. *Lt2* was not linked to either *Lt1a* or *Lt1b*, whereas the latter behaved as though allelic. The question then arose as to what events in barley metabolism were influenced by these mutations.

The enzyme aspartate kinase (AK) in barley can be separated into the three isozymic forms AKI, AKII, and AKIII. The latter two isozymes are sensitive to feedback inhibition by lysine (Rognes et al., 1983).

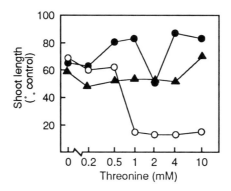

Figure 2-28. Growth of two mutants, R3004 (▲) and R3202 (●), and wild type (○) barley in a range of threonine concentrations with lysine held constant at 3 mM. (Bright et al., reprinted by permission from *Nature* vol. 299 pp. 278-279. Copyright © 1982 Macmillan Magazines Ltd.)

Analyses of the properties of these isozymes has shown that R2501 and R3202 both have an AKII with greatly decreased sensitivity to inhibition by lysine, and a normal AKIII, whereas R3004 has an unchanged AKII but an AKIII that is relatively unaffected by lysine (Table 2-30). These results, taken with the genetic data, suggested that resistance to lysine plus threonine in R2501 and R3202 was due to mutations in a gene, *lt1*, causing altered feedback sensitivity of AKII and in R3004 was due to a mutation in a gene, *lt2*, that decreases lysine inhibition of AKIII. The genes *lt1* and *lt2* were, therefore, likely to be structural loci for these two AK isozymes (Rognes et al., 1983).

The suggestion has been made that mutants having such altered regulatory feedback properties as those shown in these examples might produce cereal plants with grains of improved nutritional quality. Analyses of the soluble amino acids of seeds showed that the mutation in R2501 and R3004 caused a greater than 12-fold increase in threonine. The absence of such an accumulation in R3202 was unexplained. None of the mutants had enhanced levels of soluble lysine, probably because the activity of the first enzyme unique to lysine biosynthesis, dihydropicolinic acid synthase, is also tightly regulated by lysine ; (Table 2-31 Bright et al., 1982).

Table 2-30. Distribution of activity and lysine sensitivity of three forms of aspartate kinase (AKI, II and III) extracted from barley mutants and compared to wild type (Bomi).
[a.] Threonine-sensitive, lysine-insensitive form.
[b.] No effect of lysine.

Plant type	Distribution of aspartate kinase activity (% of total)			Ratio II/III	Inhibition by 10 mM lysine (%)		Lysine concentration (mM) required for half-maximal inhibition ($I_{0.5}^{lys}$)	
	Peak I[a]	Peak II	Peak III		Peak II	Peak III	Peak II	Peak III
Bomi (sensitive)	15	32	53	0.60	82	93	0.40	0.36
R3004 (resistant)	17	46	37	1.24	85	60	0.30	3.5
R3202 (resistant)	17	21	62	0.35	0	93	-[b]	0.32
R2501 (resistant)	9	48	43	1.12	42	90	10	0.40

(Rognes et al., *Planta 157*: 32-38, 1983.)

Table 2-31. Soluble amino acids in grains of normal barley and three mutants.

Amino Acid	Soluble content (nmol per mg nitrogen)				
	Pooled controls		R3004	R3202	R2501
	I	II			
Threonine	9±3	12± 5	116±36	16±5	147±21
Lysine	12±5	12±5	15± 3	18±5	19±8
Alanine	49±13	47±10	41± 6	68±14	19±5
Valine	18±10	28±12	17±5	38±10	19±4
Glutamine	143±25	140±36	122±20	199±39	112±22
Asparagine	167±27	181±28	122±18	237±45	131±24

(Bright et al., reprinted by permission from *Nature* vol. 299 pp. 278-279. Copyright © 1982 Macmillan Magazines Ltd.)

2-4.2.2   *Some Uses of Amino Acid Pathway Mutants in Physiological Studies*

2-4.2.2.1   Uptake of Amino Acids by Barley Plants.

From additional screens to those discussed for barley mutants with the potential to accumulate soluble lysine, two mutants were recovered that were resistant to the toxic effects of the lysine analogue S-(2-aminoethyl)-L-cysteine (*Aec*). Pure breeding lines of the mutants (called R906 and R4402 by Bright et al., 1983) were obtained and the *Aec* resistance shown to be inherited as a single, recessive, nuclear gene.

Lysine uptake by R4402 excised roots was compared with those of the wild type Maris Mink over the range 10^{-7} to 10^{-2} M lysine. Between 10^{-7} and 10^{-5} M, the mutant rate was about 30 percent of the control value, but above 10^{-5} M the two values approached each other until at 10^{-2} M the mutant rate was 90 percent of the control figure (Fig. 2-29; Bright et al., 1983).

In further tests, uptake of a number of basic, neutral, and acidic amino acids were compared in R4402 and wild-type cells. The mutant took up near normal amounts (89-108 percent) of neutral and acidic amino acids, whereas the uptake of lysine, arginine, and ornithine was impaired suggesting a specific deficiency in a basic amino acid transport system (Table 2-32; Bright et al., 1983). Lysine uptake by leaf slices of mutant and parent plants was compared to establish whether there was a loss of uptake capacity in leaf cells parallel with the loss in the roots. No significant differences in lysine or leucine uptake rates between mutants and parents was observed.

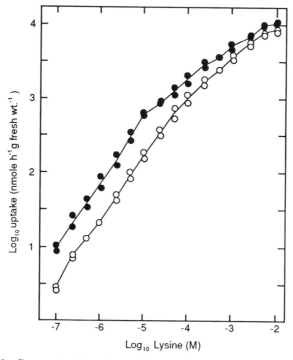

Figure 2-29. Concentration dependence of lysine uptake by roots of 4-day-old wild type (Maris Mink) (●) and R4402 (○) barley plants. (Bright et al., *Plant Physiol.* 72: 821-824, 1983; reproduced by permission of the ASPP.)

A comprehensive analysis of data on amino acid transport in higher plants has led to the proposal that two carrier systems are normally present: a general amino acid transport channel of low specificity and a more specific, basic amino acid system. The decreased uptake of lysine observed in this study was most apparent at concentrations below 10^{-5} M; at higher concentrations, the difference between the mutant and the parent decreased progressively (Fig. 2-29). Therefore, the simplest interpretation of the results is that two carrier systems for lysine exist in the barley root cell membrane and that one of these, a high-affinity system, is altered or lacking in the mutants. The missing system appears to be the one that specifically transports basic amino acids into the root cells. It also seems apparent that the *aec1* gene affects the uptake of basic amino acids into roots but has no influence on the uptake of lysine into leaf tissue (Bright et al., 1983).

2-4.2.2.2   Ethionine Resistance in Tobacco.

Cells of *Nicotiana tabacum* L. cv. Xanthi (TX1) were selected for resistance to growth inhibition by the methionine analogue ethionine;

Table 2-32. Amino acid uptake by roots of 3-day-old wild type (Maris Mink) barley and the mutant, R4402. Amino acid concentration 50μM, except for DL-ornithine (100 μM). The right column gives uptake by R4402 as percentage of uptake by Maris Mink roots.

Amino Acid	Uptake		
	Maris Mink	R4402	
	nmol/g fresh wt.h		*%*
Exp. I			
Leucine	457+14	408+17	89
Asparagine	222+71	200+14	90
Lysine	1013+82	461+35	45
Arginine	626+46	333+38	53
Exp. II			
Glutamic acid	47+ 5	52+ 6	108
Alanine	358+ 41	376+46	105
Lysine	1284+152	455+59	35
DL-Ornithine	658+35	322+30	49

(Bright et al., *Plant Physiol. 72*: 821-824, 1983; reproduced by permission of the ASPP.)

the cell line R11 resulted (Gonzales et al., 1984). An amino acid analysis of R11 revealed an increase in the cellular levels of several free amino acid pools over those in TX1 (Table 2-33). The increases were most notable for *met* and *thr*, which showed levels 110 and 18 times, respectively,

Table 2-33. Variation in the free amino acid profiles of ethionine-resistant and ethionine-sensitive tobacco suspension cultures.
[a] Expressed as nmol/g fresh weight. Met, methionine; thr, threonine; ile, isoleucine; lys, lysine; gly, glycine; ser, serine; arg, arginine; leu, leucine.

Amino Acid	TX1	R11	(R11/TX1)
Met	5[a]	511[a]	110
Thr	370	6768	18
Ile	27	146	5
Lys	47	214	5
Gly	328	1369	4
Ser	286	1466	5
Arg	50	225	5
Leu	88	270	3

(Gonzales et al., *Plant Physiol. 74*: 640-644, 1984; reproduced by permission of the ASPP.)

those of TX1. These data suggested that the resistance to the growth inhibition of ethionine exhibited by R11 was due to the overproduction of methionine. In turn, this suggested a possible alteration in one of the regulatory control enzymes in the methionine biosynthetic pathway. Methionine adenosyltransferase activity, however, was unaltered. Homoserine dehydrogenase activity and its sensitivity to feedback inhibition by *thr* was also found to be unaltered in the resistant cells. Thus, the search was extended to include AK.

Total AK activity in TX1 cells was inhibited approximately 50 percent each by *lys* and *thr*. The combination of the two resulted in almost complete inhibition in both cases, suggesting the presence in tobacco of at least two isozymes, one sensitive to *lys* and the other to *thr*. In R11, however, the percent inhibition by these amino acids was quite different; 90 pecent for *lys* and 10 pecent for *thr* (Table 2-34; Gonzales et al., 1984).

The main point to note from the data in Table 2-34 is that the total AK activity in R11 was nine times that in TX1 and that this difference was due solely to an increase in the level of activity for the lysine-sensitive isozyme in R11. A decline in total activity with increasing concentration of *lys* or *thr* should be due to the loss of activity of the corresponding feedback-sensitive isozyme, and the activity remaining at high levels should be due to the complementary isoenzyme. Therefore, the activity of the two isozymes could be calculated. By these criteria, the threonine-sensitive activities for TX1 and R11 were 0.68 and 1.04 nmol/hour/mg protein, respectively, and for the lysine-sensitive activities were 0.77 and 12.22 nmol/hour/mg protein, respectively. Clearly, the amount of the threonine-sensitive activity was about the same in both cell lines, whereas the lysine-sensitive activity of the resistant line was elevated to 16-fold that of the sensitive line. With increased levels of AK

Table 2-34. Feedback inhibition of aspartate kinase in ethionine-sensitive (TX1) and ethionine-resistant (R11) cultured tobacco cells. [a,b]. Control activity = 1.46 and 13.66, respectively.

	Inhibition	
Addition	TXl	R11
	(%)	(%)
None	0[a]	0[b]
10 mM Lys	55	92
10 mM Thr	48	10
10 mM Lys + Thr	97	100

(Gonzales et al., *Plant Physiol. 74*: 640-644, 1984; reproduced by permission of the ASPP.)

present, as in the case of R11, more aspartate would be processed into the pathway. Most of the carbon would be shunted toward *met* and *thr* at the *lys* branch-point, as *lys* stringently regulates its own synthesis at dihydrodipicolinate synthase. The carbon would then be processed by the threonine-resistant homoserine dehydrogenase to *met* and *thr* (see Fig. 2-27).

The evidence obtained in these experiments by Gonzales et al. (1984) suggested that the difference in the lysine-sensitive AK activities of the ethionine-resistant and ethionine-sensitive cell lines was probably not due to a newly induced isozyme or an alteration in enzyme structure. The question remained as to what kind of modification at the molecular level would give rise to the elevated isozyme levels. Two possibilities would be gene amplification and derepression. Indeed, Binarova et al. (1989) have also concluded that resistance of alfalfa cells either to *lys* and *thr* or to ethionine may be caused by gene amplification.

The inhibition of growth of cultured cells and of whole plants by amino acids or their analogues has been successfully exploited for the selection of mutant cell lines. As seen in the examples of those in the aspartate pathway, resistance can take several forms, although the primary manifestation would appear to be linked to the production of key enzymes in the pathway with altered properties. Uptake mutants and those having amplified genes, however, may also be isolated using selection procedures such as those discussed here. One of the consequences of resistance in many cases is the overproduction of end products of a pathway (e,g., essential amino acids such as *lys* and *met*), which has given impetus to this field of study in an attempt to improve the amino acid composition of certain crop species. It is also clear, however, that these mutants are valuable to investigate the physiological control of metabolic pathways in higher plants. Hence, many more have been isolated than are discussed here, notably those resistant to valine (for example, see Grandbastien et al., 1989).

**Case   Study   2-5:   Herbicide   Resistance   in   Plants
and Cells**

2-5.1.   Sulfonylurea-resistant and Imidazolinone-resistant Mutants

Herbicide effectiveness is judged by its ability to discriminate between a weed and a crop. Most often this discrimination has been achieved by chemically designing compounds with specificity. An alternative would be to introduce herbicide resistance into crop species. The ability to engineer such genetic changes should broaden the usefulness of existing herbicides and reduce the need to develop new ones. These requirements have led to a number of detailed investigations of the modes and sites of action of several general herbicides in crop and other model plant species. Prominent among these have been the sulfonylurea and imidazolinone herbicides (Haughn et al., 1988).

The sulfonylurea group of general herbicides is characterized by chlorsulfuron (CS) and sulfometuron methyl (SM), (Chaleff and Ray, 1984) which are the active ingredients in the Du Pont herbicides Glean and Oust (Fig. 2-30). The imidazolinones are a class of herbicides under development by the American Cyanamid Company and are the active ingredients in commercial products such as Arsenal and Scepter (Table 2-35).

In early investigations of their actions it was discovered that sulfonylurea herbicides inhibited plant growth by blocking some process of cell division. For example, excised pea roots grown in sterile culture provided a quantitative bioassay to evaluate the effects of CS on plant growth. As little as 2.8 nM of the herbicide significantly inhibited root growth, whereas 28 nM typically inhibited root growth by more than 80 percent. Casein hydrolysate at 0.1 percent (wt/vol) in the culture medium provided a significant reversal of this growth inhibition throughout the entire range of CS concentrations tested (Ray, 1984). Among the 20 L-amino acids, the only group that alleviated CS-induced inhibition of growth was that containing *val, leu*, and *ile*. When tested against various concentrations of CS, a combination of 100 µM each of *val* and *ile* completely protected the roots from growth inhibition in the presence of up to 280 nM CS (Fig. 2-31; Ray, 1984).

The results of these supplementation experiments suggested that CS acted on one of the primary steps of the isoleucine-valine (*ile-val*)

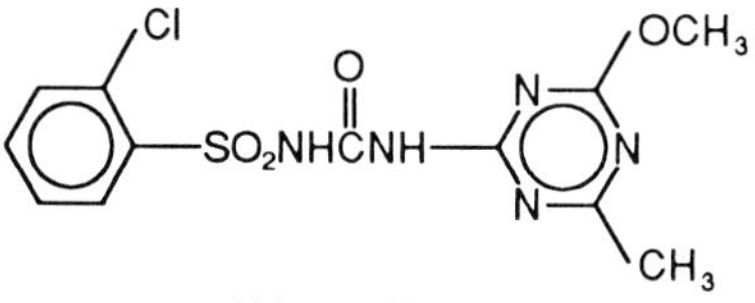

Chlorosulfuron

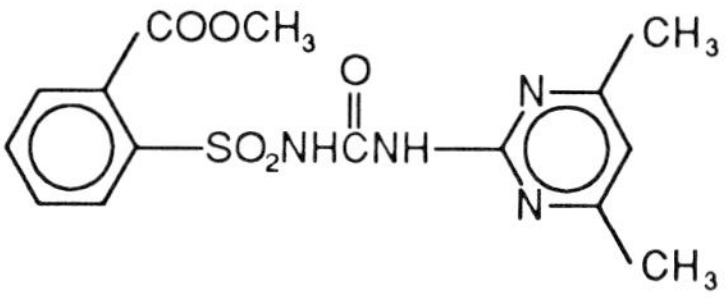

Sulfometuron methyl

Figure 2-30. Chemical structures of chlorsulfuron and sulfometuron methyl. (Chaleff and Ray, *Science 223* : 1148-1151; Copyright 1984 by the AAAS.)

Table 2-35. The chemical structures of some imidazolinones and their K_i values for acetolactate synthase extracted from corn seedlings.

AC number	R (structure)	K_i (μM)
243,997	(pyridine ring with COOH)	12
222,164	(benzene ring with COOH)	1.7
252,214	(quinoline ring with COOH)	3.4

(Shaner et al., *Plant Physiol.* 76: 545-546, 1984; reproduced by permission of the ASPP.)

pathway (Fig. 2-32). Thus, acetolactate synthase (ALS) (EC 4.1.3.18), the first common enzyme in the *ile-val* pathway, was extracted from a wide variety of plant species. CS inhibited ALS activity 50 percent (I_{50}) at very low concentrations (17 nM to 34 nM) in all species tested (Table 2-36; Ray, 1984), results that suggested ALS as the site of action of sulfonylurea herbicides, although recent evidence indicates that this in not always so (Harms et al., 1990).

Similarly, it was demonstrated that the site of action of the imidazolinone herbicides was also ALS. Three different imidazolinones were compared for their inhibitory action on ALS and all three proved to be inhibitors of the enzyme (Table 2-35). Many recent studies have confirmed that ALS is the site of action of both classes of herbicides. Such a conclusion has been confirmed through the isolation of mutants and variants that are resistant to one or other of these herbicides, examples of which are described later. A third group of herbicides has also been found to have ALS as its primary target (Subramanian and Gerwick, 1989). The action of these compounds is not discussed here.

2-5.1.1   *Isolation and Characterization of Herbicide-resistant Mutants and Variants*

One of the most definitive examples of the isolation and characterization of herbicide-resistant mutants is that exemplified by the work of Chaleff and co-workers. First, they initiated callus cultures from young leaves of a haploid plant (H1) of *Nicotiana tabacum* cv. Xanthi. Resistant cell

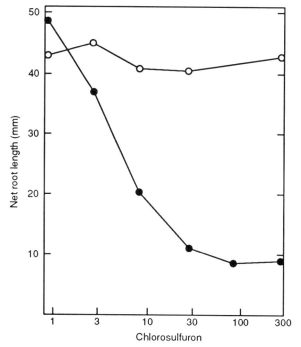

Figure 2-31. The effect of 100 μM each of valine and isoleucine on the growth of pea roots in the presence of various concentrations of CS. Control without amino acid supplement (●); with 100 μM valine and isoleucine (○). (Ray, *Plant Physiol. 75*: 827-831, 1984; reproduced by permission of the ASPP.)

lines were selected by transferring callus to medium supplemented with either CS or SM at 5.6 x 10⁻⁹ M. Plants were regenerated from cells and, as a result of crosses, resistance was shown to be conferred by a single

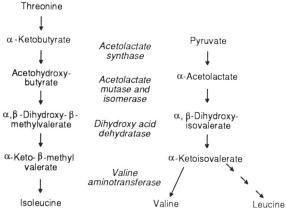

Figure 2-32. Pathway for biosynthesis of isoleucine, leucine, and valine.

Table 2-36. I_{50} values for chlorsulfuron inhibition of acetolactate synthase from various plant species.

Species	I_{50}
	nM
Pea	21.0
Wheat	18.5
Soybean	23.0
Tobacco	28.3
Green foxtail	25.8
Johnsongrass	35.9
Morning glory	24.4

(Ray, *Plant Physiol.* 75: 827-831, 1984; reproduced by permission of the ASPP.)

dominant or semidominant nuclear mutation. Linkage analyses of six mutants identified two unlinked genetic loci (Chaleff and Ray, 1984).

Phenotypic differences between callus cultures derived from several mutant isolates were revealed by their relative resistance to CS or SM (Table 2-37). For example, mutant S4 was more resistant to CS than to SM, the compound against which it was selected. Mutant C4, which was genetically linked to S4 but was of independent origin and was selected on the basis of resistance to CS, displays the same high degree of resistance to both compounds (Table 2-37; Chaleff and Ray, 1984).

Table 2-37. Relative resistance of several homozygous mutant cell lines of tobacco to sulfometuron methyl or chlorsulfuron.

Cell line	Control Weight (mg)	Sulfometuron methyl at 10 ppb Weight (mg)	Per-cent	Chlorsulfuron at 10 ppb Weight (mg)	Per-cent
H1	2573±126	32±2	1.2	33±2	1.3
S1/S1	2676±90	172±14	6.4	309±34	11.5
S4/S4	3286±191	1105±80	33.6	3171±260	96.5
S5/S5	2639±129	344±55	13.0	725±95	27.5
S6/S6	3348±159	887±92	26.5	1076±108	32.1
C3/C3	2186±165	1369±63	62.6	1896±119	86.7
C4/C4	4137±254	3473±323	83.9	3453±231	83.5

(Chaleff and Ray, *Science 223*: 1148-1151; copyright 1984 by the AAAS.)

A particular enzyme can be established as the primary (or only) site of action of a herbicide by a combination of genetic and biochemical lines of evidence that includes showing that mutants resistant to a herbicide possess an altered form of an enzyme that is insensitive to the compound and demonstrating that this form of the enzyme co-segregates with the resistant phenotype in genetic crosses. Thus, a plant heterozygous for the S4 mutation was constructed by a series of backcrosses with parent plants. Self-fertilization of this heterozygous individual yielded homozygous mutant, heterozygous and homozygous normal progeny in the ratio of 1:2:1, which is expected for segregation of a single semidominant nuclear allele. All normal sensitive segregants had a CS-sensitive form of ALS that was the same as that of the parent cell line. All individuals homozygous for the S4 mutation contained a highly resistant form of the enzyme (Table 2-38; Chaleff and Mauvais, 1984).

ALS activities in extracts from heterozygotes displayed an intermediate degree of resistance to CS (Table 2-38). The maintenance through genetic crosses of an association between the responses of the plant, the cell culture, and the ALS activity to CS provided a convincing

Table 2-38. Sensitivity to chlorsulfuron (CS) of ALS activities in progeny of a *S4*/+ heterozygote (see Table 2-37) of tobacco. The data are presented as percentage of ALS activity in the absence of CS. Callus cultures initiated from the individual isolates were tested for growth on medium supplemented with CS at 5 ppb (14 nM). Callus cultures capable of growth on this medium are designated R (resistant), S denotes sensitive callus cultures that did not grow in the presence of the herbicide.

Genotype	Callus phenotype	Uninhibited activity (percent) in the presence of chlorsulfuron		
		10 ppb	30 ppb	100 ppb
+	S	40	23	18
+/+	S	42	29	23
+/+	S	36	22	16
S4/+	R	59	50	42
S4/+	R	61	52	45
S4/S4	R	86	80	80
S4/+	R	64	53	46
S4/+	R	60	47	41
S4/S4	R	88	82	79
S4/S4	R	89	83	81

(Chaleff and Mauvais, *Science* 224: 1443-1444; copyright 1984 by the AAAS.)

argument that this enzyme is a primary major site of action of the herbicide in plants and that the production of an insensitive form of the enzyme is the basis for resistance of the mutant. What cannot be assumed is that the S4 mutation occurred in the structural gene for ALS. It is possible that the S4 mutation altered a regulatory apparatus that normally prevents expression of a more resistant ALS isozyme encoded in the tobacco genome.

Another very important point that should be stressed is that demonstration of a correlation between an altered phenotype and an altered biochemical function is especially important in assigning the basis for mutant phenotypes selected in vitro because of the great genetic variability accumulated by plant cells during maintenance in culture. Hence, detection of an altered enzyme in a variant cell line or in a regenerated plant does not, by itself, indicate that alteration of that particular activity is the basis of a new phenotype. The biochemical alteration detected and the phenotype difference observed could result from unrelated genetic events. Careful genetic and biochemical analyses, in parallel, are essential.

Conclusive evidence that ALS is the source of the resistance to these herbicides came after the cloning of a gene encoding ALS from a CS-resistant mutant of *Arabidopsis thaliana* which conferred a high level of herbicide resistance when introduced into tobacco (Haughn et al., 1988). Furthermore, the DNA sequence of the mutant gene differed from that of the wild type by a single base pair substitution (Fig. 2-33). The change at nucleotide 870, which led to the substitution of *ser* for *pro* in the ALS molecule, occurred at the same position and was identical to a mutation that confers sulfonylurea resistance in a yeast ALS gene. In the case of tobacco genes, a mutant (C3) has been found that has a single *pro-gln* replacement at amino acid 196 in one ALS gene. A S4-Hra mutant has two amino acid changes, a *pro-ala* substitution at amino acid 196 and a *trp-leu* substitution at amino acid 573 (Lee et al., 1988). Armed with the information that one or two specific amino acid changes can give rise to herbicide resistance in ALS and that ALS genes can be cloned and introduced by transgenic techniques into crop plants, it is now possible to introduce sulfonylurea resistance into crops by isolating the ALS gene from a species, mutagenizing the gene at specific sites, and transferring the mutant gene to sensitive plants.

```
                                       Pro
    Wild type  - - -   GCA  CAA  GTC  CCT  CGT  CGT   - - -


    Mutant     - - -   GCA  CAA  GTC  TCT  CGT  CGT   - - -
                                       Ser
```

Figure 2-33. Identification of the mutation that confers herbicide resistance in the *Arabidopsis thaliana* ALS gene at nucleotide 870. (Haughn et al., *Mol. Gen. Genet. 211*: 266-271, 1988.)

2-5.1.2   The Use of Herbicide Resistance for Physiological Investigations

2-5.1.2.1   Cross-resistance and Feedback Sensitivity.

A number of reports have confirmed that both sulfonylurea and imidazolinone herbicides block plant growth by inhibiting the activity of ALS (see Saxena and King, 1988). Inhibition apparently occurs through the binding of the herbicides to specific sites on the enzyme. The question then arises as to whether the two groups of herbicides bind to the same site or whether separate (or separable) sites exist on the ALS molecule, either structurally or functionally.

In an attempt to answer this question, a wide range of variants of independent origin resistant to the sulfonylurea herbicides were isolated from haploid cell cultures of *Datura* and tested for cross-resistance to some imidazolinones. The variants chosen for investigation were stable and showed high (100 to 1000-fold) resistance to the sulfonylurea herbicides CS and SM. While some also exhibited cross-resistance to imidazolinones, others showed no cross-resistance at all. Both classes of herbicide tested inhibited ALS activity isolated from wild type cells and the enzyme was found to be resistant to the sulfonylureas and to the imidazolinones in those cells showing cross-resistance to the latter (Table 2-39; Saxena and King, 1988).

Care must be taken in interpreting these results as variants isolated under conditions of cell culture can include a wide range of genetic variability in addition to that which has affected the primary target (in this case ALS) (see earlier discussion in this Case Study). This secondary variability could, for example, influence the effect herbicides have on cell growth quite independently of any change in ALS. Thus, a close correlation between growth and ALS resistance must be established. With this caution in mind, the conclusion can still be drawn from the results in Table 2-39 that the two groups of herbicides share binding sites on the enzyme. The most noteworthy observation, however, may be the *lack* of cross-resistance to imidazolinones in some of the sulfonylurea variants, suggesting the possibility of two slightly different (separable) sites of action on ALS for the two groups of herbicides. Added to this should be the observation of differential sensitivity of variants to the herbicides, which suggests the possibility of mutations having occurred at different genetic loci. The selection of additional variants against the imidazolinones as well as against the sulfonylureas will provide a spectrum of genetic material to investigate the structural-functional properties of the ALS enzyme in relation to herbicides. Indeed, more recent observations suggest that valuable information about the control of the enzyme by branched-chain amino acids can also be gained from these herbicide-resistant cell lines.

Table 2-39. Concentrations of the sulfonylurea herbicides, chlorsulfuron (CS) and sulfometuron methyl (SM), and the imidazolinones (AC252,215 and AC252,925) required to inhibit cell growth (C) and the activity of acetolactate synthase (ALS)(A) by 50 percent (I_{50}) in cell lines of *Datura* selected for resistance either to CS (the CSR lines) or to SM (the SMR line).
[a] 50 percent inhibition was not achieved even at the highest concentration used.

Cell line	I_{50}							
	CS		SM		AC252.215		AC252,925	
	C	A	C	A	C	A	C	A
WT	1.10^{-10}	$1.10^{-\varepsilon}$	5.10^{-11}	1.10^{-9}	1.10^{-8}	1.10^{-6}	3.10^{-8}	1.10^{-5}
SMR1	3.10^{-9}	1.10^{-7}	3.10^{-9}	1.10^{-7}	1.10^{-7}	1.10^{-5}	1.10^{-5}	1.10^{-4}
CSR2	3.10^{-8}	1.10^{-5}	3.10^{-8}	1.10^{-5}	3.10^{-10}	1.10^{-10}	1.10^{-9}	1.10^{-7}
CSR6	3.10^{-7}	1.10^{-4}	3.10^{-7}	1.10^{-4}	1.10^{-7}	NA^{a}	3.10^{-7}	NA
CSR10	3.10^{-8}	1.10^{-8}	1.10^{-8}	1.10^{-5}	1.10^{-6}	1.10^{-5}	1.10^{-5}	1.10^{-5}

(Saxena and King, *Plant Physiol.* 86: 863-867, 1988; reproduced by permission of the ASPP.)

For a long time (see Rathinasabapathi et al., 1990, for review), it has been known that ALS in higher plants is feedback inhibited by the branched-chain amino acids valine, isoleucine, and leucine, the end products of the metabolic pathway in which ALS is the first enzyme (see Fig. 2-32). Recent interest in the binding characteristics of herbicides to the herbicide-resistant enzyme led to an investigation as to whether these changes also brought about any modification to the binding characteristics of branched-chain amino acids to ALS.

The specific activities of ALS in extracts from wild-type *Datura innoxia* cells and four CS-resistant variants (CSR6, CSR10, CSR15, CSR18) was measured in the presence and absence of different concentrations of *val, ile*, or *leu* (Table 2-40). The presence of any of the amino acids in the assay mixture at 5 or 10 mM concentration reduced considerably the ALS activity in wild-type extracts. The CSR variants, on the other hand, had an ALS with variable levels of sensitivity to the amino acids added separately or in combination. In every case, the enzyme in the variants had lower sensitivity than that from the wild type, but in some instances only marginally so. The inhibitory effect on ALS from wild type or variants of two amino acids together (e.g., *val* and *leu*) was additive. Alanine, in contrast, an amino acid unrelated to the

Table 2-40. The effect of chlorsulfuron (CS), valine (val), leucine (leu), and isoleucine (ile) on the specific activity of acetolactate synthase from wild type (Px4) and CS-resistant variants of *Datura*.
[a.] Specific activities (nmole acetoin/mg protein/hour) of acetolactate synthase. (100 percent values) were (mean ± SE): Px4 = 179.5 ± 5.4; CSR6 = 61.8 ± 3.4; CSR10 = 63.7 ± 6.8; CSR15 = 245.8 ± 9.6; CSR18 = 166.5 ± 5.3.

	ALS activity[a] (Mean % of Control ± SE)				
Inhibitor	Px4	CSR15	CSR18	CSR10	CSR6
0	100.0±3.0	100.0±3.9	100.0±3.0	100.±3.00	100.0±3.0
$CS(10^{-8}$ M)	50.9±0.7	80.0±1.7	96.4± 4.5	92.7± 1.6	81.9±0.8
Val(5 mM)	58.8±0.6	71.2±3.9	71.3± 1.2	73.9± 2.1	77.8± 2.8
Val(10mM)	52.1±1.5	61.9±2.0	65.8±1.0	74.2±1.0	80.4±12.3
Leu(5 mM)	29.9±1.2	50.7±0.6	59.3±1.5	76.1±6.5	80.1±7.3
Leu (10 mM)	27.5±1.3	36.3±1.8	49.9±1.8	60.7±4.1	63.5±2.0
Ile(5 mM)	70.8±0.6	72.0±0.6	80.4±0.9	95.0±12.3	99.6±10.0
Ile(10 mM)	63.1±0.6	64.5±0.6	73.3±0.6	70.2±0.6	100.7±9.1
Val(5mM)+ Leu(5 mM)	22.9±1.0	43.5±1.4	61.9±2.7	60.3±2.7	69.2±2.5
Val(10mM+ Leu(10mM)	20.0±0.5	30.9±2.2	41.9±1.9	85.4±10.3	79.1±5.8
Ala(10mM)	91.9±2.5	90.2±0.6	89.8±0.9	84.2±0.5	118.1±6.3
LSD at P=0.05%	2.9	6.1	6.5	18.1	16.1

(Rathinasabapathi et al., *Plant Science 67*: 1-6, 1990; reproduced by permission of Elsevier Science Publishers.)

branched-chain pathway, did not significantly affect ALS activity from either wild-type or variant cells (Rathinasabapathi et al., 1990).

The simplest interpretation of these results (Table 2-40) was that the alterations of the CS-binding site of ALS also lead to structural or functional changes in the binding site for the amino acids, at least in the case of *val* and *leu*. The case for *ile* was less clear. The fact that these changes were, in some cases, marginal and may have no real significance suggests that different mutational events giving rise to herbicide resistance may have more or less influence on the binding site for branched-chain amino acids. Such an hypothesis requires confirmation by more stringent genetic analysis but significant changes to the feedback control of ALS by end product amino acids could affect cell metabolism detrimentally. The transfer of a gene encoding such a radically changed enzyme to, say, a crop plant could lead to the disruption of amino acid and protein metabolism in the crop and, subsequently, to a lowering of its yield. Thus, it may be necessary to pay close attention to the modifications in gene products brought about by mutation especially in the case of genes isolated from cells in culture where profound changes to an enzyme, for example, may have less deleterious effects on growth than in the case of the more stringent requirements imposed by morphogenesis in a whole plant.

An additional point of interest would be to isolate, for example, valine-resistant variants in *Datura* and test their herbicide resistance to determine whether changes to amino acid binding can, in turn, lead to changes at the herbicide binding site on ALS. A report by Relton et al. (1986) suggests that this may be so in tobacco.

2-5.1.2.2   Herbicide Resistance and Gene Amplification.

Mutations conferring resistance to antimetabolites can arise through different mechanisms, among which is a mutation allowing the production of large amounts of the target protein of the antimetabolite. In many cases, gene amplification is involved in this process. Cell lines showing gene amplification are usually selected by growth in increasing concentrations of the antimetabolite. Resistant mutants are phenotypically stable or unstable in the absence of selection pressure depending on the form of the duplicated gene sequence. For stable mutants, the repeated DNA sequence can be found within the chromosome in a region referred to as a homogeneously staining region. Unstable gene amplification often takes the form of double-minute chromosomes that lack a centromere (Xiao et al., 1987).

CSR6 is a variant isolated from cell cultures of *Datura innoxia,* which has resistance to the herbicides CS and SM (see earlier discussion in this Case Study). Cells of this variant were able to grow fully in the presence of 5×10^{-8} M CS or SM, whereas the wild type were unable to grow at 3.16×10^{-10} M (Table 2-41; Xiao et al., 1987). The DNA was extracted from isolated nuclei of wild-type cells, digested with the restriction enzymes, *Eco*R1 and *Hin*dIII, and the digest fragments separated by gel electrophoresis. These gels were then probed with ^{32}P-labeled yeast DNA containing the ILV2 gene that encodes the polypeptide of the ALS enzyme in yeast. This DNA hybridized to fragments of the *Datura* DNA in bands on gels at 5.4 and 2.1 kb (*Eco*R1 digest) or 5.5 and 4.8 kb (*Hin*dIII digest) (Xiao et al., 1987). DNA extracted from nuclei isolated from the variant CSR6, digested with the same two restriction

Table 2-41. Herbicide-resistant phenotypes of independent *Datura innoxia* cell lines. A total of 25 microcalli of each cell line were transferred to petri plates containing medium supplemented with sulfometuron methyl (SM) and incubated at 28oC for 14 days in the dark. The microcalli were then grouped by their size compared with those of wild type growing in a basal medium (MS) without SM; same as wild type (+++), 1/3 wild type (+), or no growth (-). The percentage of calli in each category is given.

| Cell line | Medium | | | | |
| | MS + 10^{-8} M SM | | | MS | |
	+ + +	+	-	+ + +	-
CSR2/472	0	20	80	100	0
SMR8/473	0	20	80	100	0
SMR1/452	44	40	16	94	6
CSR2	100	0	0	100	0
CSR3	44	36	20	100	0
CSR6	100	0	0	100	0
Px4	0	0	100	100	0

(Xiao et al., *Theor. Appl. Genet. 74*: 417-422, 1987.)

enzymes as wild-type DNA, electrophoresed and then probed with yeast ILV2 DNA, however, showed only one hybridization band, at 2.65 kb (Fig. 2-34).

This and other evidence suggested that the gene in *D. innoxia* that encodes the enzyme ALS is present in multiple copy in the CSR6 variant but not in the wild-type cells where the hybridization pattern of fragments digested by restriction enzymes is quite different. It was thought possible that the gene encoding ALS was present in tandem array in the variant cells in repeat sequences of DNA, 2.65 kb in length. After several passages of the resistant cells in the absence of selection pressure (i.e., in the absence of CS), however, the repeat sequences of the 2.65-kb fragment of DNA was no longer detectable in extracts from the nuclei of the variant. Thus, the high copy number of the ALS gene was only transiently produced and disappeared when selection pressure from the herbicide was removed. Thus, assays could not be performed to determine whether the increased copy number of the ALS gene led, in turn, to overproduction of the enzyme product in variant cells.

This mechanism may apply to other systems that confer resistance/tolerance to cellular inhibitors or environmental stresses. Herbicide resistance in other plant cell lines have been shown to result from genomic DNA amplification and to give rise to overproduction of target protein (see Xiao et al., 1987, for some examples). It may be that the

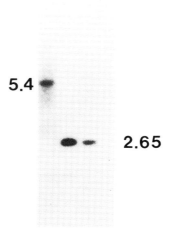

Figure 2-34. Yeast *ILV2* hybridization to *Eco*R1 and *Hind*III 2.65 kb fragments of *Datura innoxia* CSR6 DNA. Lane 1, *Eco*R1 digested wild type (Px4) DNA; lane2, EcoR1 CSR6 DNA; lane3, *Hind*III DNA; lane4, *lambda* DNA digested with *Hind*III to produce molecular markers. (Xiao et al., *Theor. Appl. Genet. 74*: 417-422, 1987.)

phenomenon of gene amplification is a common method in the plant kingdom by which organisms respond to challenges from toxic compounds. The process, whether taking a permanent or transitory form, may represent a way in which organisms can respond rapidly to an unfavorable environmental circumstance. In any case, gene amplification must be placed alongside the other ways in which plants can be modified genetically.

2-5.2.   The s-Triazine-resistant Mutants

The triazine herbicides have been in agricultural use for many years. Atrazine [2-chloro-4-(ethylamino)-6-(isopropylamino)-s-triazine] is the most widely used of the triazines, especially for corn. It rapidly degrades in corn by hydroxylation or by peptide conjugation to metabolic products that are inactive as plant growth inhibitors. Most weed species cannot rapidly metabolize atrazine and are, therefore, susceptible to the herbicide. Unfortunately, some major crops routinely used in crop rotation with corn are also incapable of rapid atrazine metabolism. As a result, soil carry-over of atrazine can cause dramatic growth reduction, in soybean, for example (Arntzen and Duesing, 1983; Hirschberg and McIntosh, 1983).

Biological research related to pesticide chemistry has focused increasingly on the molecular target sites for the active molecules. This has led to theoretical modeling of herbicide resistance based on target site modification, a development given a strong impetus by the recognition of increasing frequencies of weed species with resistance to the triazines. Since 1970, more than two dozen weed species have developed triazine resistance (Hirschberg and McIntosh, 1983; Solymosi and Lehoczki, 1989). For example, in most cases herbicide application at 5 to 10 times the normal rate will cause crop injury or death. However, 50 to 100 times the normal rate of atrazine will not control the new weed biotypes.

2-5.2.1   *The Characterization of Atrazine Resistance*

2-5.2.1.1   The Molecular Mechanism of Triazine Action.

Atrazine and simazine have long been recognized to be inhibitors of photosystem II (PSII)-dependent electron transport in chloroplasts (for a description of the photosystems of the chloroplast, see the introduction to Chapter 3). One herbicide molecule binds per PSII complex. Binding of herbicide was found to cause reversed electron flow from the second electron carrier (B) to the primary stable electron acceptor (Q) of the PSII complex. The herbicide acts by excluding (competing with) plastoquinones from their binding site at the electron carrier, Q/PQ oxidoreductase (Fig. 2-35; Arntzen and Duesing, 1983).

2-5.2.1.2   The Molecular Basis of Triazine Resistance.

In an early attempt to identify the site of action of the triazines, chloroplast thylakoids were isolated from resistant biotypes of the weed *Senecio vulgaris* L. (common grounsel). These were found to be almost totally insensitive to added atrazine in electron transport assays, although other herbicides thought to act at the same step in electron flow [for example, Diuron or DCMU: 3-(3,4 dichlorophenyl)-1,1-dimethylurea] were still active (Fig. 2-36). Measurements of radioactive herbicide binding to isolated resistant membranes revealed the complete loss of high affinity triazine binding sites, although Diuron still bound with near-normal affinity. In the case of the susceptible biotype of the weed, both Diuron and atrazine bound with comparable affinities (Table 2-42; Pfister et al., 1979). In the susceptible chloroplasts, unlabeled atrazine effectively displaced Diuron from the membranes (Fig. 2-36). With resistant chloroplasts high concentrations of unlabeled atrazine caused only slight competition against [^{14}C] Diuron.

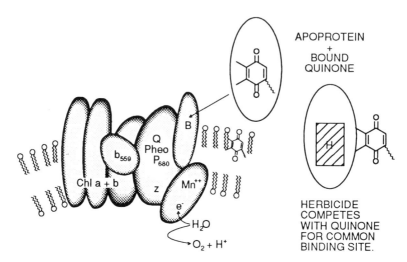

Figure 2-35. A model demonstrating the mode of action of photosystem II-directed herbicides. The structural component polypeptides are diagramatically represented as being embedded in a lipid bilayer. A polypeptide of the complex is partially exposed at the surface and serves to bind a plastoquinone reversibly (B) shown in greater detail on the right-hand side of the diagram. The quinone co-factor of B binds when oxidized but dissociates when reduced to shuttle electrons to PSI. PSII-directed herbicides competitively displace the quinone at this binding site. Subtle changes in the amino acids in the vicinity of the binding site can selectively alter herbicide and/or quinone binding affinities. (Courtesy of C.J. Arntzen and J.H. Duesing. Reprinted from *Advances in Gene Technology: Molecular Genetics of Plants and Animals,* K. Downey, R.W. Voellmy, F. Ahmad, J. Schulz (eds), Academic Press, 1983.)

It is not possible to determine from these data (Fig. 2-36; Table 2-42) whether the binding site for atrazine is totally lost in the resistant chloroplasts or if the affinity between inhibitor and the membrane is strongly diminished. The simplest conclusion to be drawn from the results of these studies was that triazine resistance was a result of subtle modification of a component of the PSII complex, the alteration resulting in a selective loss in triazine affinity without causing the loss of function of the electron carrier, as illustrated in the model described in Figure 2-37 (Pfister et al., 1979). This conclusion led to a search for the identity of the membrane component forming the binding site.

Triazine resistance is maternally inherited, suggesting that this trait is encoded by the chloroplast genome. The *psbA* gene of the chloroplast genome encodes a 32-kD polypeptide, known as the QB-binding or D1 protein, that has a special importance as it forms part of the reaction center that transports electrons from the primary electron acceptor, QA, to the secondary stable electron acceptor, QB (see Fig. 2-35). A variety of mutants have been isolated from a number of plant

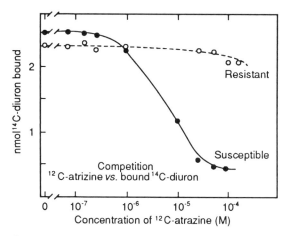

Figure 2-36. Competition between [14C]Diuron and atrazine in susceptible and resistant chloroplasts of common groundsel *Senecio vulgaris* L. Concentration of [^{14}C]Diuron: 0.5 μM. (Pfister et al., *Plant Physiol. 64*: 995-999, 1979; reproduced by permission of the ASPP.)

species that confer variable resistance to the triazine and the urea herbicides (Table 2-43; Shigematsu et al., 1989). Most triazine-resistant plants collected in the field have the same mutation in the QB protein from *ser*-264 to *gly*-264 or, in the case of the monocot *Poa annua*, from *ser*-264 to *ala*-264 (Barros and Dyer, 1988). Thus, it is now possible to present a graphic picture of how atrazine and functionally related herbicides may exert their influence on photosynthesis. This is because recently there have been substantial advances in the determination of the structure of PSII reaction centers and the role of the D1 protein in these centers.

Table 2-42. Calculated binding constants (K) and number of binding sites (X_1; chlorophylls per one bound inhibitor molecule) for Diuron and atrazine in susceptible and resistant chloroplasts of common groundsel, *Senecio vulgaris* L.

	Chloroplasts			
Inhibitor	Susceptible	Resistant		
	K	X_1	K	X_1
Diuron	1.4×10^{-8} M	420 Chl/ inhibitor	5×10^{-8} M	500 Chl/ inhibitor
Atrazine	4×10^{-8} M	450 Chl/ inhibitor	No binding detected	

(Pfister et al., *Plant Physiol. 64:* 995-999, 1979; reproduced by permission of the ASPP.)

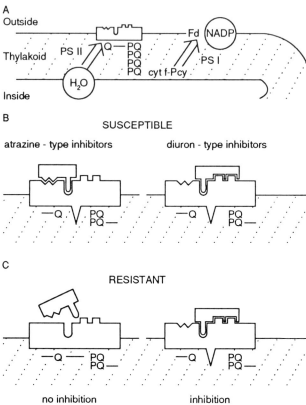

Figure 2-37. Model of inhibitor binding to chloroplast membranes. **A.** Asymmetrical orientation of photosynthetic electron transport chains; external localization of herbicide-binding component in thylakoid membrane. **B.** Existence of two different regions within the herbicide binding component: An "essential region" of the binding constituent that interacts with a common structural element of the different herbicides to produce a conformational change of the constituent, thus interrupting electron transport on the reducing side of PSII; and, different domains that are necessary for specific attachment of the inhibitors to the binding constituent. **C.** Modification of one substructure of the binding constituent in the resistant chloroplast leads to a loss of atrazine-binding capability, whereas DCMU binding to the same constituent is not substantially affected. (Pfister et al., *Plant Physiol. 64:* 995-999, 1979; reproduced by permission of the ASPP.)

The model of D1 structure currently receiving most support suggests that it has five rodlike alpha-helices (A-E) that pierce the thylakoids vertically or nearly so. The fourth (D) and fifth (E) helices and D-E

Table 2-43. Amino acid changes in the QB protein, atrazine resistance, diuron resistance in *Chlamydomonas reinhardtii, Amaranthus hybridus, Solanum nigrum, Anacystis nidulans*, and *Nicotiana tabacum* with respect to wild type.

Amino Acid Position	Codons.		Resistant Strain	Resistance	
	Wild	Mutant		Atrazine	Diuron
219	Val	Ile	*C. reinhardtii*	2.1x	17x
251	Ala	Val	*C. reinhardtii*	25x	5x
255	Phe	Tyr	*C. reinhardtii*	15x	0.6x
264	Ser	Gly	*A. hybridus* *S. nigrum*	1000x	1.4x
	Ser	Ala	*C. reinhardtii* *A. nidulans R2*	84x	10x
	Ser	Thr	*N. tabacum*	560x	40x
275	Leu	Phe	*C. reinhardtii*	Not reported	

(Shigematsu et al., *Plant Physiol. 89*: 986-992, 1989; reproduced by permission of the ASPP.)

connecting segment are aligned in the thylakoid as shown in Figure 2-38, it is suggested to make a pocket into which the quinone fits. The quinone is loosely held in this pocket and freely exchanges with a pool of plastoquinone in the adjacent thylakoid matrix. Thus, it is easily displaced by herbicides that take its place in the pocket and bind relatively tightly, disrupting electron flow. Of particular interest is the serine residue at position 264 in D1. Crystallographic studies have shown that a hydrogen bond is possible between the side chain oxygen of serine of this protein and analogues of atrazine. Mutation of this serine to either glycine or alanine abolishes this hydrogen bond and thereby decreases binding of the herbicide. From this there is the clear implication that it is the abolition of the hydrogen bond that decreases atrazine sensitivity in plants (Barros and Dyer, 1988).

The observed changes in the amino acids at points other than position 264, such as in *Chlamydomonas reinhardtii* (Table 2-43), which also convey herbicide resistance, may do so by affecting the reaction of the herbicide molecule with the D1 protein in the pocket in a more remote indirect way. This may account for the generally lower level of resistance to the herbicides seen in the case of the alga.

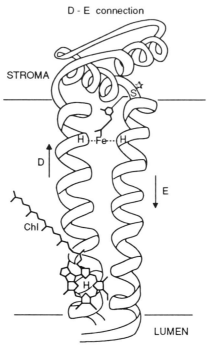

Figure 2-38. Stylized representation of the possible conformation of the quinone pocket of D1. The two transmembrane helices D and E pass through the membrane in opposite directions. Serine 264 (S*), mutation of which may result in herbicide resistance, is located in the D-E connection near the start of helix E and probably forms a hydrogen bond with the herbicide itself, shown here in the pocket near the stromal surface. (Barros and Dyer, *Theor. Appl. Genet.* 75: 610-616, 1988.)

2-5.2.2   The Use of Triazine Resistance in Physiological Studies

2-5.2.2.1   Why It Is Necessary to Use Isolines When Comparing Mutants and Wild Type.

Results from several laboratories suggested that some weed (R) biotypes resistant to triazines differ both structurally and biochemically from susceptible (S) biotypes. The R biotypes have been reported to have such changes as larger and more abundant grana lamellae, less starch, and an increase in the chlorophyll a/b ratio, among others. Such changes might be expected, or certainly explained, in cases where a mutation has decreased the flow of electrons through the light-harvesting pathway of photosynthesis. However, others have found striking differences in a number of leaf anatomical parameters between R and S biotypes (see Vaughn, 1986). But in no case were isonuclear lines used so that differences could be unequivocally linked to the mutation conferring triazine resistance.

To eliminate extraneous influences, structural and physiological comparisons ought to be made only after the mutation in question has been placed in an isogenic background with respect to the wild type from which the variant was derived. Thus, for example, comparisons were made between a pair of near isolines of canola that differed in triazine susceptibility after resistance to the herbicides was transferred from a R biotype of the weed *Brassica campestris* to the canola by repeated backcrosses. Structural and biochemical observations of these two biotypes indicated that, although the chloroplasts of the two lines showed several differences, no gross anatomical variations could be found (Table 2-44; Vaughn, 1986)

The point cannot be emphasized too strongly that because these are isonuclear lines (or nearly so) the prediction can be much more confidently made that the mutation in the 32-kD quinone-binding protein conferring triazine resistance has relatively little direct effect on the anatomy of the plant but has significant influence on certain chloroplast parameters that can be linked to the presumed role of the 32-kD protein in photosynthesis.

Table 2-44. Morphometric comparisons of chloroplast and anatomical parameters in resistant (R) and susceptible (S) biotypes of canola.
[a] Significantly different from the S biotype at the 0.05 level of confidence.

	R	S
	% volume	
Chloroplast character		
Grana lamellae	21.88[a]	15.78
Stroma lamellae	7.07	6.86
Starch	0.23[a]	16.70
Stroma (includes plastoglobuli)	70.82[a]	60.61
Anatomical character		
Epidermis	20.93	19.98
Palisade mesophyll	31.40	32.45
Spongy mesophyll	26.58	24.11
Air space (Vascular tissue was not included in these measurements)	20.27	22.19

(Vaughn, *Plant Physiol.* 82: 859-863, 1986; reproduced by permission of the ASPP.)

2-5.2.2.2   The Photosynthetic Performance of Triazine-resistant and Susceptible Biotypes.

Plant biotypes that are resistant to S-triazines under most conditions often grow less vigorously and have lower quantum yields and lower maximum rates of photosynthesis. The question arises as to whether the lower rates of light-saturated CO_2 reduction characteristic of triazine-resistant biotypes are a direct consequence of an effect on electron transfer brought about by the alteration of the QB-binding site (Ort et al., 1983). Figure 2-39 shows that the rate of CO_2 reduction in a triazine-resistant biotype of *Amaranthus hybridus* (pigweed) was lower at all levels of irradiance than the rate in a susceptible biotype. It was also confirmed that electron transport in the PSII region of the light reactions in the resistant biotype was less than in the susceptible (Table 2-45).

In chloroplasts from the susceptible plants, the rate of PSII-dependent reactions was closely comparable to the overall rate of PSI+II. It was also observed that the electron transport of PSI+II was much the same in

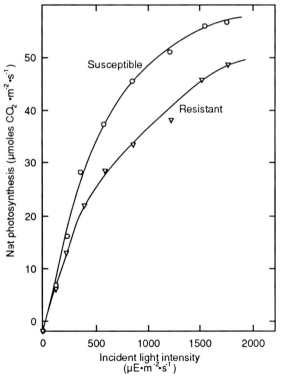

Figure 2-39.   Comparison of the dependence of CO_2-saturated photosynthesis on the incident light intensity in triazine-resistant and triazine-susceptible biotypes of *Amaranthus hybridus*. The ambient CO_2 concentration was 1500 µl/L. (Ort et al., *Plant Physiol.* 72: 925-930, 1983; reproduced by permission of the ASPP.)

Table 2-45. Comparison of electron transport activities in chloroplasts isolated from triazine-susceptible and triazine-resistant biotypes of *Amaranthus hybridus.* DAD, diiminodurene; DMMDBQ, dimethyl-methylenedioxy-*p*-benzoquinone; DHQ, duroquinol, tetramethyl-*p*-hydroquinone; MV, methylviologen.

System	Photosystem	Electron Transport Rate	
		Susceptible	Resistant
		mmol e -.mol Chl ^{-1}s $^{-1}$	
$H_2O \longrightarrow DAD_{ox}$	II	270±15	140±10
$H_2O \longrightarrow$ DMMDBQ	II	230±20	110±5
DAD $\longrightarrow$ MV	I	630±50	660±60
DHQ $\longrightarrow$ MV	I	380±30	390±40
$H_2O \longrightarrow$ MV	I + II	270±15	265±20

(Ort et al., *Plant Physiol.* 72: 925-930, 1983; reproduced by permission of the ASPP.)

both biotypes, although in resistant biotypes the rate of the PSII reactions was reduced to 50 percent that in the susceptible. The fact that chloroplasts isolated from both biotypes supported equivalent rates of whole chain electron transfer and that these rates are sufficient to support the maximum rate of CO_2 reduction seen in the susceptible biotype led Ort et al. (1983) to eliminate lower electron transfer as the direct underlying cause of lower CO_2 reduction rates in resistant biotypes.

Recent investigations using S-triazine-resistant biotypes of *Brassica napus*, on the one hand, or *Senecio vulgaris*, on the other, have confirmed the lower quantum yields and lower maximum rates of photosynthesis found in resistant *Amaranthus* by Ort et al. (1983), but have come to opposite conclusions about the connection between the observations. In the case of *Brassica*, the conclusion was that the lower quantum yields caused by modification of the QB-binding protein at the site of resistance to triazine herbicides is directly responsible for the lower maximum rates of photosynthesis in resistant biotypes (Jursinic and Pearcy, 1988); for *Senecio*, the conclusion was the same as that drawn by Ort et al. (1983), that is, that lower quantum yields alone do not adequately account for lower photosynthetic rates at saturating light levels and that other factors must be invoked to explain this observation (Ireland et al., 1988). Of course, a complicating factor in this work is the role played by rubisco activase in the connection between the light and dark reactions of photosynthesis. Simply attempting to relate quantum yield to photosynthetic rates may not be valid (see Case Studies 3-5 and 3-6).

A resolution of the cause of lower photosynthetic rates in triazine-resistant biotypes is important, especially because of the widespread desire to introduce this particular modification into crop plants. Further work must continue with near isogenic lines of herbicide-resistant plants to determine what genetic or environmental conditions might be able to compensate for this decreased photosynthetic capacity in resistant biotypes. Only by gaining this type of information will it be possible to devise strategies that allow crop plants to maintain high yields when the herbicide resistance trait has been incorporated into their genome (Jursinic and Pearcy, 1988). Evidence gathered to the present certainly suggests that productivity can be significantly affected by a triazine-resistant mutation.

2-5.2.2.3.   Triazine Resistance and Productivity.

Brassica napus cv. Regent, which had had introduced into it triazine resistance from *B. campestris*, followed by extensive backcrossing to cv. Regent, was planted out in the field in a pattern alternating with seedlings of wild type to ascertain competitive fitness. Figure 2-40 illustrates typical results obtained. All parameters measured (fresh weight, dry weight, number and weight of siliques [seed pods], and yield

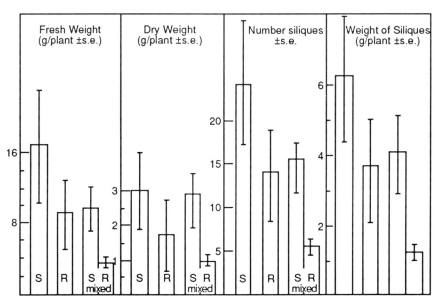

Figure 2-40. Productivity and competitive fitness of triazine-resistant *Brassica napus* at harvest. Seeds were planted in microplots either separately (separate bars) or mixed at 5-cm spacing (adjoining bars). The data from the R and S mixed microplots were further analyzed by Student's t-test and R was significantly different in all cases from S at the <0.01 level. (Gressel and Ben-Sinai, *Plant Science 38*: 29-32, 1985; reporoduced by permission of Elsevier Science Publishers.)

of seed) of the interplanted triazine-resistant biotype were all less than one-third of the susceptible biotype.

Thus, it is clear that the triazine-resistant cultivar of Regent is exceedingly less fit than the previous sensitive cultivar of this species, although the productivity is less affected. When planted alternately with the wild type, the resistant biotype competed only poorly; when planted by itself (i.e., in the absence of competition from the wild type), it performed much better (Fig. 2-40). Even so, all of these results suggest that this cultivar (this gene conferring herbicide resistance) may not be the best vehicle for the transfer of herbicide resistance to crop plants. More recently, similar investigations using isonuclear biotypes of *Senecio vulgaris* suggest that impaired chloroplast function limits growth and productivity at the whole plant level (McCloskey and Holt, 1990).

General Conclusion

The most prominent physiological "message" arising from these studies with herbicides is that care must be taken in understanding the metabolic consequences of gene modification, especially if the eventual aim is to introduce an engineered trait into an economically important species. Both in the case of those herbicides that have as their primary target the enzyme ALS and with the triazines there is clear evidence that significant indirect changes can be caused in important components of key metabolic pathways by mutations leading to herbicide resistance.

References

Arntzen, C.J. and Duesing, J.H. 1983. Chloroplast-encoded herbicide resistance. In *Advances in Gene Technology: Molecular Genetics of Plants and Animals*, K. Downey, R.W. Voellmy, F. Ahmad, and J. Schulz (eds), Academic Press, NY, pp. 273-294.

Barros, M.D.C. and Dyer, T.A. 1988. Atrazine resistance in the grass *Poa annua* is due to a single base change in the chloroplast gene for the D1 protein of photosystem II. *Theor. Appl. Genet. 75:* 610-616.

Beadle, G.W. and Tatum, E.L. 1941. Genetic control of biochemical reactions in *Neurospora. Proc. Natl. Acad. Sci. USA 27:* 499-506.

Binarova, P., Novotny, F., and Nedbalkova, B. 1989. Selection and characterization of alfalfa cell lines resistant to lysine + threonine and/or ethionine. *Biochem. Physiol. Pflanzen. 185:* 99-107.

Bright, S.W.J., Kueh, J.S.H., Franklin, J., Rognes, S.E., and Miflin, B.J. 1982. Two genes for threonine accumulation in barley seeds. *Nature 299:* 278-279.

Bright, S.W.J., Kueh, J.S.H., and Rognes, S.E. 1983. Lysine transport in two barley mutants with altered uptake of basic amino acid in the root. *Plant Physiol. 72:* 821-824.

Caboche, M. and Rouze, P. 1990. Nitrate reductase: A target for molecular and cellular studies in higher plants. *Trends in Genetics 6*: 187-192.

Carlson, J.E. and Widholm, J.M. 1978. Separation of two forms of anthranilate synthetase from 5-methyltryptophan-susceptible and -resistant cultured *Solanum tuberosum* cells. *Physiol. Plant. 44*: 251-255.

Chaleff, R.S. 1981. *Genetics of Higher Plants. Application of Cell Culture.* Cambridge University Press,NY.

Chaleff, R.S. and Mauvais, C.J. 1984. Acetolactate synthase is the site of action of two sulfonylurea herbicides in higher plants. *Science 224*: 1443-1444.

Chaleff, R.S. and Ray, T.B. 1984. Herbicide-resistant mutants from tobacco cell cultures. *Science 223*: 1148-1151.

Crawford, N.M. and Campbell, W.H. 1990. Fertile fields. *Plant Cell 2*: 829-835.

Dierks-Ventling, C. 1982. Storage protein characteristics of proline-requiring mutants of *Zea mays* (L.). *Theor. Appl. Genet. 61*: 145-149.

Dierks-Ventling, C. and Tonelli, C. 1982. Metabolism of proline, glutamate, and ornithine in proline mutant root tips of *Zea mays* L. *Plant Physiol. 69*: 130-134.

Doddema, H. and Telkamp, G.P. 1979. Uptake of nitrate by mutants of *Arabidopsis thaliana* disturbed in uptake or reduction of nitrate. II. Kinetics. *Physiol. Plant. 45*: 332-338.

Dooner, H.K. 1980. Regulation of the enzyme UFGT by the controlling element *Ds* in *bz-m4*, an unstable mutant in maize. *Cold Spring Harbor Symp. Quant. Biol. 45*: 457-462.

Dooner, H.K. 1983. Coordinate genetic regulation of flavonoid biosynthetic enzymes in maize. *Mol. Gen. Genet. 189*: 136-141.

Dooner, H.K. and Nelson, O.E. Jr. 1977. Genetic control of UDPglucose:flavonol 3-*O*-glucosyltransferase in the endosperm of maize. *Biochem. Genet. 15*: 509-519.

Dooner, H.K. and Nelson, O.E. Jr. 1979a. Interaction among *C, R* and *Vp* in the control of the *Bz* glucosyltransferase during endosperm development in maize. *Genetics 91*: 309-315.

Dooner, H.K. and Nelson, O.E. Jr. 1979b. Heterogeneous flavonoid glucosyltransferases in purple derivatives from a controlling element-suppressed *bronze* mutant in maize. *Proc. Natl. Acad. Sci. USA. 76*: 2369-2371.

Fankhauser, H., Pythoud, F., and King, P.J. 1990. A tryptophan auxotroph of *Hyoscyamus muticus* lacking tryptophan synthase activity. *Planta 180*: 297-302.

Fedoroff, N.V. 1984. Transposable genetic elements in maize. *Sci. Am. 250*: 84-98.

Fedoroff, N.V. 1989. About maize transposable elements and development. *Cell 56*: 181-191.

Feenstra, W.J., Jacobsen, E., van Sway, A.C.P.M., and de Visser, A.J.C. 1982. Effect of nitrate on acetylene reduction in a nitrate reductase deficient mutant of pea (*Pisum sativum* L.). *Z. Pflanzenphysiol. 105*: 471-474.

Gabard, J., Pelsy, F., Marion-Poll, A., Caboche, M., Saalbach, I., Grafe, R., and Muller, A.J. 1988. Genetic analysis of nitrate reductase deficient mutants of *Nicotiana plumbaginifolia*: Evidence for six complementation groups among 70 classified molybdenum cofactor deficient mutants. *Mol. Gen. Genet. 213*: 206-213.

Gavazzi, G., Nava-Racchi, M., and Tonelli, C. 1975. A mutation causing proline requirement in *Zea mays*. *Theor. Appl. Genet. 46*: 339-345.

Gonzales, R.A., Das, P.K., and Widholm, J.M. 1984. Characterization of cultured tobacco cell lines resistant to ethionine, a methionine analog. *Plant Physiol. 74*: 640-644.

Grandbastien, M.A., Missonier, C., Goujaud, J., Bourgin, J.P., Deshayes, A., and Caboche, M. 1989. Cellular genetic study of a somatic instability in a tobacco mutant: In vitro isolation of valine-resistant spontaneous mutants. *Theor. Appl. Genet. 77*: 482-488.

Gremaud, M.F. and Harper, J.E. 1989. Selection and initial characterization of partially nitrate tolerant nodulation mutants of soybean. *Plant Physiol. 89*: 169-173.

Gressel, J. and Ben-Sinai, G. 1985. Low intraspecific competitive fitness in a triazine-resistant, nearly isogenic line of *Brassica napus*. *Plant Science 38*: 29-32.

Hahlbrock, K. and Scheel, D. 1989. Physiology and molecular biology of phenylpropanoid metabolism. *Ann. Rev. Plant Physiol. 40*: 347-369.

Harms, C.T., Montoya, A.L., Privalle, L.S., and Briggs, R.W. 1990. Genetic and biochemical characterization of corn inbred lines tolerant to the sulfonylurea herbicide primisulfuron. *Theor. Appl. Genet. 80*: 353-358.

Haughn, G.W., Smith, J., Mazur, B., and Somerville, C. 1988. Transformation with a mutant of *Arabidopsis* acetolactate synthase gene renders tobacco resistant to sulfonylurea herbicides. *Mol. Gen. Genet. 211*: 266-271.

Herrmann, A., Schulz, W., and Hahlbrock, K. 1988. Two alleles of the single-copy chalcone synthase gene in parsley differ by a transposon-like element. *Mol. Gen. Genet. 212*: 93-98

Hirschberg, J., and McIntosh, L. 1983. Molecular basis of herbicide resistance in *Amaranthus hybridus*. *Science 222*: 1346-1349.

Horsch, R.B. and King, J. 1983. Isolation of an isoleucine-valine-requiring auxotroph from *Datura innoxia* cell cultures by arsenate counterselection. *Planta 159*: 12-17.

Ireland, C.R., Telfer, A., Covello, P.S., Baker, N.R., and Barber, J. 1988. Studies on the limitations to photosynthesis in leaves of the atrazine-resistant mutant of *Senecio vulgaris* L. *Planta 173*: 459-467.

Jacobsen, E. and Feenstra, W.J. 1984. A new mutant with efficient nodulation in the presence of nitrate. *Plant Sci. Lett. 33*: 337-344.

Jursinic, P.A. and Pearcy, R.W. 1988. Determination of the rate limiting step for photosynthesis in a nearly isonuclear rapeseed (*Brassica napus* L.) biotype resistant to atrazine. *Plant Physiol. 88*: 1195-1200.

King, J. 1986. Plant cells and the isolation of conditional lethal variants. *Enzyme Microb. Technol. 9*: 514-522.

King, J., Savage, A.D., and Khanna, V. 1980. *Datura innoxia* cell suspensions: A system for the isolation of mutants useful in transport studies. In *Plant Membrane Transport: Current Conceptual Issues*, R.M. Spanswick, W.J. Lucas, and J. Dainty (eds.), Elsevier/North Holland Biomedical Press, Amsterdam, pp. 623-624.

Kishinami, I. and Widholm, J.M. 1987. Auxotrophic complementation in intergeneric hybrid cells obtained by electrical and dextran-induced protoplast fusion. *Plant Cell Physiol. 28*: 211-218.

Komeda, Y., Tanaka, M., and Nishimune, T. 1988. A *th-1* mutant of *Arabidopsis thaliana* is defective for a thiamin-phosphate-synthesizing enzyme: Thiamin phosphate pyrophosphorylase. *Plant Physiol. 88*: 248-250.

Kuhn, E. and Klein, A.S. 1987. Expression of the developmental mutant *bz-m4 Derivative 6856* in maize seedlings. *Phytochem. 26*: 3159-3162.

Langridge, J. and Brock, R.D. 1961. A thiamine-requiring mutant of the tomato. *Austr. J. Biol. Sci. 14*: 66-69.

Larson, R., Bussard, J.B., and Coe, E.H. Jr. 1986. Gene-dependent flavonoid 3'-hydroxylation in maize. *Biochem. Genet. 24*: 615-624.

Larson, R. and Coe, E.H. Jr. 1977. Gene-dependent flavonoid glucosyltransferase in maize. *Biochem. Genet. 15*: 153-156.

Lee, K.Y., Townsend, J., Tepperman, J., Black, M., Chui, C.F., Mazur, B., Dunsmuir, P., and Bedbrook, J. 1988. The molecular basis of sulfonylurea herbicide resistance in tobacco. *EMBO J. 7*: 1241-1248.

Martin, C., Carpenter, R., Coen, E.S., and Gerats, T. 1987. The control of floral pigmentation in *Antirrhinum majus*. In *Developmental Mutants in Higher Plants*, H. Thomas, and D. Grierson (eds.), Cambridge University Press, Cambridge, pp. 19-52.

Marton, L., Sidorov, V., Biasini, G., and Maliga, P. 1982. Complementation in somatic hybrids indicates four types of nitrate reductase deficient lines in *Nicotiana plumbaginifolia*. *Mol. Gen. Genet. 187*: 1-3.

McClintock, B. 1984. The significance of responses of the genome to challenge. *Science 226*: 792-801.

McCloskey, W.B. and Holt, J.S. 1990. Triazine resistance in *Senecio vulgaris* parental and nearly isonuclear backcrossed biotypes is correlated with reduced productivity. *Plant Physiol. 92*: 954-962.

Mendel, R.R., Marton, L., and Muller, A.J. 1986. Comparative biochemical characterization of mutants at the nitrate reductase/molybdenum cofactor loci *cnxA, cnxB* and *cnxC* of *Nicotiana plumbaginifolia*. *Plant Science 43*: 125-129.

Mol, J.N.M., Stuitje, A.R., and van der Krol, A. 1989. Genetic manipulation of floral pigmentation genes. *Plant Molec. Biol. 13*: 287-294.

Muller, A.J. 1983. Genetic analysis of nitrate reductase-deficient tobacco plants regenerated from mutant cells. Evidence for duplicate structural genes. *Mol. Gen. Genet. 192*: 275-281.

Oostindier-Braaksma, F.J. and Feenstra, W.J. 1973. Isolation and characterization of chlorate-resistant mutants of *Arabidopsis thaliana*. *Mutat. Res. 19*: 175-185.

Ort, D.R., Ahrens, W.H., Martin, B., and Stoller, E.W. 1983. Comparison of photosynthetic performance in triazine-resistant and susceptible biotypes of *Amaranthus hybridus*. *Plant Physiol. 72*: 925-930.

Pfister, K., Radosevich, R., and Arntzen, C.J. 1979. Modification of herbicide binding to photosystem II in two biotypes of *Senecio vulgaris* L. *Plant Physiol. 64*: 995-999.

Racchi, M.L., Gavazzi, G., Monti, D., and Manitto, P. 1978. An analysis of the nutritional requirements of the *pro* mutant of *Zea mays*. *Plant Sci. Lett. 13*: 357-364.

Racchi, M.L., Gavazzi, G., Dierks-Ventling, C., and King, P.J. 1981. Characterization of proline-requiring mutants in *Zea mays* L. *Z. Pflanzenphysiol. 101*: 303-311.

Ranch, J.P., Rick, S., Brotherton, J.E., and Widholm, J.M. 1983. Expression of 5-methyltryptophan resistance in plants regenerated from resistant cell lines of *Datura innoxia*. *Plant Physiol. 71*: 136-140.

Rathinasabapathi, B., Williams, D., and King, J. 1990. Altered feedback sensitivity to valine, leucine and isoleucine of acetolactate synthase from herbicide-resistant variants of *Datura innoxia*. *Plant Science 67*: 1-6.

Ray, T.B. 1984. Site of action of chlorsulfuron. Inhibition of valine and isoleucine biosynthesis in plants. *Plant Physiol. 75*: 827-831.

Reddy, A.R., Britsch, L., Salamini, F., Saedler, H., and Rohde, W. 1987. The A1 (Anthocyanin-1) locus in *Zea mays* encodes dihydroquercetin reductase. *Plant Sci. 52*: 7-13.

Relton, J.M., Wallsgrove, R.M., Bourgin, J.P., and Bright, S.W.J. 1986. Altered feedback sensitivity of acetohydroxyacid synthase from valine-resistant mutants of tobacco (*Nicotiana tabacum*). *Planta 169*: 46-50.

Rognes, S.E., Bright, S.W.J., and Miflin, B.J. 1983. Feedback-insensitive aspartate kinase isoenzymes in barley mutants resistant to lysine plus threonine. *Planta 157*: 32-38.

Sahi, S.V., Saxena, P.K., Abrams, G.D., and King, J. 1988. Identification of the biochemical lesion in a pantothenate-requiring auxotroph of *Datura innoxia* P. Mill. *J. Plant Physiol. 133*: 277-280.

Saux, C., Lemoine, Y., Marion-Poll, A., Valadier, M.H., Deng, M., and Morot-Gaudry, J.F. 1987. Consequences of absence of nitrate reductase activity on photosynthesis in *Nicotiana plumbaginifolia* plants. *Plant Physiol. 84*: 67-72.

Savage, A.D., King, J., and Gamborg, O.L. 1979. Recovery of a pantothenate auxotroph from a cell suspension culture of *Datura innoxia* Mill. *Plant Sci. Lett. 16*: 367-376.

Saxena, P.K., Mii, M., Crosby, W.L., Fowke, L.C., and King, J. 1986. Transplantation of isolated nuclei into plant protoplasts. *Planta 168*: 29-35.

Saxena, P.K. and King, J. 1988. Herbicide resistance in *Datura innoxia*. Cross-resistance of sulfonylurea-resistant cell lines to imidazolinones. *Plant Physiol. 86*: 863-867.

Saxena, P.K., Hammerlindl, J., Crosby, W.L., and King, J. 1990. Introduction of resistance to kanamycin into the protoplasts from a pantothenate-requiring auxotrophic cell line of *Datura innoxia* P. Mill. via direct gene transfer. *Plant Science 70*: 105-114.

Shaner, D.L., Anderson, P.C., and Stidham, M.A. 1984. Imidazolinones. Potent inhibitors of acetohydroxyacid synthase. *Plant Physiol. 76*: 545-546.

Shigematsu, Y., Sato, F., and Yamada, Y. 1989. The mechanism of herbicide resistance in tobacco cells with a new mutation in the QB protein. *Plant Physiol. 89*: 986-992.

Shimamoto, K. and King, P.J. 1983. Isolation of a histidine auxotroph of *Hyoscyamus muticus* during attempts to apply BUdR enrichment. *Mol. Gen. Genet. 189*: 69-72.

Solymosi, P. and Lehoczki, E. 1989. Characterization of a triple (atrazine-pyrazon-pyridate) resistant biotype of common lambsquarters (*Chenopodium album* L.). *J. Plant Physiol. 134*: 685-690.

Stewart, G.R. and Orebamjo, T.O. 1979. Some unusual characteristics of nitrate reduction in *Erythina senegalensis* DC. *New Phytol. 83*: 311-319.

Subramanian, M.V. and Gerwick, B.C. 1989. Inhibition of acetolactate synthase by triazolopyrimidines. In *Biocatalysis in Agricultural Biotechnology*, J.R. Whitaker and P.E. Sonnet (eds), American Chemical Society Symposium Series #389, American Chemical Society, Washington, DC, pp. 277-288.

Sung, Z.R. 1979. Relationship of indole-3-acetic acid and tryptophan concentrations in normal and 5-methyltryptophan-resistant cell lines of wild carrots. *Planta 145*: 339-345.

Sung, Z.R., Smith, R., and Horowitz, J. 1979. Quantitative studies of embryogenesis in normal and 5-methyltryptophan-resistant cell lines of wild carrot. The effects of growth regulators. *Planta 147*: 236-240.

Suter, M., Schnebli, V., and King, P.J. 1988. The development of a negative selection system for the isolation of plant temperature-sensitive auxin auxotrophs. *Theor. Appl. Genet. 75*: 869-874.

Tonelli, C. 1985. D-proline effect on L-proline-requiring mutants in *Zea mays* (L.). *Plant Cell Physiol. 26*: 1205-1210.

Toriyama, K., Kameya, T., and Hinata, K. 1987. Selection of a universal hybridizer in *Sinapis turgida* Del. and regeneration of plantlets from somatic hybrids with *Brassica* species. *Planta 170*: 308-313.

Vaughn, K.C. 1986. Characterization of triazine-resistant and -susceptible isolines of canola (*Brassica napus* L.). *Plant Physiol. 82*: 859-863.

Wallsgrove, R.M., Risiott, R., King, J., and Bright, S.W.J. 1986a. Biochemical characterization of an auxotroph of *Datura innoxia* requiring isoleucine and valine. *Plant Science 43*: 109-114.

Wallsgrove, R.M., Risiott, R., Negrutiu, I., and Bright, S.W.J. 1986b. Biochemical characterization of *Nicotiana plumbaginifolia* auxotrophs that require branched-chain amino acids. *Plant Cell Rep.3*: 223-226.

Wang, X-M., Scholl, R.L., and Feldmann, K.A. 1986. Characterization of a chlorate-hypersensitive, high nitrate reductase *Arabidopsis thaliana* mutant. *Theor. Appl. Genet. 72*: 328-336.

Warner, R.L. and Kleinhofs, A. 1981. Nitrate utilization by nitrate reductase-deficient barley mutants. *Plant Physiol. 67*: 740-743.

Warner, R.L., Kleinhofs, A., and Narayanan, K.R. 1985. Genetics, biochemistry and physiology of nitrate reductase-deficient mutants of barley. In *Exploitation of Physiological and Genetic Variability To Enhance Crop Productivity*, J.E. Harper, L.E. Schrader, and R.W. Howell (eds.), American Society of Plant Physiologists, Waverley Press, Baltimore, MA, pp. 23-30.

Widholm, J.M. 1972a. Tryptophan biosynthesis in *Nicotiana tabacum* and *Daucus carota* cell cultures: Site of action of inhibitory tryptophan analogs. *Biochim. Biophys. Acta 261*: 44-51.

Widholm, J.M. 1972b. Cultured *Nicotiana tabacum* cells with an altered anthranilate synthetase which is less sensitive to feedback inhibition. *Biochim. Biophys. Acta 261*: 52-58.

Wienand, U., Weydemann, U, Niesbach-Klosgen, U., Peterson, P.A., and Saedler, H. 1986. Molecular cloning of the *c2* locus of *Zea mays*, the gene coding for chalcone synthase. *Mol. Gen. Genet. 203*: 202-207.

Wray, J.L. 1986. The molecular genetics of higher plant nitrate assimilation. In *A Genetic Approach to Plant Biochemistry*, A.D. Blonstein, and P.J. King (eds.), Chapter 5, Springer-Verlag, New York, pp. 101-157.

Wray, J.L. 1988. Molecular approaches to the analysis of nitrate assimilation. *Plant, Cell Environ. 11*: 369-382.

Xiao, W., Saxena, P.K., King, J., and Rank, G.H. 1987. A transient duplication of the acetolactate synthase gene in a cell culture of *Datura innoxia. Theor. Appl. Genet. 74*: 417-422.

Mutants Affecting the
                  Chloroplast and Photosynthesis

Photosynthesis is, essentially, the only path through which energy can
be introduced into the living world. Because of this, great emphasis has
been placed on understanding the complex processes that comprise
photosynthesis. Mutants have been one of the tools used to dissect and
analyze the photosynthetic apparatus. An array of genotypes has been
generated in a variety of ways and a number of classes of these mutants
have been of significant value in biochemical and physiological studies
of both the light and dark reactions of photosynthesis. Case Studies 3-
1 to 3-4 deal with mutants that have defects in some component(s) of the
light reactions. Case Studies 3-5 and 3-6 give examples of mutants with
lesions in the dark metabolic pathways.

Inside the Light Reactions

The chloroplast has a double membrane, the envelope, that controls
molecular traffic into and out of the organelle. Inside the chloroplast is
another membrane system, the thylakoids, in which the light energy
conversion and electron transport processes occur (Fig. 3-1). The
thylakoids are highly organized, flattened tubes or sacs containing the
photosynthetic pigments that are, in certain regions, stacked to form
grana. Other, longer thylakoids connect one granum to another through
the chloroplast matrix, the stroma (Salisbury and Ross, 1985).
    The pigments associated with photosynthesis are present in the
thylakoid membranes. The main pigments are the chlorophylls a and b,
but these are accompanied by the carotenoids: the carotenes (orange in
color) and the xanthophylls (yellow). All of the pigments are embedded
in and attached to specific proteins and are organized into physically
separable photosystems (Fig. 3-2).
    The photosynthetic electron transport chain can be conveniently
divided into three regions: the photosystem I (PSI) and II (PSII)
complexes, organized around their respective reaction centers, and the
cytochrome b_6-f complex. In addition, thylakoids contain two complexes

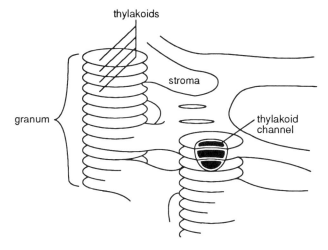

Figure 3-1. The arrangement of the internal membranes of a chloroplast. (From *Plant Physiology*, Third Edition, by Frank B. Salisbury and Cleon W. Ross © 1985 by Wadsworth, Inc. Reprinted by permission of the publisher.)

not directly involved in electron transport: the light-harvesting chlorophyll a/b protein complexes (LHCI and LHCII) and the coupling factor or adenosine triphosphate (ATP) synthetase (Fig. 3-2).

When isolated chloroplasts are treated with certain mild detergents, PSI and PSII are removed from thylakoids and can be separated by polyacrylamide gel electrophoresis (PAGE). Analysis of the separated green bands shows that PSI contains mainly chlorophyll a, small amounts of chlorophyll b, and some β-carotene, attached to several proteins. Also present are two Fe-S proteins similar to ferredoxin. PSII, too, contains chlorophylls a and b, but in different proportions to

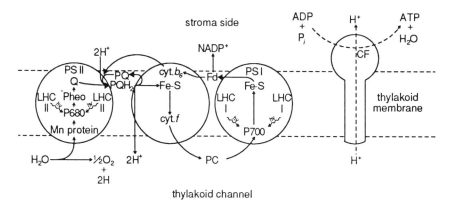

Figure 3-2. Light-driven electron transport in a chloroplast thylakoid. (From *Plant Physiology*, Third Edition, by Frank B. Salisbury and Cleon W. Ross © 1985 by Wadsworth, Inc. Reprinted by permission of the publisher.)

those in PSI, and β-carotene. In addition, PSII has its primary electron acceptor, a colorless chlorophyll a lacking Mn^{2+} called pheophytin (Pheo), closely associated with which is a quinone, Q. Finally, PSII contains one or more proteins with bound Mn^{2+} that are probably involved in H_2O oxidation (Salisbury and Ross, 1985).

In addition to the photosystems, two other major green bands can be separated from chloroplasts by electrophoresis. In each band both chlorophylls a and b, as well as xanthophylls, can be found, but little or no β-carotene. These are the light-harvesting complexes that function with PSI (LHCI) and PSII (LHCII) to absorb light energy and transfer it to the appropriate reaction centers (P680 or P700) of the appropriate photosystem. The distinct locations of PSI and PSII raises problems of cooperation between them as they act in concert to transfer electrons from H_2O to $NADP^+$. Two mobile electron carriers provide the necessary connection: a copper-containing protein, plastocyanin, and a group of quinones, mainly plastoquinone (PQ), which move laterally and vertically within the fluid membrane. Action and cooperation between the two photosystems requires another complex of two cytochromes, b_6 and f, along with another Fe-S protein, as well as ferredoxin that transfers electrons from PSI to $NADP^+$ (Fig. 3-2).

A final, separate component of thylakoids necessary for photophosphorylation is a protein complex called the coupling factor (CF) complex. This can either hydrolyze ATP or synthesize it in a reverse reaction favored by light-driven electron transport (Salisbury and Ross, 1985). The CF complex is composed of a spherical headpiece (the CF_1 particle) attached to a stalk piece (the CF_0 particle) that extends across the thylakoid membrane (Fig. 3-2).

The chloroplast may be biochemically complex but is genetically relatively simple. By no means all of the proteins known to occur in the chloroplast are manufactured there. The nuclear genome must, therefore, play a major role in chloroplast biogenesis. Gene cloning techniques are being used to study the encoding of chloroplast proteins by nuclear genes as well as the mechanisms by which these proteins are transported across the chloroplast envelope, proteolytically processed, targeted to their appropriate locations, and in some cases, assembled into multiprotein complexes (Taylor et al., 1987).

The more straightforward approaches of molecular biology are not always adequate to dissect the more complex processes of chloroplast biogenesis and to identify the functions of specific proteins. A complementary approach would be to identify and exploit mutations that disrupt one or more of these complex mechanisms. This would allow not only the biochemical identification of the particular processes affected but also an investigation of the regulatory interactions between the nuclear and plastid genomes. One example of the complexity of the latter problem is to consider that the plastid has a number of developmental states other than the mature chloroplast. These states change several times during the development of a given cell. Correlated with changes in developmental status are major changes in the expression of nuclear and plastid genes (Taylor et al., 1987).

In Case Studies 3-1 to 3-4, consideration is given to a number of cases where mutation has disrupted events in the light reactions of photosynthesis. Phenotypes to be discussed range from the visible, a permanent or temporary delay in the synthesis of chlorophyll, to the visualizable, by the use of particular light filters, to those that bring about no obvious change in the plant and that can be detected only by applying biochemical tests.

Case Study 3-1: The Chlorophyll b-deficient Mutants

One type of mutation in higher plants and algae, which has been widely studied, results in the reduced expression of chlorophyll b. This pigment comprises about one-fourth of the chlorophyll in most higher plants and is bound in pigment-protein complexes associated with both PSI and PSII. Two major phenotypes are expressed in this group; those plants with no detectable chlorophyll b and those with a reduced amount of the pigment. Among the former are the *chlorina* mutants of barley (Markwell et al., 1986). Their color may vary from pale green to yellow-green. Most differ from the normal by one recessive gene. One of the many such mutants isolated from barley had little or no chlorophyll b and became known as the *chlorina f2* mutant (Table 3-1; Highkin, 1950).

The total chlorophyll content of the *chlorina f2* mutant did not differ significantly from normal (Table 3-1) and the mutation was not lethal as it is in many cases. The mutant was capable of completing its life cycle and, therefore, useful to study the role of chlorophyll b in the photometabolism of higher plants. One of its characteristics observed in early experimentation was that, although the mutant plant had photosynthesis that proceeded at the same rate as normal plants, as soon as the plants started growing autonomously there was a decrease in growth rate of the mutant (Fig. 3-3; Highkin and Frenkel, 1962).

Table 3-1. Spectrophotometric determinations of pigment contents of normal and *chlorina* stocks of barley (expressed as mg/g dry wt).

Stocks	Chl. a	Chl. b	Total chl.	% chl. a	Ratio a/b
Normal (Average)	6.19±0.335	2.35±0.256	8.55±0.0183	72.5	2.66±0.425
Chlorina f2	8.10±0.071	0.34±0.114	8.43±0.840	96.1	23.83±0.479

(Highkin, *Plant Physiol. 25*: 294-306, 1950; reproduced by permission of the ASPP.)

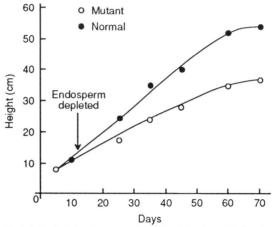

Figure 3-3. Total height of normal and *chlorina f2* barley plants as a function of age, depicting the decrease in growth rate of the mutant plants against that of the normal plants after depletion of the endosperm. (Highkin and Frenkel, *Plant Physiol. 37*: 814-820, 1962; reproduced by permission of the ASPP.)

Furthermore, the decrease in growth of the mutant was accompanied by a decrease in the content of sucrose and reducing sugars. Because chlorophyll b was known not to be essential for photosynthesis, the reason for the lowered growth rate and other parameters remained obscure, but led to a deeper investigation of the nature of the lesion giving rise to the *chlorina f2* phenotype.

3-1.1   Characterization of the *chlorina f2* Mutant of Barley

A clue as to the role played by chlorophyll b in the chloroplast was brought to light by the application of electron microscopy. Goodchild et al. (1966) demonstrated that there were fewer lamellae per granum and number of grana per chloroplast in the *chlorina f2* mutant, although the structure of the lamellae did not appear affected by the lack of chlorophyll b (Table 3-2). The mutant chloroplast gave the impression of being more disorganized than the wild type but it appeared that chlorophyll b was not an essential structural component of the grana.

With the advent of sodium dodecyl sulfate(SDS)-gel electrophoresis, it became possible to examine in detail the polypeptide constituents of thylakoid membranes. The *chlorina f2* mutant of barley became an interesting plant with which to study the effect of the total absence of chlorophyll b on the composition of all the known pigmented components of the membrane. Electrophoresis of SDS extracts of the photosynthetic membranes of normal barley plants showed the pattern for higher plants usually found using the methods available at the time (Fig. 3-4).

Table 3-2. Average numbers of lamellae and grana[a] in normal and *chlorina f2* chloroplasts of barley.

[a] A granum is defined as a group of lamellae containing two or more single lamellae.

[b] Highly signifigant, P<0.001.

Total lamellae Chloroplast	Grana/chloro plast	Lamellae granum	Single lamellae chloroplast	Grana/chloro -plast with 8 or more lamellae
Normal				
260.67	49.44	5.07	12.83	7.17
Mutant				
176.59	33.94	3.90	46.41	1.18
Difference				
84.08[b]	15.50[b]	1.17[b]	-33.58[b]	5.99[b]

(Goodchild et al., *Exptl. Cell Res. 43*: 684-688, 1966.)

The zone with the lowest electrophoretic mobility was the P700-chlorophyll a-protein (PSI); the zone of intermediate mobility was the photosystem II chlorophyll-protein (PSII); and the fastest migrating zone was free pigment. Analysis of the membranes from the *chlorina f2* mutant revealed a total absence of any pigment electrophoresing in the location of PSII chlorophyll-protein. All the other protein bands were present and no extra bands appeared in the mutant (Fig. 3-4).

These results (Fig. 3-4) showed that the expression of the *chlorina f2* mutation not only eliminated chlorophyll b from the plant but also changed other characteristics of the thylakoid membrane. Thus, the impression was gained that the absence of chlorophyll b was not due simply to a defective enzyme in the pathway of pigment biosynthesis. An equally likely primary site for expression of the mutation appeared to be the protein of the PSII chlorophyll-protein complex. Also arising from this experiment and other previous work (see Thornber and Highkin, 1974) was the realization that the absence of the PSII chlorophyll-protein complex did not impair PSII activity, as the barley mutant could still live photoautotrophically. Therefore, the decision was taken to name the complex the light-harvesting chlorophyll a/b protein (LHC). The previous name implied that it was essential for PSII activity. Investigations using mutants, such as *chlorina f2,* led to the recognition that the complex provides light energy to PSII, but that it is not essential for its function.

During the next few years after these early studies, more and finer detailed analyses of the proteins in wild type and *chlorina f2* mutants of barley were performed. For example, using a high resolution,

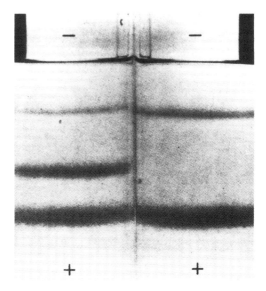

Figure 3-4. Polyacrylamide gel electrophoresis of sodium dodecylsulfate extracts of normal (*Hordeum vulgare* cv. Lyon) (left) and the *chlorina f2* mutant (right) photosynthetic membranes. The direction of electrophoresis is from the cathode to the anode. (Thornber and Highkin, *Eur. J. Biochem. 41*: 109-116, 1974.)

discontinuous SDS-PAGE system it was possible to show that the mutant chloroplast membrane protein completely lacked a 25-kD peak, which accounted for about 18 percent of the chloroplast membrane protein in the normal plant, and also had other significant reductions in polypeptides at 27.5 and 20-kD (Henriques and Park, 1975). Because of these drastic reductions in membrane proteins, Henriques and Park (1975) found difficulty in explaining other results they obtained by the freeze-fracture technique.

Freeze-fracture splits a membrane along an internal plane, yielding two complementary faces. Application of this technique to chloroplast lamellae led to the visualization of these membranes as particulate structures with the particles arranged in a repeatable way. Surprisingly, deletion of polypeptides in the *chlorina f2* mutant had no visible effect on the arrangement or size of these particles in the thylakoid membrane (Henriques and Park, 1975). This observation was unexpected because the particles in question were known to be comprised largely of protein and because the missing polypeptides accounted for a large fraction (about 20 percent) of membrane protein. Henriques and Park (1975) hypothesized that the particles contained a multitude of different polypeptides and that the removal of only a few did not lead to any detectable morphological alterations. They also hypothesized that some of the protein was not in the particles but differently arranged in the structural framework of the photosynthetic membrane. Therefore, absence of it did not so profoundly affect particle structure. Later studies, however, have provided a quite different explanation for the interesting observations by Henriques and Park (1975). Some of these investigations are discussed later.

Bellemare et al. (1982) made a number of significant observations regarding the occurrence and processing of thylakoid membrane proteins in wild-type and *chlorina f2* barley during their attempt to understand the possible roles of chlorophylls a and b in the assembly of the light-harvesting complex (LHC). The availability of the mutant totally lacking chlorophyll b provided an opportunity to define the role of the pigment in LHC biogenesis.

By the early 1980s, LHC was thought to be made up of three to five polypeptides with which were associated both chlorophyll a and b. The two major polypeptides of the complex were known to be synthesized on free cytoplasmic polysomes as soluble precursors, p15 and p16, that were four to five kD larger than their mature products. Then the precursors were reported to be imported into intact chloroplasts by an ATP-dependent mechanism, cleaved to their mature sizes, and inserted into thylakoid membranes where they were thought to recruit chlorophylls a and b to form the LHC (see Bellemare et al., 1982).

By two-dimensional PAGE, it was found that, in addition to the 25-kD (p15) polypeptide found by others to be missing, the *chlorina f2* mutant also lacked, or was greatly deficient in, polypeptides 14 (27-kD), 20 (22-kD), 24 (21-kD), and 25 (20-kD), and contained reduced amounts of p16 (24-kD) (Bellemare et al., 1982). Polypeptides 14, 15, and 16 are major components of the LHC of PSII (LHCII) whereas the others are constituents of the PSI chlorophyll antennae (LHCI). Although the mutant membranes were greatly deficient in these polypeptides, the surprising discovery was made that p14, 15, 20, and 25 were imported upon incubation of the mutant translation products with wild-type chloroplasts, suggesting that their mRNAs were present in the mutant at levels comparable to those of the wild type. Also, intact chloroplasts isolated from the *f2* mutant imported, processed, and assembled in membranes these same polypeptides if presented with either mutant or wild-type translation products. More specifically, although the mutant was devoid of chlorophyll b and had a greatly reduced amount of p15, the precursor p15 was still imported, converted into mature form, and integrated into the thylakoid membrane (Bellemare et al., 1982). Thus, the conclusion must be drawn that chlorophyll b is unnecessary for these biosynthetic events.

These (Bellemare et al., 1982) and other results suggested that the *chlorina f2* mutant was not deficient in the ability to synthesize and process LHC proteins but, rather, that the proteins failed to accumulate because of their rapid turnover. The hypothesis then arose that the pigment chlorophyll b may be required for the stabilization and accumulation of certain polypeptides in thylakoids. This idea has gained acceptance through recent experiments in which the chloroplasts of the *f*2 mutant were solubilized with digitonin and fractionated by PAGE with sodium deoxycholate in the running buffer. Using this procedure, rather than SDS for solubilization, a chlorophyll a protein analogous to the major LHC from wild-type chloroplasts was recovered (Table 3-3) (Duranton and Brown, 1987). No significant differences between the units from the wild type and the mutant could be found in any respect except when the stabilities of the units were compared. The lack of chlorophyll b appeared to affect the normal organization of chlorophyll a and protein in such a way as to render the complex more unstable (not shown in Table 3-3).

Thus, although the precise lesion in the *chlorina f*2 mutant remains unknown, the variant has already provided valuable information on the role of at least one chlorophyll pigment in the biogenesis of light-harvesting components of the chloroplast. It has also been used in a number of physiological studies in addition to those detailed previously. Some of these investigations are illustrated in the following pages.

3-1.2   Other Physiological Investigations Using the *chlorina* *f*2 Mutant of Barley

3-1.2.1   *The Stoichiometry of the Photosystems*

Studies of the structural and functional organization of higher plant chloroplasts suggests that the stoichiometry of the thylakoid membrane complexes and the light-harvesting antennae is adjusted and optimized in response to prevailing environmental conditions. For example, an imbalance in the absorption of light between PSII and PSI and, therefore, an imbalance in the rate of electron transport between the

Table 3-3. The weight ratios of protein to chlorophyll a in the lamellae and various fractions of wild type and *chlorina f*2 mutant barley.

	Wild-type		Mutant
lamellae	4.5		5.0
PS1	3.5-4.0	M1	4.0-4.5
PSII	4.8	M2	4.8-5.0
LHC	2.2	M3	2.0
		M4	2.0

(Duranton and Brown, *Photosyn. Res. 11*: 141-151, 1987; reprinted by permission of Kluwer Academic Publishers.)

two photosystems, might be expected to trigger increased biosynthesis or activity in connection with the photosystem representing the rate-limiting step. Such a mechanism would allow changes in the stoichiometry between the two photosystems to occur in response to changing light quality during the growth of a plant. Such a phenomenon can be investigated through the use of the *f2* mutant (Ghirardi et al., 1986).

The balance of electron transport between the three main complexes in the thylakoid membrane (PSII, cytochrome b_6-f, and PSI) could be altered by specific mutations. The *chlorina f2* mutant of barley has a less stable LHCII, in particular. As a consequence, the proportion of the appressed thylakoid membrane area is substantially reduced compared with that of the wild type (see Section 3-1.1). This reduction was reflected in a concomitant lowering of chlorophyll on a per-component basis in mutant chloroplasts (Table 3-4; Ghirardi et al., 1986). The stoichiometry ratio of electron transport components were Q/PQ/cytochrome f/P700 = 1.8:9.4:1.1:1.0 in the wild type and 3.0:10.8:1.0:1.0 in the mutant chloroplasts. Thus, the relative concentrations of PQ, cytochrome b_6-f, and PSI remained substantially unaffected by the mutation. The mutant, however, had much higher amounts of PSII reaction center complexes compared with the wild type as reflected in a Q/P700 ratio of 3.0 for the mutant and 1.8 for the wild-type (Table 3-4).

The hypothesis arising from these observations is that the mutation reduced the light-harvesting capacity of PSII by preventing the stabilization or assembly of LHCII (Ghirardi et al., 1986). Hence, the elevated relative concentration of PSII in the mutant may be thought of as an attempt by the plant to restore the balance of light absorption

Table 3-4. Concentration of electron transport components in wild-type and *chlorina f2* mutant barley chloroplasts.

Component ratio	Wild-type	*Chlorina f2* mutant
Chl a/Chl b	2.9	∞
Chl/Q	332±26	104±19
Chl/P-700	595±33	313±11
Chl/PQ	63± 7	29± 6
Chl/Cyt f	541±64	325±39
Q/P-700	1.8	3.0
PQ/P-700	9.4	10.8
Cyt f/P-700	1.1	1.0

(Ghirardi et al., *Biochim. Biophys. Acta 851*: 331-339, 1986; reproduced by permission of Elsevier Science Publishers.)

between PSII and PSI, a problem precipitated by the increased instability of LHCII. Of course, such a response suggests the existence of a chloroplast mechanism capable of detecting and correcting an electron transport imbalance between the photosystems in higher plants, the precise nature of which is not known.

3-1.2.2   Cation-induced Grana Stacking

As discussed in Section 3-1.2.1, green plants have the ability to adapt to changes in the spectral quality of ambient light. In particular, they have the capacity to detect and correct an imbalance in the rates of excitation of PSI and PSII. This capacity is seen, for example, when plants exposed to light that preferentially stimulates PSII are suddenly exposed to excess light that preferentially stimulates PSI (see Burke et al., 1979). Initially, the plants use the new light regime inefficiently, but within 5 min the efficiency of photosynthesis rises as the plant alters the organization of its photosynthetic apparatus to make better use of illumination.

One hypothesis put forward in connection with this ability of plants to adjust to changing illumination was that changes in the cationic composition of the stroma, especially Mg^{2+} ions, were alone responsible for these alterations in the organization of the chlorophyll-protein complexes of the thylakoid. From in vitro experiments it was concluded that high salt concentrations promoted the excitation energy transfer to PSII, whereas low salt promoted excitation energy transfer to PSI (Burke et al., 1979). The component of the thylakoid suggested as the mediator of these adjustments was the LHC. Because the *f2* mutant of barley has a modified assembly of LHCII, Burke et al. (1979) used it to test this particular hypothesis.

Isolated chloroplasts from the chlorophyll-b-less and wild-type barley incubated in either 10 mM Na^+, or in 10 mM Na^+ plus 3 mM Mg^{2+}, or in 10 mM Na^+ plus 10 mM Mg^{2+} were examined with the electron microscope. Wild-type thylakoids suspended in 10 mM Na^+ (Fig. 3-5A) lost all grana stacking and appeared as long, parallel membrane sheets. Distinct regions of membrane appression (grana) were maintained in the wild-type membranes incubated in either 3 or 10 mM Mg^{2+} (Fig. 3-5, B,C). Membranes of the *f2* mutant chloroplasts appeared unstacked and highly vesiculated when incubated in 10 mM Na^+ (Fig. 3-6A). When incubated in 3 mM Mg^{2+}, the mutant membranes remained largely unstacked with only a few regions appressed (Fig. 3-6B). More distinct regions of appression were evident when the mutant membranes were incubated in 10 mM Mg^{2+} (Fig. 3-6C; Burke et al., 1979).

These results (Fig. 3-5) were in agreement with the hypothesis of cation regulation of chloroplast membrane function. The barley mutant chloroplasts required higher concentrations of divalent cations to maintain grana stacks in isolated chloroplasts than did the wild-type

membranes, possibly because of the deficiency in the LHCII assembly.
The results were also consistent with the suggestion that LHC mediates
cation-induced grana stacking and excitation energy distribution

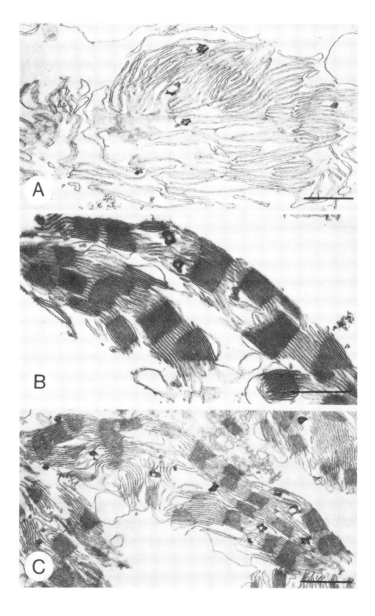

Figure 3-5. Electron micrographs of thylakoids isolated from wild-type
barley and resuspended in (A) 10 mM NaCl; (B) 10 mM NaCl plus 3 mM
MgCl$_2$; and (C) 10 mM NaCl plus 10 mM MgCl$_2$. (Burke et al., *Plant Physiol.*
63: 237-243, 1979; reproduced by permission of the ASPP.)

between PSI and PSII. More recent experimentation, however, some of which is discussed later, suggests that, in addition to stromal cation concentrations, other factors are also involved in thylakoid organization.

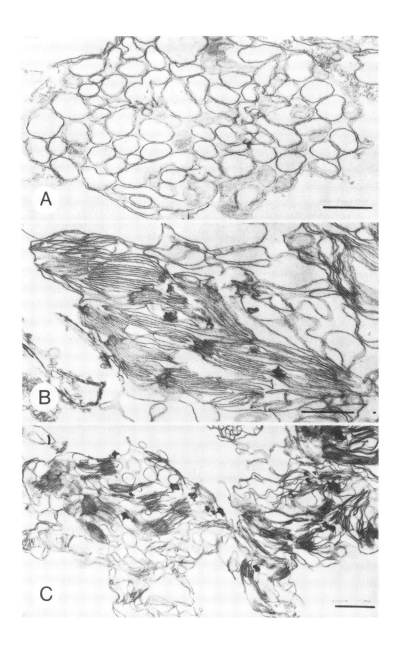

Figure 3-6. Legend as in Fig. 3-5 but for the *chlorina f2* mutant. (Burke et al., *Plant Physiol. 63*: 237-243, 1979; reproduced by permission of the ASPP.)

3-1.2.3 The Protein Phosphorylation Model of Thylakoid Organization

Although in vitro experiments, such as those discussed in Section 3-1.2.2, suggested that cations in the stroma control the assembly of thylakoid components, it was not established that the salt concentrations of the stroma in vivo could be lowered sufficiently to promote excitation energy transfer to PSI to the extent observed, and it was also unclear how an imbalance in the rates of excitation of PSI and PSII could influence the ionic composition of the stroma. Thus, recently, another model has emerged in which adaptive changes in chlorophyll-protein organization of the thylakoid are explained in terms of a reversible phosphorylation of the LHC (Bennett et al., 1984). The protein phosphorylation model acknowledges the importance of stromal cations but explains the adaptive changes in terms of alterations to the surface charge of the membrane through reversible phosphorylation of a surface-exposed segment of the LHC. The *chlorina f2* mutant of barley not only lacks chlorophyll b, but is also severely deficient in the organization and assembly of LHC polypeptides. It provides a system with which to examine the role of the LHC in the phosphorylation process and to assess the possible involvement of other non-LHC proteins in the phosphorylation model.

Both the cation model and the protein phosphorylation model take account of the lateral heterogeneity of thylakoids, that is, PSI is located almost exclusively in the exposed (intergranal) regions of the thylakoid system of lamellae, whereas most PSII and LHC are found in the appressed (granal) regions. Both models also recognize that the lateral heterogeneity is maintained by high cation concentrations in the stroma. They differ, however, in their predictions about the mobility of chlorophyll-protein complexes in vivo. According to the cation model, as illustrated in Section 3-1.2.2, changes in the cationic composition of the stroma affect interactions among PSI, PSII and LHC largely by increasing or decreasing lateral heterogeneity. High salt concentrations increase membrane appression, promote lateral heterogeneity and increase the separation of PSI from PSII, and LHC, thereby promoting excitation energy transfer from LHC to PSII. Low salt concentrations abolish lateral heterogeneity and enable PSI, PSII, and LHC to intermingle, thus promoting excitation energy transfer from LHC to PSII to PSI. The LHC phosphorylation model envisions the movement of LHC units between the appressed and exposed regions of the thylakoid system depending on their phosphorylation. The phosphorylated LHC is found mainly in the exposed regions where it can transfer excitation energy to PSI. Nonphosphorylated LHC is found principally in the appressed regions where it can excite PSII (Bennett et al., 1984).

The previous description is of the mechanism to correct an imbalance in the distribution of excitation energy between the two photosystems. The question of how the two models compare with respect to the mechanism by which the imbalance is sensed in the first place is a matter of importance. The ionic model suggests that any imbalance in favor of PSI increases the Mg^{2+} ion concentration of the stroma, whereas an imbalance in favor of PSII decreases it. What is not clear from the model is how these changes are brought about; how they are triggered. In contrast, the protein phosphorylation model is claimed to be more precise on this point. The protein kinase that phosphorylates LHC is said to be sensitive to the redox state of the plastoquinone (PQ) pool. Overexcitation of PSII reduces the pool, activates the kinase, promotes LHC phosphorylation and movement from the appressed to the exposed regions, and thereby increases excitation energy transfer to PSI at the expense of PSII. Conversely, overexcitation of PSI oxidizes the pool, inactivates the kinase, leads to net dephosphorylation of the LHC by a phosphatase that is always active in the chloroplast and to movement of LHC from the vicinity of PSI to the appressed regions where it serves as a light-harvesting complex for PSII. What is not yet clear (Bennett et al., 1984) is how the activity of the kinase is regulated by the redox state of the PQ pool.

Using [^{32}P]orthophosphate feeding experiments with wild type and the *chlorina f2* mutant, the phosphorylation model was tested by Bennett et al. (1984). When [^{32}P]orthophosphate was supplied to leaves detached from the mutant, no thylakoid polypeptide was phosphorylated. Heavy labeling of both LHC and non-LHC proteins occurred in wild-type leaves under the same conditions and in intact chloroplasts of *f2* supplied in vitro with the label. The results obtained *in organello* indicated that the mutant contained potentially active protein kinase in the thylakoids that was active in vitro but not in vivo. The absence of chlorophyll b in the mutant is crucial to explaining these results. Bennett et al. (1984) hypothesize that it is the preferential absorption of light by chlorophyll b that switches on the kinase in the wild type. In the mutant it is impossible to find a wavelength that will preferentially excite PSII. With the majority of the chlorophyll a located in PSI in the mutant, PSII should always receive less excitation than PSI. Thus, under normal circumstances the PQ pool should be oxidized and the kinase be inactive, as observed.

To test this hypothesis, attempts were made to find conditions under which the PQ pool of the *f2* mutant would become reduced in vivo and lead to kinase activation. Two different treatments were used for this purpose. In one, *f2* leaves were allowed to take up [^{32}P]orthophosphate in the presence of 1 mM iodoacetamide, a potent inhibitor of the Calvin cycle. Under these circumstances, $NADP^+$ regeneration would be inhibited, electron flow through PSI would cease, and the PQ pool should become reduced and activate the kinase. In the other treatment, *f2* leaves were allowed to take up [^{32}P]-orthophosphate under a low photon fluence rate of 50 μmol/m²/s for 4 hours and then exposed to saturating white light (2,000 μmol/m²/s) for 5 minutes before harvest. Under high photon fluence rates Calvin cycle enzymes limit photosynthesis and the majority of the components of the electron transport chain become reduced, including the PQ pool that, in turn, should activate the kinase. The results in Figure 3-7 show that kinase activation was, indeed, observed under both of these experimental conditions.

The results in Figure 3-7 support the hypothesis that the kinase of the mutant is inhibited in vivo because the activity of the Calvin cycle, combined with the unusual distribution of chlorophyll between the photosystems, maintains a rapid turnover of PSI and keeps the PQ pool oxidized. This hypothesis also explains why the kinase is activated *in organello*. Isolated chloroplasts that are capable of CO_2 fixation rapidly consume endogenous CO_2 and must be supplied with HCO_3 for continued Calvin cycle activity. In the absence of exogenous CO_2, the cycle is inhibited, the PQ pool is reduced, and the kinase is activated.

Thus, if the assumptions made by Bennett et al. (1984) are valid, it can be said that predictions based on the phosphorylation model were borne out in these experiments. The results also emphasize the responsiveness of the proposed thylakoid protein phosphorylation system in vivo to changes in the activity of the Calvin cycle and to changes in

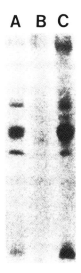

A B C

Figure 3-7. In vivo activation of the thylakoid protein kinase of the chlorina f2 mutant of barley. Detached leaves of the mutant were allowed to take up [^{32}P]orthophosphate at a photon flux of 50 μmol/m²/s. In one case (A), the radioisotope solution also contained 1 mM iodoacetamide. After 4 hours, thylakoids were isolated immediately (A,B) or after the leaves had been exposed for 5 minutes to a photon flux of 2,000 μmol/m²/s (C). The Figure shows a radioautograph of gels on which phosphorylated proteins were separated. (Bennett et al., *Advances in Photosynthetic Research 3*: 99-106, 1984; reprinted by permission of Kluwer Academic Publishers.)

light intensity. It should be said, however, that the phosphorylation model is only one of the hypotheses under investigation and is by no means accepted as the mechanism by which interactions between PSI and PSII occur. The control of photosynthetic pathways appears to be exceedingly complex (see also Case Study 3-5). What is clear is that the *f2* mutant can be a valuable tool in the search for an explanation of the adjustments known to occur in thylakoid activity in response to changing environmental conditions.

Case Study 3-2:   The Virescent Mutants

A common class of mutations in large-seeded plants are those that cause a delay in greening. Many gene mutations are known to affect the plastid pigments of seedlings. Some give rise to entirely white plants, lacking both chlorophyll and carotenoid pigments, which can be maintained only by growing them on full nutrient media. Others lack chlorophylls only and range in color from cream white to bright yellow. Among the latter are the virescent mutants that, after a lapse of time, begin to green and often, eventually, become indistinguishable from normal plants. They survive early stages of seedling growth through the use of stored nutrients in the large seeds from which they come. They also comprise a diverse group of mutants both genotypically and phenotypically.

3-2.1   Characterization of Virescent Mutants

Virescent mutants are characterized by a lag in chlorophyll accumulation in young leaves although in most cases the chlorophyll biosynthetic pathway is unimpaired.   An example is the rate of chlorophyll accumulation in dark-grown peanut seedlings: In the normal plant chlorophyll accumulation is complete after an exposure to continuous illumination for 48 to 72 hours. In contrast, there is a lag period of 72 hours in the virescent mutant leaves, followed by a phase of rapid chlorophyll accumulation. The chlorophyll content of the two types of leaves is nearly equal after 120 hours in the light (Figure 3-8; Benedict and Ketring, 1972). Benedict and Ketring (1972) also noted a similar delay (Fig. 3-8) in the synthesis of protein in mutant leaves. They speculated that the low level of protein synthesis during the lag period in virescent leaves might limit the synthesis of factors essential for the development of both cell and chloroplast constituents.

Investigations of the reason for the delay in appearance of the chlorophylls in virescent leaves led most directly to an examination of the assembly of granal components in the chloroplasts of the mutants. Early studies using cotton suggested that many characteristics of these

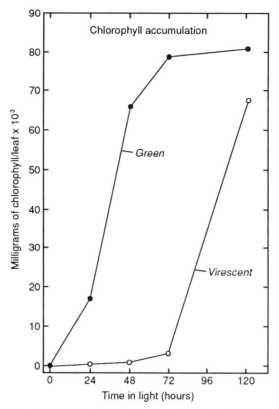

Figure 3-8. The light-induced formation of chlorophyll in virescent and wild-type peanut leaves. (Benedict and Ketring, *Plant Physiol.* 49: 972-976, 1972; reproduced by permission of the ASPP.)

mutants seemed to derive from the aberrant structure of their chloroplasts (Benedict and Kohel, 1970; Benedict et al., 1972). Examination of the chloroplasts in the mutant leaves showed a general lack of grana associated with the period of low chlorophyll content. As the mutant leaves matured, grana developed to the same extent as the grana in the chloroplasts of the wild-type leaves. Thus, the high photosynthetic rate characteristic of mutant leaves during greening seemed to be directly related to the lower chlorophyll content during that stage of delayed development. It was thought to be the result of smaller photosynthetic units in the mutant leaves (Benedict and Kohel, 1970; Benedict et al., 1972).

More recent studies with the *vir-c* mutant of tobacco has led to a greater understanding of the relationship between the structure and function of virescent mutants. Unlike most other virescent mutants, *vir-c* is maternally inherited and is presumed, therefore, to be associated with the chloroplast genome. The existence of both nuclear and cytoplasmic mutations producing the same phenotype supports the view that virescent mutants comprise a diverse group in which each distinct mutant locus disturbs a specific factor regulating chlorophyll content.

The *vir-c* mutant of tobacco has a delayed granal stack formation in young leaves. Wild-type ("Samsun") leaves showed a steady increase in the number of thylakoid profiles per granum from an immature, through a half-expanded, to a fully expanded stage of development (Table 3-5; Archer and Bonnett, 1987). In *vir-c* leaves, there was no change in the thylakoid profiles until the fully expanded stage of development was reached when the number of profiles was the same as in "Samsun." A reduction in thylakoid membranes would then also reduce chlorophyll accumulation sites. These observations linked the reduction in granal thylakoids specifically with the delay in chlorophyll accumulation and suggested that the mutation affected the timing of thylakoid development. Poor development of grana is the most frequently reported, but not the only structural aberration in virescent mutants.

This phenotypic pattern suggested that some component necessary for chlorophyll accumulation and thylakoid development was absent or nonfunctional in young and in mutant leaves at an early stage of development but was supplied later in leaf growth, permitting recovery. If several different proteins are required for the normal sequence of thylakoid stacking, defects in any one of them would lead to a reduction in thylakoid stacking and an inability to accumulate photosynthetic pigments. Specific effects on photosynthetic activity would depend on the role of the defective protein. The virescent phenotype may result from mutations within the structural gene products of similar function late in leaf development. Alternatively, they may be regulatory mutations controlling the timing of gene expression and, as such, may provide a genetic tool to understand the regulation of chloroplast development.

In pursuit of such specific aberrations in thylakoid development, the total protein content in chloroplasts near the half-expanded stage of

Table 3-5. Thylakoid profiles in grana of *vir-c* tobacco mutants and *Nicotiana tabacum* L. var. "Turkish Samsun" chloroplasts at three stages in leaf development. Grana were counted in 6 to 10 different chloroplasts for each datum point. Data for half-expanded leaves are different at the 99 percent level of significance.

	Leaf Stage		
	Immature	Half expanded	Fully expanded
vir-c	5.8	5.7	18.4
Samsun	5.6	10.1	16.3

(Archer and Bonnett, *Plant Physiol.* *83*: 920-925, 1987; reproduced by permission of the ASPP.)

development was examined. The *vir-c* chloroplasts contained 2.9 to 4.3 times less thylakoid protein than "Samsun" (Table 3-6; Archer et al., 1987). Stromal protein levels were not as severely affected as were those of the thylakoid. When the proteins synthesized by the mutant chloroplasts were examined using a light-driven, in vitro protein synthesis, however, the amounts of thylakoid and stromal proteins synthesized in a 30-minute period by the chloroplasts isolated from half-expanded leaves of "Samsun" and *vir-c* were similar, suggesting that mutant chloroplasts were capable of carrying out translation at normal rates, at least for short periods. When the in vitro-synthesized proteins were analyzed in detail by PAGE, a single polypeptide at about 37.5-kD was found to be missing in samples taken from half-expanded leaves of *vir-c*, but to be present in fully mature mutant leaves and in wild-type leaves at all stages of development.

The absence of the 37.5-kD protein in developing leaves of the *vir-c* mutant suggested that it played a role in the mutant phenotype. The polypeptide could be important in granal stack formation or could be involved in the assembly of the thylakoid protein complexes. Decreased ability to assemble could account for the lower amount of protein in the mutant chloroplasts noted here (Table 3-6), and in the case of other virescent mutants, as there is evidence that unassembled protein components of chloroplast protein complexes may be degraded. Thus, this mutant was thought to be of value in understanding the control of thylakoid protein complex assembly as well as of chloroplast gene expression (Archer et al., 1987). That virescent mutants are of value in a variety of physiological investigations, including thylakoid assembly, is shown in the discussion that follows.

Table 3-6. Protein content in chloroplasts from half-expanded leaves of *vir-c* and *Nicotiana tabacum* "Samsun".

	Samsun	*vir-c*	Samsun/ *vir-c*
	$ng/L \times 10^5 chloroplasts$		
Thylakoid protein			
Experiment 1	667	155	4.3
Experiment 2	406	139	2.9
Stromal protein			
Experiment 1	558	426	1.3
Experiment 2	358	165	2.2

(Archer et al., *Plant Physiol. 83*: 926-932, 1987; reproduced by permission of the ASPP.)

3-2.2   The Use of Virescent Mutants in Physiological Investigations

3-2.2.1   Thylakoid Biogenesis

The biogenesis of photosynthetic membranes in eukaryotes follows a programmed differentiation that allows efficient conversion of absorbed light to biochemical energy (see Fig. 3-11). PSI activity and PSI-dependent phosphorylation of adenosine diphosphate (ADP) can be detected before PSII reaction center activity, which, in turn, is followed by appearance of water-splitting activity. Insertion of the chlorophyll a/b LHC into thylakoids lags behind the electron transport chain and may continue for some time after the electron carriers have achieved final activities. It is probably advantageous for the developing chloroplasts to delay assembly of the major light-harvesting apparatus until electron transport is adequately high. Excessive photon capture may lead to irreversible damage. It is not known at what levels (transcription, translation, assembly, activation of assembled complexes) the differentiation sequence is controlled, nor is it clear what signals direct the process. Mutants where the differentiation program has been altered should be useful in probing the molecular mechanisms involved. Among these are the virescent mutants.

The mutant v^*-424 of *Zea mays* shows virescent-type development. As can be seen from the results in Table 3-7 (Polacco et al., 1985), most

Table 3-7. Estimation of some thylakoid polypeptides from the v^*-424 mutant of *Zea mays*.

[a] Leaf sections were taken from expanding third leaves to emerge and were measured from the juncture of leaves with the sheath.

[b] Arbitrarily set to 1.0. CPI = major chlorophyll protein complex of PSI; CF_I = ATPase of coupling factor; LHCP = polypeptides of the light-harvesting complex.

Leaf Section[a]	Chla/b Ratio	Polypeptide Band:Mutant/Normal					
		CPI	α-CF_I	β-CF_I	PSII	34-KD	LHCP[b]
cm							
0-4	2.0	0.23	0.41	0.54	0.24	0.50	1.0
4-8	2.1	0.33	0.53	0.33	0.35	0.55	1.0
8-12	2.3	0.37	0.43	0.39	0.36	0.60	1.0
12-tip	2.4	0.38	0.55	0.59	0.39	0.70	1.0

(Polacco et al., *Plant Physiol.* 77: 795-800, 1985; reproduced by permission of the ASPP.)

of the components of the thylakoid measured were reduced in early greening of v^*-424. These analyses showed values for most non-LHC polypeptides that were 0.23 to 0.55 normal levels when thylakoids containing equivalent amounts of LHCP were compared (Table 3-7). Clearly, the assembly of the major chlorophyll a/b LHC was not delayed relative to most other thylakoid protein complexes during thylakoid biogenesis.

Thus, although it is probably advantageous for developing chloroplasts to delay assembly of the major light-harvesting apparatus until electron transport is adequately high, the phenotype of v^*-424 demonstrates that the normally late appearance of LHC, at least, is not obligatory for thylakoid differentiation. Because LHC is the only major thylakoid complex whose polypeptides are synthesized on cytoplasmic ribosomes, if the v^*-424 allele can be shown not to perturb chloroplast protein synthesis and LHC turnover, the primary lesion is most likely an alteration in the timing mechanism that makes LHC a late event of chloroplast development.

3-2.2.2   *Photosynthetic Capability*

As already indicated, virescent mutations, although similar in phenotype, have varied effects on photosynthesis. These mutants have higher, lower, or similar rates of photosynthesis to those in the equivalent normals from which they were derived. These effects may reflect alteration of a different photosynthetic component in each virescent mutant. One example is that of the *vir-c* mutant of tobacco in which the photosynthetic rate in half-expanded leaves saturated at a higher light intensity than in the wild type "Samsun." The half-saturation for *vir-c* was 2.7×10^{16} quanta/cm^2/s, compared with 0.94×10^{16} quanta/cm^2/s for "Samsun." At a light intensity that saturated photosynthesis in both plants, the photosynthetic rates were very similar when expressed on the basis of dry weight, or on the basis of unit leaf volume (Table 3-8; Archer and Bonnett, 1987).

The photosynthetic rates at low-light intensity were noticeably different, with the mutant photosynthesizing at a much lower rate than "Samsun" (Table 3-8). Coupled with the requirement for a higher light intensity to saturate the light reactions, the poorer photosynthetic rate at low-light intensities suggested that the light collecting ability of *vir-c* was reduced. Such a conclusion correlates well with the observations by electron microscopy (see earlier discussion in this Case Study) that significantly fewer thylakoids per granum occur in half-expanded leaves of *vir-c* than in "Samsun." It further suggested that if several different proteins were required for the normal sequence of thylakoid stacking, defects in any one of them could lead to a reduction in thylakoid stacking and an inability to accumulate photosynthetic pigments. Specific effects on photosynthetic activity would, then, depend on the

Table 3-8. Photosynthetic rates for half-expanded leaves of *vir-c* and "Samsun" genotypes of *Nicotiana tabacum* L.
[a.] 2.5 x 10^{17} quanta/cm²/s.
[b.] 7 x 10^{15} quanta/cm²/s.

	Saturating Light[a]	
	Rate/mg dry weight	Rate/cm³ leaf tissue x 10^4
	μmol of O evolved / h	
vir-c	0.606±0.110	2.259±0.765
"Samsun"	0.477±0.115	2.302±0.273

	Low Light[b]	
	Rate/mg dry weight	Rate/cm³ leaf tissue x 10^4
vir-c	0.081±0.074	0.457±0.312
"Samsun"	0.231±0.061	1.164±0.714

(Archer and Bonnett, *Plant Physiol. 83*: 920-925, 1987; reproduced by permission of the ASPP.)

role of the defective protein. For instance, in contrast to the *vir-c* mutant of tobacco, in the case of virescent corn, less light was required to saturate the light reactions than in the wild type indicating that photosynthetic units were larger, or more numerous, or both. The nuclear virescent mutant of cotton photosynthesized at rates very similar to the wild type, whether light intensity was low or high, whereas the virescent mutant of peanut had a lower photosynthetic rate (Archer and Bonnett, 1987).

Thus, virescent mutations, although similar in phenotype, have varied effects on photosynthesis. These effects may reflect alteration of a different photosynthetic component in each mutant and allow this series of mutants to be used to study the correlation between photosynthetic rates and the details of chloroplast development (Archer and Bonnett, 1987).

3-2.2.3   Influence of the Light Reactions on the Dark Reactions

One model of leaf photosynthesis put forward proposed that electron transport is a key factor in regulating carbon assimilation (however, see Case Study 3-5). It was suggested that whole-chain electron transport regulated maximum carboxylation velocity through the regeneration of ribulose bisphosphate (RuBP). The hypothesis was based on the consideration that in the photosynthetic carbon reductive cycle two molecules of phosphoglyceric acid are consumed to produce one molecule of RuBP. In this process, two molecules of NADPH and three molecules of ATP are consumed. Because two electrons are required to reduce one molecule of NADP⁺, then if substrate is not limiting it is possible to relate electron transport to the rate of phosphoglyceric acid production and carboxylation velocity. To test the practical application of this model, a comparison was drawn between a virescent mutant of *Triticum durum* (var. Cappelli) and its wild type.

Certain photosynthetic parameters in the flag leaves of field-grown plants were measured and compared. Electron transport rates were higher in the wild type than in the mutant (Fig. 3-9; Tomarchio et al., 1983). When expressed on a leaf area basis, the plastids from the wild type were more than two-fold more active in electron transport than those from the mutant. In contrast, the RuBP content per unit leaf area was similar in the two lines (Fig. 3-9).

At this point, the assumption had to be made that the amount of RuBP measured was sufficient to saturate the ribulose bisphosphate carboxylase (rubisco) present. If the enzyme constituted 50 percent of soluble proteins, the reported RuBP content appeared to be sufficient to saturate the carboxylase activity found. For example, comparing the results obtained in a single data set, May 18, the measured carboxylase activity was 2.7 and 2.9 μmol/dm² hour for the wild type and mutant, respectively (Fig. 3-9B). According to the model, the electron transport rate needed to satisfy the NADPH requirement at the carboxylation velocity derived from the in vitro activity was 3.5 mEq/dm² hour for the wild type and 3.7 for the mutant. But it was apparent that these values could be achieved only by the wild-type, which showed a value of uncoupled electron transport of 6.1 mEq/dm² hour whereas the mutant had a rate of only 2.5 mEq (Fig. 3-9A). This discrepancy would be even

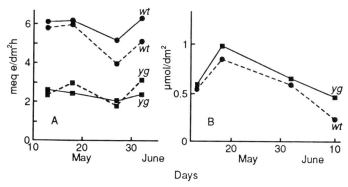

Figure 3-9. Seasonal trend of electron transport rate (A) of wild-type *Triticum durum* var. Cappelli (●) and a virescent mutant (■) measured through O_2 evolution (-) and O_2 uptake (---). (B) RuBP content of wild type (●) and mutant (■). (Tomarchio et al., *Plant Physiol.* 73: 192-194, 1983; reproduced by permission of the ASPP.)

greater if additional electron transport needed to satisfy the ATP requirement was considered (4.0 and 4.3 mEq/dm^2 hour, respectively).

On the basis of these results (Fig. 3-9), the discrepancy observed between the mutant and the wild type in photosynthetic electron transport rates, and the energetic requirements for carboxylation could not be explained. It may be that photosynthetic electron transport is not the sole energy-dependent determinant for RuBP regeneration in chlorophyll-deficient mutants when light is saturating. Alternative mechanisms exist in the interaction of photosynthesis, photorespiration, and respiration, for instance (but the role of rubisco activase seems not to be taken into account here; see Case Study 3-5.)

3-2.2.4   *Photosynthetic Enzymes in a Virescent Mutant of Maize*

Enzymes of the C-4 pathway of photosynthesis are located in mesophyll and bundle sheath cells. Whether all the enzymes are nuclear-encoded is unknown. Also, it is not known whether the light-dependent increase in levels of various enzymes of the C-4 cycle and the reductive pentose-P pathway in etiolated leaves is dependent on chloroplast development. To investigate these gaps in knowledge, Edwards and Jenkins (1988) used a virescent mutant of maize ($v16/v16$) that at 20°C is deficient in 70s (chloroplast) ribosomes, is chlorotic, and does not develop normal thylakoids, but which at 30°C develops normal green leaves. With this material, Edwards and Jenkins (1988) were able to probe the capacity for induction of C-4 cycle and reductive pentose-P cycle enzymes in the light under a condition where synthesis of chloroplast proteins on 70s ribosomes should be restricted.

The rate of photosynthesis, chlorophyll content, and activities of several photosynthetic enzymes were determined 13 days after planting. There was no measurable chlorophyll in leaves of $v16/v16$ grown at 20°C, whereas seedlings grown at 30°C had relatively high levels of the pigment. The rate of photosynthesis on a leaf area basis was similar in wild type (Oh43) and $v16/v16$ grown at 30°C, and in Oh43 grown at 20°C. In contrast, there was no photosynthesis with seedlings of $v16/v16$ grown at 20°C. The rate of respiration in the light (minus 0.6 mol/m^2/s) was similar to the dark respiration measured on a separate leaf (minus 0.69 mol/m^2/s), indicating there was no assimilation of CO_2 in $v16/v16$ grown at 20°C (Table 3-9; Edwards and Jenkins, 1988).

The activities of rubisco of the reductive pentose-P pathway and four enzymes of the C-4 pathway were also measured in the same leaf samples used for the photosynthetic assays (Table 3-9). On a leaf area basis the activity of rubisco in $v16/v16$ grown at 20°C was extremely low, only 2 percent of the activity of Oh43 grown at the same temperature. The activities of the enzymes of the C-4 pathway, however, were relatively much higher. In comparison to Oh43 grown at 20°C, $v16/v16$ had 22 percent as much pyruvate, inorganic phosphate (Pi) dikinase, 35

Table 3-9. The rate of photosynthesis and activity of certain photosynthetic enzymes from leaves of a virescent mutant (*v16/v16*) and wild-type (Oh43) seedlings of maize grown at 20°C and 30°C.

The ratio D/C represents the ratio of activity in *v16/v16* grown at 20°C to the activity in Oh43 grown at 20°C.

- = No chlorophyll detected.

Measurement	Oh4 30°C (A)	*v16/v16* 30°C (B)	Oh4 20°C (C)	*v16/v16* 20°C (D)	D/C
		(mg m^{-1})			
Chl	315	221	270	0	
		(μmol m^{-2} s^{-1})			
Photosynthesis rate	17.1	17.5	15.4	-0.6	
Rubisco	21.5	25.4	18.9	0.4	0.02
NADP-malic enzyme	65.7	52.8	42.4	19.8	0.47
PEP carboxylase	38.8	49.0	35.1	23.8	0.68
Pyruvate, Pi dikinase	7.3	5.9	6.3	1.4	0.22
NADP-malate dehydrogenase	52.5	51.6	31.9	11.3	0.35
		(μmol mg Chl^{1} min^{1})			
Photosynthesis rate	3.3	4.8	3.4	-	
Rubisco	4.1	4.2	4.2	-	
NADP-malic enzyme	12.5	14.3	9.4	-	
PEP carboxylase 7.4	7.4	13.3	7.8	-	
Pyruvate, Pi dikinase	1.4	1.6	1.4	-	
NADP-malate dehydrogenase	10.0	14.0	7.1	-	

(Edwards and Jenkins, *Aust. J. Plant Physiol.* 15: 385-395, 1988; reproduced by permission of the CSIRO Editorial and Publishing Unit.)

percent as much NADP-malate dehydrogenase, 47 percent as much NADP-malic enzyme, and 68 percent as much phospho*enol*pyruvate (PEP) carboxylase (Table 3-9). Also, the photosynthesis rate and enzyme activities with *v16/v16* grown at 30°C tended to be higher on a chlorophyll basis than in Oh43, probably due to the mutant having less chlorophyll per leaf area. The results with Oh43, on either a leaf area or chlorophyll basis, were similar in seedlings grown at 20°C and 30°C (Table 3-9; Edwards and Jenkins, 1988).

The low activity of rubisco in the mutant grown at 20°C was consistent with the requirement of 70s ribosomes for its synthesis, whereas the high activities of enzymes of the C-4 cycle, including those that are chloroplastic, suggests their synthesis is nuclear encoded. Other results suggest the chloroplast may exert some control on the synthesis of some nuclear-encoded enzymes of the C-4 cycle and reductive pentose-P pathway. The lower levels of some of these enzymes in *v16*/ *v16* than in Oh43 when grown at 20°C may indicate that the stage of development of the chloroplast can act as a signal for the expression of certain nuclear genes (Edwards and Jenkins, 1988). Whether or not this is the case, the investigation demonstrated the value of such ribosomal mutants to study the interaction between the chloroplast and the cytosol.

Conclusion

Earlier it was pointed out that the virescent mutants are phenotypically similar but genotypically heterogeneous. In the main, they are variants with deficiencies in the assembly of thylakoid components. Their main value in physiological terms is that they allow investigation of the effects caused by disruption of the protein assembly in the chloroplast. The four examples given provide some notion of the range of physiological phenomena that can be addressed through the use of this class of mutants. There are undoubtedly many more besides and no lack of variant material in many plant species with which to work.

**Case Study 3-3:   The High Chlorophyll Fluorescence
                      (hcf) Mutants**

One of the most elegant and effective methods of isolating mutants with defects in photosynthesis is based on chlorophyll fluorescence. The basic concept is that any lesion that inhibits electron flow through the electron transport chain will result in more of the absorbed energy being reemitted as chlorophyll fluorescence. By irradiating with light of less than 640 nm and viewing colonies through a filter that transmits only wavelengths greater than 650 nm those colonies with defects can be readily identified by their leaf fluorescence. In contrast to the identification of mutants on the basis of chlorophyll-deficiency (see Case Studies 3-1 and 3-2), this approach permits the isolation of mutants with normal chloroplast development. Because most of the *hcf* mutants are lethal, variant lines must be isolated by screening segregating families and then maintaining them as heterozygotes. Mutant seedlings in the *hcf* class contain normal, or nearly normal, amounts of chlorophyll but die when seed reserves are exhausted. Lethality is due to a deficiency of one or more activities in photosynthetic electron transport, ATP generation, or carbon reduction (Somerville,

1986). Because chlorophyll and carotenoid pigments occur in the chloroplast, mutations that directly or indirectly block chlorophyll biogenesis can often be identified by pigment deficiencies. Unfortunately, many of these pigment-deficient mutations exhibit extensive pleiotropism, limiting their value in studying the regulation of chloroplast development. With most of these mutations it is not possible to determine if the lesion identifies a rate-limiting step in chloroplast biogenesis or a gene that plays a truly regulatory role. *hcf* mutants also exhibit pleiotropic effects, but in many cases these are more focused. It is the focused nature of these pleiotropies that makes them attractive for the study of chloroplast biogenesis.

3-3.1   Characterization of Some *hcf* Mutants of Maize

As mentioned, the mutants isolated on the basis of the *hcf* phenotype fall into many classes, including mutants with defects in PSI, PSII, photophosphorylation, CO_2 fixation, and most other functions related to photosynthesis. According to Somerville (1986), more than 31 loci that give rise to the *hcf* phenotype have been assigned to chromosome arms of maize, a species in which most of these mutants have been identified. A few examples are given of the kinds of mutants that have been isolated using the *hcf* technique, and their characterization.

When photosynthetic electron transport is blocked by chemical inhibitors or by genetic mutation, a higher level of visual leaf fluorescence is observed. Such a phenomenon is shown in Figure 3-10 (Miles and Daniel, 1973) in which a variegated leaf of *Oenothera* (half mutant and half normal) can be seen to be fluorescing on the right mutant side. Application of the photosynthesis electron transport inhibitor DCMU to the normal side of the same leaf provokes this side to fluoresce also (Fig. 3-10).

Using this high fluorescence technique, Miles and Daniel (1974) were able to isolate for the first time in higher plants three recessive, seedling-lethal mutants of *Zea mays*, designated *hcf1, hcf2,* and *hcf3.* When CO_2 fixation was measured, the high fluorescent plants were seen to fix little or none of it. The rate of photosynthesis in milligrams of CO_2 fixed per square centimeter of leaf per hour was 18 for wild type, 0.92 for *hcf1,* 0.33 for *hcf2,* and 0 for *hcf3.*

Chloroplasts of *hcf1* plants evolved O_2 at a normal rate with either potassium ferricyanide or DCIP (2,6-dichlorophenol-indophenol) as the electron acceptor; however, only very small amounts of O_2 could be measured with *hcf2* or *hcf3* chloroplasts (Table 3-10). Furthermore, when photoreduction of these electron acceptors as well as methyl viologen was measured, *hcf1* reduced all three dyes, but *hcf2* and *hcf3* chloroplasts showed zero to low rates. Because *hcf1* could carry on photoreduction and O_2 evolution, it has a functional PSII. Conversely, *hcf2* and *hcf3* did not perform these reactions and could be considered

DCMU   Control

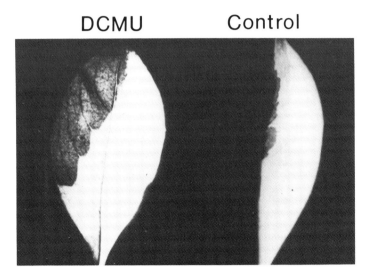

Figure 3-10. Fluorescence of an *Oenothera* leaf which is mutant along the right side of the midrib and normal on the left. The left picture is of the same leaf after submerging the petiole in DCMU for 30 minutes. DCMU = 3-(3',4'-dichlorophenyl)-1,1-dimethylurea. (Miles and Daniel, *Plant Science Letters 1*: 237-240, 1973; photograph courtesy of Dr. C.D. Miles; reproduced by permission of Elsevier Science Publishers.)

devoid of PSII. When methyl viologen (MV) reduction was measured with ascorbate-reduced DCIP as the electron donor to bypass photosystem II, all high fluorescent mutants carried on good rates of electron transport (Table 3-10). Because all mutants reduced MV from the electron donor system, they must have functional PSI up to the point at which MV accepts electrons from the chain. To test the remaining portion of the PSI electron transport chain from the MV point, the photoreduction of $NADP^+$ was measured. PSII involvement was eliminated by using ascorbate-reduced DCIP as an electron donor. In this reaction, *hcf2* reduced $NADP^+$ at a near normal rate, whereas *hcf1* showed a rate of 50 percent of wild type (Table 3-10). Taken together, these data suggest that *hcf1* is limited in PSI and has an altered NADP reductase enzyme system, whereas *hcf2* and *hcf3* are blocked in PSII (Miles and Daniel, 1974). To test this hypothesis, the characterization of the *hcf3* mutant was followed further.

Especially noticeable in a preliminary analysis of the thylakoid lamellae of *hcf3* was the great depletion in a major lamellar polypeptide with an apparent M_r of 32-kD as well as the near or total absence of the cytochrome b_{559} found in the wild type (Leto and Miles, 1980). Both of these components are known to be integral parts of PSII. In a later,

Table 3-10. Photosynthetic electron transport reactions in wild type and high fluorescence mutants of maize.
DCIP = 2,6-dichlorophenol-indophenol.

Reaction	Electron Acceptor	Electron Donor	*hcf1*	*hcf₂*	*hcf₃*	Wild Type
			O_2 μmoles / hr-mg chl			
O_2 evolution	$K_3Fe(CN)_6$		157	4	7	186
	DCIP		226	40	4	212
Photoreduction	$K_3Fe(CN)_6$		82	0	0	113
	DCIP		85	11	0	122
	Methyl viologen	Ascorbate-DCIP	138	177	217	175
	NADP	Ascorbate-DCIP	25	57		63

(Miles and Daniel, *Plant Physiol. 53*: 589-595, 1974; reproduced by permission of the ASPP.)

more detailed study, the thylakoid membrane polypeptides of mutant *hcf3* and normal sibling seedlings of maize were separated by PAGE. The mutant was shown to be missing, or having a great reduction in, polypeptides with apparent molecular masses of 49, 45, 34, 32, 16, 12, and 10-kD. In addition, a value of 3.9 Mn atoms per 400 chlorophylls (i.e., per PSII and PSI reaction center) was obtained from wild-type membranes. In contrast, thylakoids of *hcf3* contained only 0.7 Mn atoms per 400 chlorophylls, a decrease of 82 percent (Metz and Miles, 1982). This reduction in Mn binding in *hcf3* suggested that most or all of the water-splitting capacity had been lost. The missing 49- and 45-kD polypeptides are homologous with those of the chlorophyll a/b-protein centers. The 34-kD polypeptide, much reduced in *hcf3*, is probably involved in water photolysis in other systems. The 32-kD band is a rapidly turning-over, photoinducible polypeptide that contains the binding site for the triazine herbicides. It is probably directly involved in electron transport on the reducing side of PSII between the primary acceptors and the PQ pool (Case Study 2-5). The function of cytochrome b_{559} is also in PSII and it co-migrates in the lithium dodecyl sulfate-PAGE used in these experiments (Metz and Miles, 1982) at 10-kD, a band that is nearly completely absent in *hcf3*.

Thus, it can be said that the mutation in *hcf3* has resulted in the loss of all or nearly all of the components that can be associated with PSII. That a single-site, nuclear mutation can result in the loss of a group of polypeptides, several of which have been shown to be synthesized on chloroplast ribosomes, suggests that PSII exists as a physiological unit whose presence in the membrane is regulated independently of the other major complexes of the thylakoids.

Other clear examples of *hcf* mutants are those having to do with a completely different complex than that described in *hcf3*. A thylakoid complex that contains cytochrome f and b_{563}, an iron-sulfur center, and one or two unidentified polypeptides has been shown to mediate electron transport between PSII and PSI. The complex operates as a functional unit and is likely subject to the same types of regulatory controls that govern the production of PSII and PSI complexes. Two nonallelic, nuclear recessive mutants, *hcf2* and *hcf6*, appear to lack the entire cytochrome f/b_{563} complex (Table 3-11; Metz et al., 1983).

The H_2O-MV results in Table 3-11 indicated that both mutants were incapable of whole chain electron transport from water to methyl viologen. PSII activity was normal or enhanced in *hcf6*, but somewhat reduced in *hcf2* (Table 3-11, H_2O-PD/FeCN). Thylakoids from both mutants showed high rates of O_2 uptake in the diaminodurene ascorbate-MV photoreaction (Table 3-11, DAD/Asc-MV), indicating that PSI is present and functional. Both mutants were shown to lack completely both cytochrome f and b_{563} but at the same time to have no deficiency of cytochrome b_{559} (Table 3-12). Both mutants also seemed to be missing a 17.5-kD polypeptide identified as part of the cytochrome f/b_{563} complex in spinach (Metz et al., 1983).

It seemed clear from these results (Tables 3-11 and 3-12) that *hcf2* and *hcf6* represented examples in which single-site, nuclear mutations resulted in the loss of an entire thylakoid membrane complex and were

Table 3-11. Light-dependent electron transport reactions of chloroplasts isolated from mutants *hcf2* and *hcf6* and normal siblings of maize.

Rates for the mutants are given as the range of percentages relative to the wild-type control rates. Numbers in parentheses are typical electron transport rates, μmol O_2/mg Chl/hour.

[a] MV = methyl viologen; DAD/Asc = diaminodurene ascorbate; PD = phenylene diamine; FeCN = potassium ferricyanide.

Sample	Electron Transport Rate		
	H_2O-MV[a]	DAD/Asc-MV	H_2O-PD/FeCN
	% control		
Wild type	100(120)	100(433)	100(184)
hcf-2	0	130-200	40-60
hcf-6	0	120-160	90-120

(Metz et al., *Plant Physiol. 73*: 452-459, 1983; reproduced by permission of the ASPP.)

Table 3-12. Estimated cytochrome concentrations of thylakoid membranes isolated from mutants *hcf2* and *hcf6* and normal siblings of maize.

Sample	Cyt f	b_{563}	b_{559}
	nmol Cyt / μmol Chl		
Wild type	1.8	2.6	3.2
hcf-2	0	0	2.9
hcf-6	0	0	4.2

(Metz et al., *Plant Physiol. 73*: 452-459, 1983; reproduced by permission of the ASPP.)

analogous to *hcf3* where the PSII complex was missing. The mechanism involved, however, is not at all clear. Cytochrome f and b_{563} are both synthesized on chloroplast DNA. Thus, it may be that the absence of one polypeptide of nuclear origin (the 17.5-kD band) caused the whole complex to be unstable and not to accumulate in the thylakoid membrane.

An important advantage of the *hcf2* and *hcf6* mutants is that they provide a method to study photosynthetic membranes with the cytochrome f/b_{563} complex removed, thus allowing investigations of the function of cytochrome b_{559} in the native state without interference from the other two cytochromes. Also, comparison can be made between these mutants and chemical inhibition of electron transport at the cytochrome f/b_{563} complex (Metz et al., 1983).

3-3.2   The Use of *hcf* Mutants in Other Physiological Studies

3-3.2.1   *Assembly of Coupling Factor Components*

It should be remembered that plastids of higher plants can differentiate into a number of forms including photosynthetically active chloroplasts, starch-storing amyloplasts, and carotenoid crystal-containing chloroplasts. Thus, plastid metabolism requires the products of hundreds of genes, only some of which are found within the plastid. Most plastid proteins are products of the nuclear-cytoplasmic system; they must be imported from outside the organelle. Understanding the mechanisms for integration of nuclear and organellar gene expression, particularly those for multimeric components, is a central problem of eukaryotic cell biology. A number of the *hcf* nuclear gene mutants of maize, as was shown in the preceding Section, have been shown to be deficient in one or more components of the photosynthetic apparatus. Maize plants with nuclear gene mutations that affect the levels of protein products of specific plastid genes thus provide opportunities for learning about the relationships beween activities of nuclear and plastid genomes in the development of chloroplasts.

The nuclear gene maize mutant *hcf* *-38 is deficient in α and β subunits of CF_1. These subunits are products of plastid genes in maize. The levels of several peptides, in addition to the subunits of CF_1, are strikingly lower in the mutant (see Table 3-13 for a few selected examples; Kobayashi et al., 1987).

Other measurements, however, suggested that the amounts of mRNAs for the polypeptides of the CF_1 subunits were not less than 60 percent of those in wild-type plants (Table 3-14). Finally, intact chloroplasts of the mutant and its wild-type sibling were labeled with [^{35}S]methionine followed by a chase with an excess of cold methionine to determine whether *hcf* *-38 plastids could synthesize subunits of CF_1. Results confirmed those obtained in the case of the mRNAs. The α and β subunits were synthesized in nearly comparable amounts by wild-type and mutant chloroplasts. Degradation of the subunits was not observed in the mutant chloroplasts during a 90-minute chase with cold methionine. Thus, the nuclear gene mutation seemed to affect the abundance of the α and β subunits of CF_1 posttranslationally (Kobayashi et al., 1987).

These data (Tables 3-13 and 3-14) led to the conclusion that, although the subunits, together with the other polypeptides of CF_1, were made and appeared as assembled elements of thylakoid membranes, CF_1 units failed to accumulate in the mutant plants beyond about 10 percent of the normal level. In asking the question as to why this was

Table 3-13. Ratios of thylakoid proteins of mutant *hcf*-38 and its wild-type sibling of maize. Aliquots of thylakoid protein preparations of equal chlorophyll content from mutant *hcf*-38 and from its wild type sibling were subjected to polyacrylamide gel electrophoresis and stained with Coomassie brilliant blue. The area under each peak (determined by densitometric tracing) was used for the calculation. Mutant/Wild = ratio of the area under each peak: mutant/wild-type x 100.

Peptides	Mutant/Wild
α-CF_1	10
ß-CF_1	10
49-kD (PSII)	20
45-kD (PSII)	10
16-kD (PSII)	10

(Kobayashi et al., *Plant Physiol. 85*: 757-767, 1987; reproduced by permission of the ASPP.)

Table 3-14. Levels of transcripts of several chloroplast genes in mutant *hcf*-38 and its wild type maize sibling.

Genes	Mutant/Wild
	Based on total RNA
α subunit, CF_1	0.8 ± 0.4
β, ε subunit, CF_1	0.7 ± 0.4
Q_B protein (PG 32)	0.9 ± 0.2
Large subunit, rubisco	1.0 ± 0.3
16 S rRNA	1.2 ± 0.1

(Kobayashi et al., *Plant Physiol. 85*: 757-767, 1987; reproduced by permission of the ASPP.)

so, several possibilities presented themselves for consideration and investigation. There was the possibility of instability of the complex between CF_1 and CF_0 because of a mutation in one of the nuclear-coded polypeptides of the two subunits that resulted in unstable associations, a mutation in one of the CF_0 peptides that resulted in too few attachment sites for CF_1, a seemingly unrelated change in some other thylakoid component, or changes in specific proteolytic activity not yet identified. Whatever the reason, mutations of this kind represent a way to approach the problem of how the production, assembly, and insertion of multimeric components of the photosynthetic apparatus is controlled in the cell (Kobayashi et al., 1987).

3-3.2.2 The Use of hcf Mutants to Study the Assembly of Chloroplast Components

As we have already seen, chloroplast morphogenesis is under complex temporal and spatial constraints and is photoregulated. It requires the coordination of nuclear and plastid genomes. Proplastids can potentially develop into one of several organelles (eg., chromoplast, leucoplast, chloroplast) depending on their organ or tissue location. All may contain the same genome. Maize is a C-4 plant and produces two chloroplast types that are separately located in either bundle sheath or mesophyll cells. A major difference between these two chloroplast populations is a marked reduction of LHCII in bundle sheath thylakoids and its predominance in mesophyll thylakoids where it accounts for 50 percent of the total thylakoid protein and of the chlorophyll (see Polacco et al., 1987).

LHCII is made up of chlorophylls a and b, carotenoids, and polypeptides (LHCP). Regulation of formation of any of these components could potentially modulate LHCII assembly. All LHCP are believed to be encoded by a nuclear, multigene family, the *cab* genes. Figure 3-11 diagrams the proposed timing of initial activity and/or assembly of major thylakoid complexes during proplastid development, including the components of LHCII (Polacco et al., 1987).

During the development of both proplastids and etioplasts into chloroplasts, LHCII assembly lags behind the appearance of active reaction centers (Fig 3-11). Polacco et al. (1987) have devised a screening protocol that has permitted the isolation of several allelic nuclear mutations at the *v24* (a virescent) locus in maize. These mutations were observed to alter the timing of LHCII accumulation (see also Case Study 3-2). The protocol depended on: (1) the availability of a large number of delayed greening mutants where the early stages of chloroplast differentiation were moved from within the tightly rolled leaf sheath to the exposed leaf lamina stage of development, and (2) the use of chlorophyll fluorescence as an indicator of the functional status of the photosynthetic membrane. Mutations at the *v24* locus slowed chloroplast morphogenesis and exhibited abnormal leaf chlorophyll fluorescence kinetics; a high initial yield (f_o) with a reduced variable component (f_v, Fig. 3-12). High fluorescent yield (*hcf*) was the result of premature LHCII assembly. Aberrant fluorescence was developmentally conditional and could not be induced in mature tissue (called the *hcfv* trait, Fig. 3-12, line A). In progenies of these mutants a second *v* phenotype was found showing delayed greening identical to the original *hcfv* trait but lacking the *hcf* property. LHCII assembly was late in these low-fluorescing, virescent (*lcfv*, Fig. 3-12, line C) seedlings.

Further genetic analysis of *hcfv* and *lcfv* progeny showed that these allelic traits are modified by an unlinked modifying locus *Mof** (modifier of fluorescence) which has two codominant alleles (*Mof**-1, *Mof**-2). *Mof**-1 is defined as the allele responsible for the *hcfv* trait associated with early assembly of LHCII, whereas *Mof**-2 is responsible for the *lcfv*

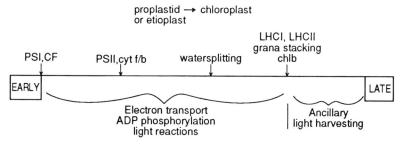

Figure 3-11. Thylakoid morphogenesis. Timing of initial activity and/or assembly of major thylakoid complexes during proplastid development in maize. (Polacco et al., *Dev. Genet. 8*: 389-403; copyright © 1987 Wiley-Liss; reprinted by permission of Wiley-Liss, A Division of John Wiley and Sons, Inc.)

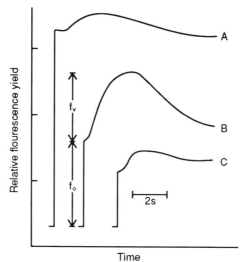

Figure 3-12. Leaf chlorophyll fluorescence kinetics for mutants. Pale green, exposed leaf regions are compared for three phenotypes: A. *hcfv* (*v24-576 v24-576 Mof˙-1 Mof˙-1*); B. normal; C. *lcfv* (*v24-576 v24-576 Mof˙-2 Mof˙-2*). (Polacco et al., *Dev. Genet. 8*: 389-403; copyright © 1987 Wiley-Liss; reprinted by permission of Wiley-Liss, A Division of John Wiley and Sons, Inc.)

trait (Fig 3-12). Analyses of total leaf RNA were carried out. When probed for the genes that encode the LHC proteins (*cab*), the *Mof* ˙-1 *Mof*-1 genotype coincided with a fourfold increase in mRNA compared with *Mof* ˙-2 *Mof* ˙-2.

These investigations, and other experiments not included in this account, have led to the working hypothesis: (1) that *v24* delays some process so that levels of *cab* transcripts are limiting for LHCII assembly in tissue fully exposed to light, and (2) that *Mof* ˙ regulates either *cab* transcription or turnover of *cab* mRNA. What is clear is that the regulation of chloroplast biogenesis is under complex control and that the provision of mutants such as these can be used to dissect effectively the regulatory processes.

3-3.2.3   *Future Prospects From the Use of* hcf *Mutants*

In principle, a set of nonallelic *hcf* mutations that are all deficient in the accumulation of one specific complex should identify nuclear gene products that function in the synthesis and assembly of that complex. The fact that all polypeptides of a given complex seem to be reduced to an equal extent in *hcf* mutants suggests that thylakoid complexes cannot be stably assembled unless all component polypeptides are available. The absence of all of the component polypeptides of one complex might, therefore, be a pleiotropic consequence of a mutation that prevents the synthesis or accumulation of only one component polypeptide. Mutations in the following kinds of genes might cause a *hcf* phenotype (Taylor et al., 1987):

1. A gene coding for a component polypeptide of the affected complex,

2. A nuclear gene regulating the transcriptional or translational activity of nuclear genes coding for chloroplast proteins,

3. A nuclear gene regulating the expression of chloroplast genes. Such regulation might be transcriptional, translational, or might involve processing of multigene transcripts. It is likely that such a mutation would affect the accumulation of polypeptides comprising more that one thylakoid complex,

4. A gene whose product is involved in the transport of nuclear-encoded chloroplast proteins from their site of synthesis in the cytosol, across the chloroplast envelope, and then to their appropriate location within the chloroplast,

5. A gene coding for a processing protease responsible for removal of the aminoterminal peptides of nuclear-encoded chloroplast proteins,

6. A gene coding for a protein involved in the assembly of thylakoid multiprotein complexes or the assembly of multisubunit stromal enzymes. The accelerated turnover of PSII polypeptides in *hcf3* chloroplasts is consistent with a mutation of this last class (Taylor et al., 1987).

The range of pleiotropic effects of *hcf* mutation on the accumulation of chloroplast RNAs and proteins indicates that the mutations identify nuclear genes whose products are involved in a number of different steps in chloroplast development. Several *hcf* mutations have been generated with maize transposable elements, and it is now possible, in principle, to clone the gene using the transposon as a tag. This strategy should provide invaluable insights into the mechanisms of chloroplast development once it becomes clear which of the many *hcf* mutants are the most interesting.

Case Study 3-4:   Chloroplast Lipid Mutants

In the present model of glycerolipid metabolism in higher plants, two pathways contribute to the synthesis of chloroplast glycerolipids in leaf cells (Fig. 3-13; Kunst et al., 1988).

The chloroplast is the sole site of de novo fatty acid synthesis and the main products of this process are C-16:0-ACP and C-18:1-ACP (see legend to Fig. 3-13 for abbreviations). These fatty acids either enter the "prokaryotic pathway" of lipid biosynthesis through acylation of glycerol-3-phosphate (G3P) within the chloroplast or are exported as coenzyme A thioesters to enter the "eukaryotic pathway" associated with the endoplasmic reticulum (Fig. 3-13). Most of the enzymes of the prokaryotic

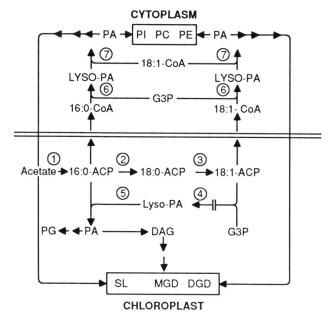

Figure 3-13. An abbreviated scheme for lipid biosynthesis in the leaves of a species of plant containing hexadecatrienoic acid (the C-16:3 species). The steps identified by numbers are as follows: 1, fatty acid synthesis; 2, elongation; 3, desaturation; 4 and 6, acyl transfer to *sn*-glycerol 3-phosphate; 5 and 7, acyl transfer to lysophosphatidic acid. The symbols used for various components are: G3P, *sn*-glycerol 3-phosphate; DAG, diacylglycerol; PA, phosphatidic acid; lyso-PA, 1-acyl-*sn*-glycerol 3-phosphate; PI, phosphatidylinositol; PC, phosphatidylcholine; PE, phosphatidylethanolamine; PG, phosphatidylglycerol; MGD, monogalactosyldiacylglycerol; DGD, digalactosyldiacylglycerol; SL, sulfolipid; 16:0, hexadecanoyl (C-16:0); 18:0, octadecanoyl (C-18:0); 18:1, *cis*-9-octadecenoyl (C-18:1); CoA, coenzyme A; ACP, acyl carrier protein. (Kunst et al., *Proc. Natl. Acad. Sci. USA 85*: 4143-4147, 1988.)

pathway are located in the inner membrane of the chloropast envelope where they catalyze the synthesis of phosphatidylglycerol (PG), monogalactosyldiacylglycerol (MGD), digalactosyldiacylglycerol (DGD), and sulfolipid (SL), the major glycerolipids of the thylakoid membranes. The eukaryotic pathway is responsible for the synthesis of the glycerolipids such as phosphatidylcholine (PC), phosphatidylethanolamine (PE) and phosphatidylinositol (PI), which are found primarily in extrachloroplast membranes.

In the majority of higher plants PG is the only product of the prokaryotic pathway, and the remaining chloroplast lipids are synthesized entirely by the eukaryotic pathway. These species are known as "18:3" plants. In a number of species, including *Arabidopsis thaliana,* however, both pathways contribute to the synthesis of MGD,

DGD, and SL (Fig 3-13; Kunst et al., 1988). These species characteristically contain substantial amounts of hexadecatrienoic acid (C-16:3), which is found only at the *sn*-2 position of galactolipid molecules produced by the prokaryotic pathway. These species have been termed "16:3" plants to distinguish them from 18:3 plants whose galactolipids contain predominantly β-linolenate.

Thus, chloroplast thylakoid membranes contain a unique lipid composition that is ubiquitous in higher plants and differs considerably from that of other cellular membranes. The pigments of photosynthesis and the major acyl lipids represent approximately 25 to 50 percent, respectively, of the total chloroplast lipid. This specialized acyl lipid composition has led to the suggestion that lipids play a vital role in photosynthesis. To investigate the roles of various lipids in chloroplast function, C.R. Somerville and co-workers have isolated a series of mutants of *Arabidopsis thaliana* (L.) Heynh. with specific alterations in leaf membrane fatty acyl composition. In this Case Study, some of these mutants, and the conclusions drawn from them, are described and discussed as this group of mutants represents the most complete investigation to date of the involvement of lipids in physiological processes in higher plants.

The discussion of particular mutants and their use in investigating phenomena associated with lipids in chloroplasts is organized differently from other Case Studies in this Chapter. Rather than having a separate description of the isolation and characterization of the mutants followed by a discussion of their use in dissecting physiological processes, each mutant will be discussed in turn along with the way in which it has been used in further investigations. In most cases, the characterization and physiological aberrations of these mutants have been investigated together.

3-4.1   The Characterization and Physiology of Chloroplast Lipid Mutants

To investigate the functional significance of membrane lipids, a series of mutants of *Arabidopsis thaliana* have been isolated that are deficient in particular membrane fatty acids. The mutants were isolated without selection by direct analysis of the leaf fatty acid composition of individual M_2 plants. The M_2 population was created by mutagenizing approximately 20,000 wild-type seeds (the M_1 generation) and permitting the resulting plants to self-fertilize. A leaf was removed from individual M_2 plants and the fatty acid composition determined by gas-liquid chromatography. From among approximately 2,000 plants examined in this way 89 were found with anomalies in fatty acid composition. After further selections in subsequent generations, at least seven of the original 89 lines were found to have stably inherited changes in fatty acid composition at five genetic loci (Browse et al., 1985). Details are given of the ways in which three of these mutants have been used so far in uncovering the functions of lipids in higher plants.

3-4.1.1 The Role of 3-trans-Hexadecenoic Acid in the Thylakoid Membrane

One of the mutant lines, designated JB60, lacked detectable levels of the unusual fatty acid, 3-*trans*-hexadecenoic acid (*trans*-C-16:1), and had a corresponding increase in palmitic acid (C-16:0), from which the *trans*-C-16:1 is derived, but was otherwise indistinguishable from the wild type in fatty acid composition (Table 3-15; Browse et al., 1985). The gene was inherited as a single nuclear mutation at a locus designated *fadA* (fatty acid desaturation).

The fatty acid *trans*-C-16:1 is esterified only to position 2 of PG. When the PG of the wild type and *fadA* mutant were compared it was found that the decrease in *trans*-C-16:1 was compensated by a proportional increase in the C-16:0 content of PG. These observations suggested that the mutant was deficient in activity for a desaturase that specifically converts C-16:0 at position 2 of PG to *trans*-C-16:1. However, no such enzyme has yet been discovered and other explanations could not be ruled out (Browse et al., 1985).

The ubiquitous presence of *trans*-C-16:1 in the thylakoid membranes of all higher plants and green algae has led to much speculation about its role in membrane organization and function. Because *trans*-C-16:1-PG is not present in etiolated tissue but accumulates during light-

Table 3-15. Fatty acid composition of leaves from mutant and wild-type *Arabidopsis*.
Each value is the mean $\pm$ standard error of independent measurements made on 10 leaves of each line.

Fatty acid	Percentage of total	
	Wild type	Mutant JB60
$C_{16:0}$	15.8 ± 0.4	18.3 ± 0.3
trans-$C_{16:1}$	1.8 ± 0.2	0
$C_{16:2}$	0.7 ± 0.1	0.8 ± 0.1
$C_{16:3} + C_{18:0}$	12.6 ± 0.8	11.6 ± 0.6
$C_{18:1}$	2.7 ± 0.4	2.7 ± 0.5
$C_{18:2}$	18.9 ± 0.8	18.9 ± 0.5
$C_{18:3}$	47.4 ± 1.5	47.2 ± 0.9

(Browse et al., *Science* 227: 763-765; copyright 1985 by the AAAS.)

induced chloroplast development and occurs only in chloroplast membranes, it has been inferred that this lipid has a specific role associated with the light reactions of photosynthesis. Recently, attention has been focused on an apparent association of this lipid with the light-harvesting chlorophyll a/b protein complex, (LHCP) which also accumulates in thylakoid membranes during light-induced chloroplast development and is thought to have an important role in the formation of the appressed membranes of the grana. Although several parameters of photosynthesis having to do with light-harvesting complexes were measured, no differences between the wild type and JB60 could be detected. The conclusion had to be drawn that the role of *trans*-C-16:1 was not a major one in the light reactions of photosynthesis in normal environmental conditions. The role would appear to be more subtle or is manifested only during particular developmental or environmental conditions. Only one clue arose as to a possible role for the fatty acid (Browse et al., 1985).

Recent models for the native structure of LHCP have indicated that it is made up of three structurally equivalent subunits. The reported presence of approximately one molecule of bound *trans*-C-16:1-PG per LHCP oligomer is suggestive of a possible specific role for the lipid in maintaining the oligomeric state of the protein. The availability of a mutant lacking *trans*-C-16:1 provided a means to investigate whether this particular lipid played a key role in LHCP structure.

When thylakoid chlorophyll-protein complexes are solubilized in SDS and electrophoresed they separate into a characteristic pattern of about six major chlorophyll-containing bands. The LHCP oligomer band can be either increased or decreased by varying the cation content of the solubilizing buffers. As the concentration of cations is increased, the proportion of LHCP monomer in the extracts from the thylakoid membranes is increased and that of the oligomer correspondingly decreased. The solubilization of thylakoid membranes from *Arabidopsis* wild type and the *fadA* mutant in solutions of SDS containing very low concentrations of NaCl revealed that the thylakoids of the *fadA* mutant contained normal levels of the LHCP oligomer (Fig. 3-13; McCourt et al., 1985). As the NaCl concentration was raised from 0 to 100 mM the amount of LHCP in both the mutant and wild type decreased from about 7 percent to zero. The membranes from the mutant, however, were much more sensitive to salt-induced dissociation of LHCP oligomer than those of the wild type. The concentration of NaCl, which gave 50 percent reduction in LHCP concentration, was about 13 mM in the mutant compared with 37 mM in the wild-type (Fig. 3-14).

The conclusion from results such as those in Figure 3-14 is that *trans*-C-16:1-PG is not required for LHCP formation but in some way stabilizes the oligomer so that it is less susceptible to SDS-mediated dissociation. How this observation might be linked to a specific role for *trans*-C-16:1-PG in vivo has not yet been ascertained. It seems that the lipid could be an element of the fine-tuning mechanisms that have

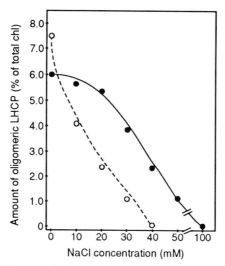

Figure 3-14. Effect of NaCl concentration on the proportion of chlorophyll in LHCP oligomer in mutant (○) and wild type (●) *Arabibopsis*. (McCourt et al., *Plant Physiol.* 78: 853-858, 1985; reproduced by permission of the ASPP.)

evolved to optimize the efficiency of photosynthetic electron transport. Through the use of tools such as is represented by the *fadA* mutant, it is possible to investigate such subtle physiological effects with some precision.

3-4.1.2   Lipid Desaturation

The lipids of chloroplast membranes contain an unusually high proportion of polyunsaturated fatty acids. Depending on the plant species, linolenic (C-18:3) or a combination of linolenic and hexadecatrienoic (C-16:3) may account for more than 80 percent of the total fatty acids of this organelle. The synthesis of these fatty acids occurs by sequential desaturation through the agency of desaturase enzymes. For example, the synthesis of α-linolenate in higher plants occurs by the sequential desaturation of stearate. The first double bond is inserted by a soluble chloroplast enzyme; the second and third are introduced only after the fatty acid has been incorporated into a glycerolipid molecule. Similar sequential events control the synthesis of the C-16:3 fatty acids.

Although the broad outline of the pathway of lipid desaturation exists, many uncertainties remain. It is not yet known how many distinct desaturases exist or which events occur outside as well as inside the chloroplast. Given the predominance of trienoic fatty acids in leaf lipids and their suggested importance to photosynthesis and other plant functions, it would be interesting and valuable to understand the operation and control of the fatty acid desaturases. Each desaturation reaction seems, however, to involve the interaction of several membrane-bound components that have not yet been characterized. Such a complex system could benefit from the isolation and study of mutants with specific alterations in fatty acid composition (Browse et al., 1986).

The mutant JB1 of *Arabidopsis thaliana* contains reduced levels of both C-16:3 and C-18:3 fatty acids and corresponding increased levels of C-16:2 and C-18:2 precursors due to a single, recessive, nuclear mutation at a locus that has been designated *fadD* (fatty acid desaturation, gene "D") (Table 3-16). The effect was also shown to be temperature sensitive; when mutant plants were grown at 18°C the fatty acid composition was similar to that of the wild type; at 26°C, however, the composition was that seen in Table 3-16.

The simplest hypothesis to explain the results in Table 3-16 and the temperature sensitivity observed is that a specific structural or regulatory component of a *n*-3 desaturase has been made thermolabile by the *fadD* mutation and that the desaturase controlled by the *fadD* gene acts on acyl chains with no apparent specificity for the length of the chain (16- or 18-carbon), its point of attachment to the glycerol backbone (*sn-1* or *sn-2*) or for the lipid head group. It is also probable that as both intra- and extrachloroplastic fatty acids are equally affected

Table 3-16. Fatty acid composition of chloroplast and extrachloroplast membranes from wild-type and mutant *Arabidopsis* grown at 23°C.

	Chloroplasts		Extrachloroplast Membranes	
			Measured	
	WT	JB1	WT	JBI
Fatty acid composition (mol %)				
C 16:0	16.4	13.7	34.5	36.2
C 16:1	3.0	3.5	1.6	1.7
C 16:2	0.7	12.1	0	2.2
C 16:3	14.0	6.1	3.3	0.5
C 18:0	1.2	1.4	3.6	5.3
C 18:1	3.5	4.8	4.8	6.0
C 18:2	12.8	22.2	25.2	28.4
C 18:3	47.7	35.6	26.0	18.0

(Browse et al., *Plant Physiol. 81*: 859-864, 1986; reproduced by permission of the ASPP.)

by the mutation that C-18:3 formed inside the chloroplast is reexported to other cellular sites.

Thus, it would appear possible to examine the phenomenon of lipid desaturation through the isolation of mutants with specific deficiencies in the polyunsaturated fatty acids. Indeed, recently two other desaturation mutants have been isolated from *Arabidopsis* (Browse et al., 1989; Kunst et al., 1989a) and more can be expected. If the technique can also yield temperature sensitive mutants of the kind reported in the example detailed here, then an especially powerful tool is available for studies of lipid biosynthesis in higher plants.

3-4.1.3   *What Effect Does the Reduction in Amount of Specific Trienoic Acids Have on Chloroplasts?*

The use of fatty acid-deficient mutants, such as *fadD*, provides a tool for investigating the role of the missing or deficient trienoic fatty acids in photosynthesis or in the economy of the chloroplast. An *fadD* mutant (designated as JB101, a derivative of JB1), for example, was inferred to have an approximately 20 percent reduction in the amount of chloroplast lipid per cell (McCourt et al., 1987). No changes, however, could be found in any of the photosynthetic parameters measured. The mutation had no major effect on CO_2 fixation, electron transport, or fluorescence. The most striking differences between the mutant and the wild type were on some of the morphometric measurements of chloroplasts (Table 3-17). Granal width increased by about 20 percent in the mutant, which caused a reduction in the total length of stromal thylakoids and in the number of grana per plastid. Grana and stroma membranes were reduced in amount to approximately 73 and 64 percent of wild type levels, respectively. The most striking effect was a 45 percent reduction in the cross-sectional area of plastids in the mutant. This decrease in chloroplast size was not accompanied by a proportional decrease in the amount of chloroplast per unit fresh weight. However, at 19°C there was no difference between the number of chloroplasts per cell in the mutant (32.4$\pm$10) and wild type (34.6$\pm$11), at 27°C the mutant and wild type had averages of 57.9$\pm$25 and 40.1$\pm$15 chloroplasts per cell, respectively (McCourt et al., 1987).

The simplest hypothesis arising from these observations is that the *fadD* gene product controls only the *n*-3 desaturase and that the amount of trienoic acid directly regulates chloroplast volume. The increased number of chloroplasts in the mutant implies that the chloroplasts are smaller because they divide more frequently. The forces responsible for determining chloroplast size and organization are entirely obscure and it is not possible to evaluate this hypothesis from these investigations. But more mutants with reduced trienoic acid composition would be useful in evaluating the possible relationship revealed here between chloroplast structure and trienoic acid content.

Table 3-17. Morphometric analysis of chloroplasts from mutant JB101 and wild-type (WT) *Arabidopsis*.
a. These values were derived from others in the table.

	WT	JB101
Grana/plastid	54.1±13.0	37.2± 8.2
Thylakoids/granum	5.5±3.3	4.8±2.5
Granal width (μm)	0.40±0.03	0.48±0.03
Stroma thylakoids/plastid	67.0±20.6	55.7±14.2
Stroma thylakoid length (μm)	0.26±0.03	0.20±0.03
Total grana (μm/plastid)	119.0	85.7
Total stroma (μm/plastid)[a]	17.4	11.1
Total thylakoids (μm/plastid)[a]	136.4	96.8
Grana/stroma[a]	6.8	7.7
Surface area (μm²/plastid)	9.7± 2.0	5.3±1.6

(McCourt et al., *Plant Physiol. 84*: 353-360, 1987; reproduced by permission of the ASPP.)

The conclusion to date from this work is that a high content of trienoic fatty acids normally found in chloroplast lipids may be an important factor regulating organelle biogenesis, but it is not required to support normal levels of the photosynthetic activities associated with thylakoid membranes (McCourt et al., 1987).

3-4.1.4   A Deficiency in Plastid Glycerol-3-Phosphate Acyltransferase Activity Alters Chloroplast Structure and Function in Arabidopsis

As already indicated in this Case Study, in C-16:3 plants such as *A. thaliana*, both the prokaryotic and eukaryotic pathways are involved in lipid biosynthesis. The majority of the higher plants, the C-18:3 plants, however, use the prokaryotic pathway only for the synthesis of phosphatidylglycerol. Because the C-18:3 plants have abandoned the prokaryotic pathway for the synthesis of chloroplast glycerolipids, the question arises as to the role of this pathway in C-16:3 plants (Kunst et al., 1989b).

Somerville and co-workers (Kunst et al., 1988) described the isolation and biochemical characterization of a class of mutants of *Arabidopsis* that lacked activity for the first enzyme of the prokaryotic pathway, glycerol-3-phosphate acyltransferase, due to a single nuclear mutation at the *act1* locus (Table 3-18). Both the mutants JB25 and LK8 lacked activity of this enzyme. Therefore, this mutation effectively converted a C-16:3 plant into a C-18:3 plant and provided an opportunity to explore the functional significance of the prokaryotic pathway by comparing an isogenic mutant and wild-type plants.

Kunst et al. (1988) discovered that as a consequence of the deficiency in the prokaryotic pathway, the *act1* mutants showed specific alterations in the composition of leaf membrane lipids. These changes included greatly reduced levels of C-16:3 acyl groups and a corresponding increase in 18-carbon fatty acids. The effects these and other changes had on chloroplast structure and function in the *act1* mutants were then investigated (Kunst et al., 1989b). A morphometric analysis was compiled of chloroplasts from the wild type and the allelic mutant lines JB25 and LK8 (Table 3-19). Although the total amount of membrane was the same in mutant and wild-type chloroplasts, the average number of thylakoid membranes per granum was reduced from 6.2 in the wild

Table 3-18. Enzyme activities in chloroplast or whole-cell extracts of mutant and wild type *Arabidopsis*.
ND = not determined.

Enzyme	Activity.pmol/min per mg of protein	
	Chloroplast	Whole cell
Gro*P* acyltransferase		
Wild type	11.0	116.0
JB25	0.4	5.7
Wild type + JB25	ND	64.04
LK8	ND	3.9
Monoacyl-Gro *P* acyltransferase		
Wild type	170	3.4
JB25	100	3.5
LK8	ND	3.4
Rubisco		
Wild type	400	570
JB25	490	550
P-*enol* pyruvate carboxylase		
Wild type	0.224	10.2
JB25	0.239	12.6

(Kunst et al., *Proc. Natl. Acad. Sci. USA.* 85: 4143-4147, 1988.)

Table 3-19. Morphometric analysis of chloroplasts from mutant lines and wild type *Arabidopsis*.

Measurement	Wild-Type	JB25	LK8
Grana/plastid	54.5± 6.6	90.0±7.2	87.6±7.6
Thylakoids/granum	6.2±37	3.7±1.6	3.9±1.7
Granal width (μm)	0.40±0.04	0.4±0.03	0.4±0.04
Stroma length (μm)	0.20±0.01	0.2±0.03	0.2±0.02
Stroma/plastid (μm)	103.9±12	99.8±10	96.1±13
Appressed membrane/plastid (μm)	114.1	97.0	104.2
Nonappressed membrane/plastid (μm)	40.7	52.8	52.9
Total membrane (μm)	154.8	149.8	157.1
Appressed/nonappressed membrane	2.8	1.8	2.0
Surface area (μm ^{2}/plastid)	9.9±2	10.8±1	10.7±1

(Kunst et al., *Plant Physiol. 90*: 846-853, 1989b; reproduced by permission of the ASPP.)

type to 3.8 in the mutant. This change was accompanied by a 65 percent increase in the number of grana per chloroplast. As a result, the amount of nonappressed membrane was about 30 percent higher in the mutants. The net effect of these changes was a decrease in the overall ratio of appressed to nonappressed membrane from 2.8 to about 1.9 in the mutants (Table 3-19; Kunst et al., 1989b). The striking similarity of these effects in the two independently isolated *act1* mutants confirms that these ultrastructural effects are specifically due to the changes in lipid composition resulting from loss of glycerol-3-phosphate acyltransferase activity.

These changes in structure, however, had only minor effects on the functions of chloroplast and did not significantly affect growth or development in the mutants. In this regard, as a single mutation can lead to the loss of the prokaryotic pathway, the question arises as to why some species retain the pathway and others do not. Under the conditions of the experiments reported by Kunst et al. (1989b), there were no apparent physiological disadvantages associated with the loss of the prokaryotic pathway. All that can be said, until further information becomes available, is that the effects noted are consistent with other evidence suggesting that membrane lipid composition is an important determinant of chloroplast structure but has relatively minor direct effects on the function of the membrane proteins associated with photosynthetic electron transport (Kunst et al., 1989b).

Conclusion

Thus, through the use of an array of lipid mutants, a picture seems to be emerging of the role of lipids, at least in chloroplast membranes, as being one of structural rather than functional significance. The isolation of more variants will help to clarify further this role and help to determine whether this view of the lipids should be modified.

The Synthesis or Dark Reactions of Photosynthesis

The sequence of reactions involving CO_2 fixation and formation of the complex carbohydrates produced in photosynthesis is often referred to as the carbon reduction cycle. The first stable compound formed in the cycle is 3-phosphoglyceric acid (3-PGA), which is itself produced from a five-carbon sugar, ribulose-1,5-bisphosphate (RuBP) (Fig 3-15). The cycle occurs in the stroma of chloroplasts and consists of a complex series of reactions that generate the end products of photosynthesis (notably sucrose and starch), as well as the regeneration of RuBP for the continuation of the cycle. The enzyme that catalyzes the formation of 3-PGA from RuBP is commonly called ribulose bisphosphate carboxylase, equally commonly abbreviated as rubisco (Salisbury and Ross, 1985).

$$CO_2 + \text{D-ribulose-1,5-bisphosphate} + H_2O \longrightarrow 2\ \text{3-phospho-D-glycerate}$$

Rubisco is functional in all photosynthetic organisms with the exception of a few photosynthetic bacteria. It is important not only because of the essential reaction it catalyzes but also because it seems to be by far the most abundant protein on the earth. Chloroplasts contain approximately half of the total protein in leaves and about one-fourth to one-half of their total protein is rubisco, which is important, therefore, in our diet and of other animal species. In Case Study 3-5, examples are discussed of attempts to investigate the role of rubisco in photosynthesis through genetic modifications of the enzyme.

This same rubisco protein also catalyzes an additional reaction involving molecular oxygen:

$$O_2 + \text{D-ribulose-1,5-bisphosphate} \longrightarrow \text{3-phospho-D-glycerate} + \text{2-phosphoglycolate}$$

This enzymic activity is called ribulose bisphosphate oxygenase and it catalyzes the first reaction in the process of photorespiration, which

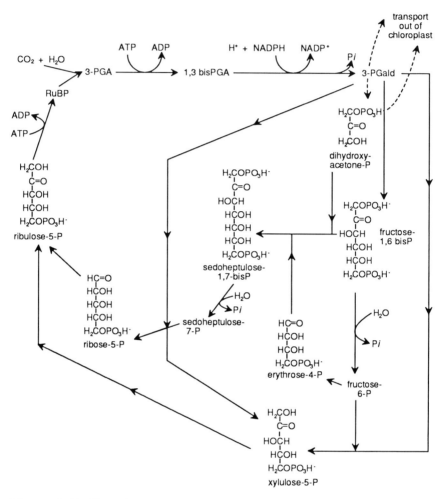

Figure 3-15. Reactions of the carbon reduction cycle.

many plant physiologists regard as the most important metabolic constraint on plant productivity (Fig. 3-16; Ogren, 1984).   In photorespiration, the O_2- and light-dependent release of CO_2 from the leaves of certain species, the oxygenation of RuBP, produces one molecule each of PGA and 2-phosphoglycolate. Two molecules of the latter are then metabolized to one each of CO_2 and PGA by a complex series of reactions occurring in the separate organelles - the chloroplast, the peroxisome, and the mitochondrion. Because CO_2 and O_2 are mutually competitive for the rubisco enzyme, photorespiration in intact plants can be manipulated simply by exposing them to atmospheres containing various CO_2/O_2 ratios. Photorespiration is inhibited by a higher than normal ratio of CO_2/O_2 and stimulated by a lower than normal ratio. Because net CO_2 fixation increases by about 50 percent when photorespiration is suppressed, the development of plant lines lacking this activity could increase plant productivity considerably. Some examples of such mutants are considered in Case Study 3-6.

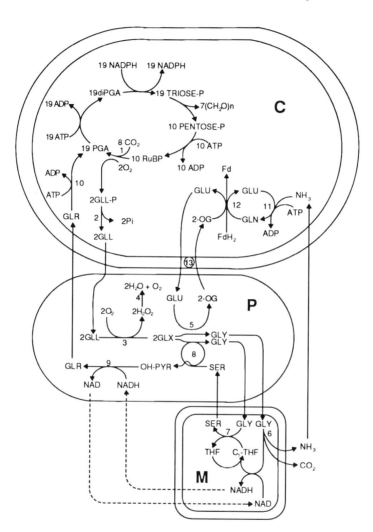

Figure 3-16. Approximate stoichiometry of C-3 photosynthesis and photorespiration in air at 25°C. Reactions occur in the chloroplast (C), peroxisome (P), and mitochondrion (M). Enzymes in the pathway are: 1, Ribulose bisphosphate carboxylase/oxygenase; 2, glycolate-P phosphatase; 3, glycolate oxidase; 4, catalase; 5, glutamate-glyoxylate aminotransferase; 6, glycine decarboxylase; 7, serine transhydroxymethylase; 8, serine-glyoxylate aminotransferase; 9, hydroxypyruvate reductase; 10, glycerate kinase; 11, glutamate synthase; 12, glutamine synthetase; 13, chloroplast dicarboxylic acid transporter. (Ogren, reproduced, with permission, from the *Annual Review of Plant Physiology*, vol. 35, © 1984 by Annual Reviews Inc.)

Case Study 3-5:   Mutations Affecting Rubisco Activity

Fraction I protein (or rubisco, see introduction) is an oligomer of large and small subunits of greatly differing molecular weights. The large subunit is encoded in the chloroplast genome, and is synthesized inside the chloroplast, whereas the small subunit is encoded in the nuclear genome, and is synthesized by cytoplasmic ribosomes as a higher molecular weight precursor. This precursor crosses the chloroplast envelope with removal of the extra sequence, and combines with the large subunit in the stroma to form the holoenzyme, rubisco (Fig. 3-17; Ellis, 1979).

The most famous paradox about rubisco is that the purified enzyme has an affinity for CO_2 that is far too low to account for the affinity of either leaves or isolated intact chloroplasts for this substrate. Much recent interest has centered on an understanding of this in vivo to in vitro discrepancy.

3-5.1   The Chlorophyll-Rubisco Connection

It has been understood for some time that light has a stimulatory effect on the development of rubisco activity in chloroplasts.

Some early reports suggested a correlation between the rate of synthesis of rubisco and the greening of leaves, that is, the formation of chlorophyll. An experimental approach to this problem was provided through the use of virescent leaves in which chlorophyll development showed a pronounced lag. If rubisco development paralleled that of

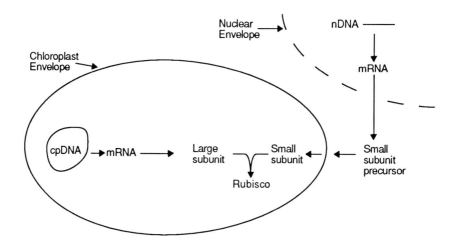

Figure 3-17. Interaction of genetic systems in the synthesis of rubisco. (Modified from Ellis, *Trends in Biochemical Science* 4: 241-244, 1979; reproduced by permission of Elsevier Publications Cambridge.)

chlorophyll, then there should also have been a lag in the appearance of the former in greening virescent leaves. It was found, however, that the development of rubisco in virescent leaves of cotton paralleled that in normal leaves. On a chlorophyll basis the development of rubisco in virescent leaves far exceeded that in green leaves (Fig. 3-18; Benedict and Kohel, 1969).

Thus, through the use of such mutants it seemed to be quite clear that chlorophyll development was not linked to rubisco synthesis. Another explanation of the stimulatory effect of light on rubisco activity was required.

3-5.2 Rubisco Activase

Other early research on activation was driven by an anomaly in the kinetics of rubisco activity: In chloroplasts and leaves of higher plants the $K_m(CO_2)$ was in the range of atmospheric CO_2 (about 10 μM), but the $K_m(CO_2)$ of isolated rubisco was 10 times higher. Later, it came to be recognized that an enzyme-Mg^{2+}-inorganic carbon complex was required for full rubisco activity, and that this complex formed just before the carboxylation of RuBP. In addition to activation by CO_2 and Mg^{2+}, rubisco is activated by preincubation with many phosphorylated

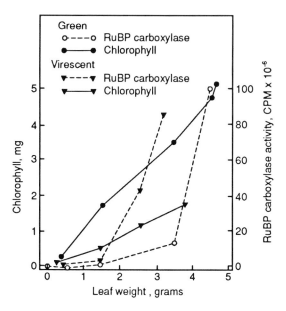

Figure 3-18. The synthesis of chlorophyll and rubisco in virescent and green cotton leaves. Green leaves: (○) Rubisco (●) chlorophyll; virescent leaves: (▽) Rubisco (▼) chlorophyll. (Benedict and Kohel, *Plant Physiol. 44*: 621-622, 1969; reproduced by permission of the ASPP.)

metabolites, such as the phosphorylated sugars and other intermediates of the Calvin cycle. The activation effect attributable to light could be linked in this scheme to the influence light was thought to have on the concentration of Mg^{2+} in the stroma of the chloroplast and of its pH (see Case Study 3-1 for other effects of cations and phosphorylated compounds in the chloroplast). Thus, the removal of H^+ from the stroma when chloroplasts are illuminated causes the movement of Mg^{2+} into the stroma to replace the protons. This causes the pH of the stroma to become more alkaline, a condition that also enhances rubisco activation. A major problem with attributing the regulation of rubisco activity solely to Mg^{2+} and pH effects is that the purified enzyme rapidly loses activation under assay conditions that mimic those thought to exist in the chloroplast (Ogren et al., 1986).

Investigation of the activation of rubisco was greatly enhanced by the isolation of mutants of *Arabidopsis thaliana* that lacked activation of the enzyme in vivo. The method used to isolate these mutants was that used to generate lines with defects in the photorespiratory pathway (see Case Study 3-6) (Somerville et al., 1982). The method also yielded mutants with a reduced affinity of the carboxylation reaction of rubisco for CO_2, because the selection process depended on photosynthetic performance at high CO_2 concentrations. Following the usual pattern, one line isolated using this selection technique was green and healthy in appearance when grown under high CO_2 but became chlorotic after several days of illumination in standard atmospheric conditions. The effect of light and CO_2 concentration on rapidly extractable rubisco activity in intact plants is seen in Figure 3-19.

Prior to illumination, both mutant and wild type had similar levels of rubisco activity. Following illumination, wild type activity increased, whereas activity in the mutant decreased. High CO_2 (1 percent) stimulated the activity observed in wild-type extracts but did not prevent the decline in the mutant. Thus, the mutant either lacked the mechanism of activation or light brought about the production of an inhibitor of the enzyme. Whatever the mechanism, nothing prevented the bicarbonate-stimulated activation of the enzyme from the mutant to normal levels of activity once it was removed from the cell (Fig 3-20; Somerville et al., 1982).

Thus, it seemed that something in the chloroplast milieu was specifically preventing bicarbonate activation of the enzyme in vivo, even at CO_2 concentrations 30-fold greater than found in air. No regulatory substance, sugar phosphate or other, could be found to explain the different behavior of rubisco in the mutant and the wild type. Nor did the mutation (designated *rca*) give rise to any difference in Mg^{2+} levels in chloroplasts or pH values important in photosynthesis. Another enzyme specific regulatory agent appeared to be involved.

The nature of the unidentified component that activated rubisco was first revealed by two-dimensional gel electrophoresis of chloroplast stromal proteins of wild type and the activation-deficient mutant of

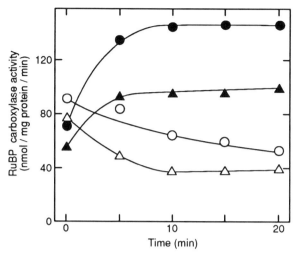

Figure 3-19. Effect of light and CO_2 concentration on the activation state of rubisco in intact plants of wild-type and mutant *Arabidopsis thaliana*. Plants were illuminated in air ($\triangle$,$\blacktriangle$) or in 1 percent CO_2, balance N_2 ($\bigcirc$,$\bullet$), and at intervals quickly homogenized and extracts assayed for rubisco activity. Wild type ($\bullet$,$\blacktriangle$); mutant ($\bigcirc$,$\triangle$). (Somerville et al., *Plant Physiol.* 70: 381-387, 1982; reproduced by permission of the ASPP.)

Arabidopsis. Two polypeptides of about 47 and 50-kD were missing in the mutant. The absence of the 50-kD polypeptide co-segregated with the physiological phenotype (the 47-kD polypeptide was masked by other proteins). Thus, it was concluded that the missing component was a soluble chloroplast protein called rubisco activase (Salvucci et al., 1985).

Additional evidence for the involvement of a protein in rubisco activation was obtained from the development of a reconstituted chloroplast assay system that included washed chloroplast thylakoids, rubisco, RuBP, and a partially purified extract of chloroplast proteins. In this system, chloroplast extracts from spinach and wild-type *Arabidopsis* stimulated rubisco activity in the light, but the extract of mutant chloroplasts did not. Stimulation of activity was also observed in lysed spinach and *Arabidopsis* wild-type chloroplasts, but not in lysed chloroplasts from the mutant (Table 3-20; Salvucci et al., 1985; Ogren et al., 1986).

3-5.3 The Coordinate Control of Light and Dark Reactions of Photosynthesis

The detection of a rubisco activase in plants provided a mechanism to explain the coordinate regulation of the flow of photons through the electron transport chain with the activity of rubisco and, in turn, with

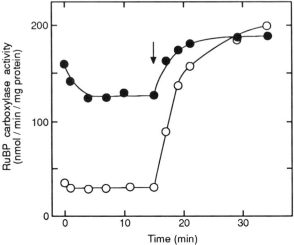

Figure 3-20. Bicarbonate activation of rubisco in crude extracts from wild-type and mutant *Arabidopsis thaliana*. The arrow indicates the point at which 10 mM (final concentration) bicarbonate was added to assay mixtures. Wild type (●), mutant (○). (Somerville et al., *Plant Physiol. 70*: 381-387, 1982; reproduced by permission of the ASPP.)

CO_2 fixation so that a balance is maintained between the light and dark reactions of photosynthesis.

Rubisco activation in leaves was stimulated by increasing light intensity but not by increasing CO_2 concentration. As shown in Figure 3-21, the activation of rubisco by light in both the wild type and the *rca* mutant closely matched the steady state photosynthetic rate of the plants (Salvucci et al., 1986). In the dark, the activation state of rubisco was similar in the mutant and wild-type plants, increasing as a function of CO_2 concentration over the range of 0 to 2,000 μL of CO_2/L. In light, the rubisco activation state in both the mutant and the wild-type was relatively insensitive to changes in the external CO_2 concentration between 100 and 2,000 μL of CO_2/L (Fig. 3-22).

The extent of light activation, however, differed substantially between the two strains. Although the wild-type activation state in the light was considerably higher than in the dark at CO_2 concentrations below 2,000 μL CO_2/L and in CO_2-free air, the activation state of the mutant was lower in the light than in the dark except in CO_2-free air where activation states in the light and dark were similar (Fig 3-23). At 2,000 μL CO_2/L, rubisco activation level in the mutant in the light was only 30 percent of wild type.

By modulating the first reaction in carbon reduction, rubisco activase ultimately controls the balance between the utilization and regeneration of RuBP and hence the rate of net photosynthesis under steady-state conditions. In illuminated leaves, the steady-state RuBP concentration is saturating for rubisco regardless of the CO_2 concentration or light

Table 3-20. Light activation of rubisco in a reconstituted chloroplast system and in lysed chloroplasts.

Experiment	Source of chloroplasts	Source of extract	Rubisco activity (1) nmol CO_2 mg rubisco protein[1] min[1]; (2) nmol CO_2 min[-1]		
			Dark	Light	Light-dark
(1)	spinach	none	32	36	4
	spinach	spinach	40	523	483
	spinach	A. thaliana (w.t.)	36	119	83
	spinach	A. thaliana (mutant)	34	37	3
(2)	spinach	(lysed)	24	84	60
	A. thaliana (w.t.)	(lysed)	13	33	20
	A. thaliana (mutant)	(lysed)	16	18	2

(Ogren et al., *Phil. Trans. R. Soc. Lond. B. 313*: 1986; Salvucci et al., *Photosyn. Res. 7*: 193-201, 1985.)

intensity. RuBP forms a complex with inactivated rubisco and only very slowly activates in the presence of CO_2 and Mg^{2+}. Rubisco activase uses rubisco-RuBP as a substrate, and as rubisco activase is apparently driven by electron transport through PS1 in illuminated thylakoids (Campbell and Ogren, 1990), the amount of active rubisco is directly determined by the light intensity received by the chloroplast. This mechanism ensures that the RuBP concentration is always saturating and is sufficient to maintain the maximum possible rates of photosynthesis whatever the ambient light intensity, CO_2 concentration, and other environmental parameters.

In contrast to previous models regarding light and CO_2 limitations of photosynthesis, which considered that energy supply directly regulated RuBP synthesis and utilization (see, for example, Case Study 3-2.2.3), the use of this mutant and the consequent discovery of rubisco activase has led to the idea that the activation level of rubisco and the rate of RuBP synthesis are coordinately regulated by processes in the thylakoid membrane to maintain a saturating level of RuBP and, thus, maximum rates of photosynthesis (Salvucci et al., 1986). Recent evidence suggests that this regulation is accomplished in more than one way in different species of plants. In addition to activase, tight binding inhibitors are

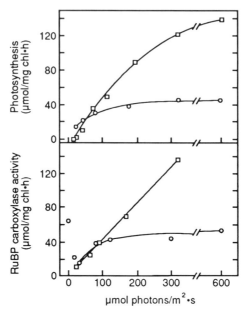

Figure 3-21. Response of net photosynthesis and rubisco activation level (initial activity) to light intensity in *Arabidopsis* wild type (□) and *rca* mutant (○). (Salvucci et al., *Plant Physiol. 80*: 655-659, 1986; reproduced by permission of the ASPP.)

also a widespread mechanism for regulation of rubisco (Kobza and Seemann, 1988; Salvucci, 1989). The isolation and characterization of this mutant of *Arabidopsis thaliana*, however, provided a major breakthrough in understanding rubisco activation in situ.

Case Study 3-6:   Photorespiratory Mutants

Photorespiration, the O_2- and light-dependent release of CO_2 from the leaves of certain species, is inseparable from photosynthetic CO_2 fixation, yet has no acknowledged function.   The relationship between photosynthesis and photorespiration is determined by the kinetic properties of the bifunctional enzyme ribulose-1,5-bisphosphate carboxylase/oxygenase (rubisco) for which CO_2 and O_2 are mutually competitive substrates. Carboxylation of ribulose bisphosphate (RuBP) produces two molecules of phosphoglycerate (PGA), whereas oxygenation of RuBP produces one molecule each of PGA and 2-phosphoglycolate (see introduction). Two molecules of phosphoglycolate are subsequently metabolized to one each of CO_2 and PGA by a complex series of reactions occurring in three separate organelles: the chloroplast, the peroxisome, and the mitochondrion (Ogren, 1984). According to this scheme (Fig. 3-

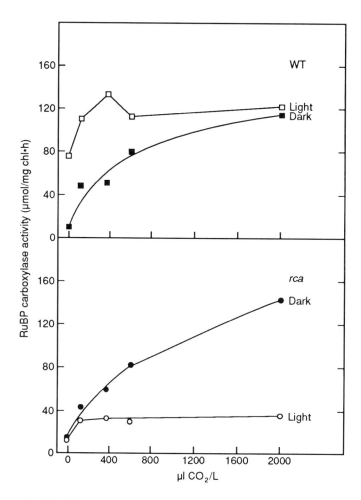

Figure 3-22. The response of rubisco activation level in the light and dark to external CO_2 concentration in *Arabidopsis* wild type (□,■) and *rca* mutant (○,●). (Salvucci et al., *Plant Physiol. 80*: 655-659, 1986; reproduced by permission of the ASPP.)

16), photorespiratory release of CO_2 results from the mitochondrial decarboxylation of glycine. Because CO_2 and O_2 are mutually competitive, photorespiration in intact plants can be manipulated simply by exposing them to atmospheres containing various CO_2/O_2 ratios. Photorespiration is inhibited by a higher than normal ratio of CO_2/O_2 and stimulated by a lower than normal one.

Control of photorespiration, an energy-wasting process that characterizes C-3 plant species, represents, in principle, a major contribution to increased crop productivity. Net CO_2 fixation can be increased by about 50 percent in C-3 plants when photorespiration is inhibited by reducing the O_2 concentration around the leaf from 21 to

1%. The key is to prevent rubisco acting as an oxygenase and thus removing potential precursor of photosynthesis to the photorespiratory pathway. In turn, this will necessitate altering the enzyme so that the carboxylation/oxygenation ratio is increased under normal atmospheric conditions.

To pursue this problem systematically a two-step strategy has been devised for recovery of plants with reduced RuBP oxygenase activity. First, the plant is genetically altered so that the presence of the oxygenase is a lethal character when the plant is grown in air. Then, revertants that grow normally in air may have reduced oxygenase activity because of secondary mutations that alleviate the effect of the first mutation. Therefore, it was possible to undertake the isolation of mutants with defects in the photorespiratory pathway on the premise that such mutants would not be viable under normal atmospheric conditions (350 μL/L CO_2/21 percent O_2/balance N_2) but viable under conditions that limit photorespiration (eg., 1 percent CO_2 / 21 percent O_2/balance N_2) (Somerville and Ogren, 1982).

Thus, mutants with defects in photorespiratory metabolism have been isolated in *Arabidopsis* and barley by exploiting the observation that the flow of carbon into the photorespiratory pathway can be prevented without any deleterious effects by placing plants in air containing high levels (1%) of CO_2. Thus, mutants with defects in photorespiratory metabolism were isolated as plants that were normal in appearance when grown in air containing high CO_2 but that turned chlorotic within several days after being transferred to standard atmospheric conditions because of the disruption of photorespiratory metabolism (Somerville and Ogren, 1982).

3-6.1   The Isolation of Photorespiratory Mutants in *Arabidopsis*

In the case of *Arabidopsis thaliana*, wild-type seed was mutagenized with ethyl methane sulfonate, brought to maturity in normal cultural conditions, and bulk collected. The M_2 generation was grown to the four-leaf stage in air enriched to 1 percent CO_2. Plants lacking vigor or showing chlorosis were removed. The CO_2 concentration was reduced to 0.03 percent and plants that remained normal were discarded after 4 days. The remainder were returned to 1 percent CO_2 for 4 days and those that failed to recover were discarded. A 500-fold reduction in population size resulted from these manipulations. The remaining plants were taken to maturity in 1 percent CO_2 without further screening (Somerville and Ogren, 1979).

A similar screening applied in the M_3 generation, to individual lines established from the M_2, resulted in the retention of 40 probable mutants from an initial population of about 40,000 M_2 plants. Because intermediates of the photorespiratory pathway are rapidly labeled with $^{14}CO_2$, many of the mutants were readily characterized by simply labelling the plants with $^{14}CO_2$, examining the distribution of label in the photorespiratory metabolites and performing confirming enzyme assays. This approach (Somerville and Ogren, 1979) permitted the characterization of mutants deficient in activity for seven enzymes associated with photorespiration. Others have since been added to this number through the use of the same technique. A number of major points regarding photorespiration have been settled using these mutants, some of which are detailed in the remainder of this Case Study.

3-6.1 The Characterization and Physiology of Some Photorespiratory Mutants

3-6.2.1 Source of Photorespiratory Glycolate

Although glycolate has long been recognized as an important precursor of photorespiratory CO_2, it was not until the discovery that rubisco could act also as an oxygenase that a convincing mechanism for glycolate metabolism became apparent. Attempts to quantify the flow of carbon through phosphoglycolate (P-glycolate) failed, however, probably because the chloroplast has a very active phosphatase. The isolation of a mutant lacking P-glycolate phosphatase provided unequivocal evidence that P-glycolate is the precursor of glycolate and photorespiratory CO_2 (Somerville and Ogren, 1982).

In an atmosphere containing 21 percent O_2 and 352 $\mu l/L$ CO_2, the photosynthetic rate of one of the *Arabidopsis* mutants initially approached that of the wild-type rate but rapidly declined to zero (Fig. 3-23; Somerville and Ogren, 1979). This reduction could be completely reversed by a 15 to 20 minutes dark treatment suggesting that inhibition was caused by a reversibly altered metabolic pool rather than photooxidative damage. The probable site of the lesion was deduced by 2 minutes labeling with $^{14}CO_2$ in normal atmospheric conditions when 20 percent of the ^{14}C incorporated was found in phosphoglycolate but was barely detectable in the wild type (Table 3-21). The increased labeling of the phosphoglycolate pool was accompanied by a large decrease in the labeling of glycolate, serine, and glycine, all products of photorespiration derived from phosphoglycolate.

These results suggested that the mutant was deficient in phosphoglycolate phosphatase, a chloroplast enzyme. This was confirmed by assaying the enzyme in crude leaf extracts (Table 3-22). The F_1 results suggested further that a single recessive nuclear mutation was responsible. F_1 progeny were phenotypically normal and had restoration of one-half of the phosphatase activity.

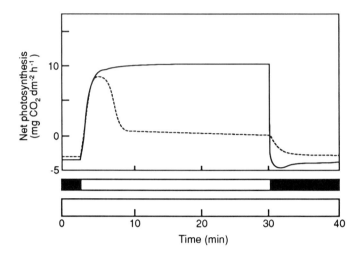

Figure 3-23. Net CO_2 exchange of wild-type and mutant *Arabidopsis* in an atmosphere with 352 μL/L CO_2, 21 percent O_2, balance N_2. The response of the wild-type plants is indicated by the solid line, and that of the mutant by the broken line. The open or closed bar in the Figure represents light or dark conditions, respectively. (Somerville and Ogren, reprinted by permission from *Nature* vol. 280 pp. 833-836. Copyright © 1979 Macmillan Magazines Ltd.)

3-6.2.2   *The Source of Photorespiratory CO_2*

Two reactions could account for the release of photorespiratory CO_2: the enzymic decarboxylation of glycine and the nonenzymic oxidation of glyoxylate to CO_2 and formate. This issue was clarified by using a mutant lacking serine hydroxymethyltransferase (*stm* mutant). In this mutant, glycine was not decarboxylated and accumulated as a metabolically inactive endproduct of photorespiratory carbon flow. During the first few minutes of illumination the mutant did not evolve CO_2. After several minutes, however, photorespiration began and increased until it approached that of the wild type. The mechanism implied by these observations was that during photorespiration, amino groups cycle between the amino acids glutamate, glycine, and serine and the α-keto acids, glyoxylate, oxoglutarate, and hydroxypyruvate. If a large percentage of these mobile amino groups accumulate in glycine, then glyoxylate cannot be aminated and, being unstable, would decarboxylate spontaneously. This was confirmed in the mutant when an exogenous source of NH_3 suppressed CO_2 release in the mutant but had no effect in the wild-type (Somerville and Ogren, 1982), as described later.

In photorespiration, glycine is converted to stoichiometric amounts of CO_2, NH_3 and the C-1 group of $N5$, $N10$-methylene tetrahydrofolate

Table 3-21. Products of $^{14}CO_2$ assimilation by wild-type and mutant *Arabidopsis.* Values indicate percentage of total water-soluble counts recovered, and represent the mean of two independent experiments.

	Wild-type	CS119
Basic fraction	21.4	11.0
Glycine	10.5	0.5
Serine	8.4	0.9
Alanine	2.0	9.0
Neutral fraction	6.1	14.8
Acid-1 fraction	36.7	23.3
Glycolate	2.7	0.9
Acid-2 fraction	8.8	29.1
Phosphoglycerate	6.3	5.6
Phosphoglycolate	0.7	19.6
Acid-3 fraction	24.4	17.6
Recovery	97.4	95.8

(Somerville and Ogren, reprinted by permission from *Nature* vol. 280 pp. 833-836. Copyright © 1979 Macmillan Magazines Ltd.)

(see Fig. 3-16). There is an absolute requirement for tetrahydrofolate (THF) in this reaction, which must be regenerated from glycine + C-1-THF = serine + THF. The mutants were recognized by their ability to

Table 3-22. Phosphoglycolate phosphatase activity in wild-type and mutant leaf extracts.
[a] Enzyme activity is expressed as μmol/minute/mg protein.

Strain	Phosphoglycolate phosphatase activity [a]	Rubisco activity [a]
Wild type	0.22	0.30
CS119	0.01	0.28
F1(WTxCS119)	0.08	0.23
Mixture	0.11	0.29

(Somerville and Ogren, reprinted by permission from *Nature* vol. 280 pp. 833-836. Copyright © 1979 Macmillan Magazines Ltd.)

accumulate high levels of [^{14}C]glycine (47 percent of label versus 6 percent in the wild type) in 10 minutes of photosynthesis in normal O_2 levels and by having a low serine level. Although the serine transhydroxymethylase level was 15 percent that in the wild type, that in the mitochondria was judged, on biochemical evidence, to be absent. Although the mutants were unable to metabolize glycine, they exhibited a substantial rate of light and O_2-dependent release of CO_2 (Fig. 3-24; Somerville and Ogren, 1981).

These mutants (Fig. 3-24) had a rate of photorespiration as much as 30 percent that in the wild type. This suggests CO_2 is arising by a mechanism other than by glycine decarboxylation. The source of this CO_2 might be as follows: continued exposure to photorespiratory conditions could result in all readily transferable amino groups becoming trapped in glycine due to continued glyoxylate synthesis and amination. Under these circumstances, the demand for NH_3 imposed by glyoxylate transamination might far exceed the ability of the plant to provide reduced nitrogen from NO_3^- reduction, therefore, the rate of glycine synthesis would be expected to decline eventually due to depletion of suitable amino donors. Then, if glyoxylate cannot be converted to glycine it may be directly decarboxylated to give CO_2. Provision of NH_3 artificially to mutant leaves should, if the above is the case, cause a reduction in the evolution of CO_2 whereas no effect should be found in the case of wild-type leaves. The results in Figure 3-25 support this conclusion (Somerville and Ogren, 1981). Leaf fragments of the wild type released CO_2 into a CO_2-free gas stream in a typical response that was unaffected by the presence of 10 mM NH_3 in the medium (Fig. 3-25B). Leaf fragments of the mutant line released CO_2 at about 50 percent of the wild-type rates when placed on unsupplemented medium

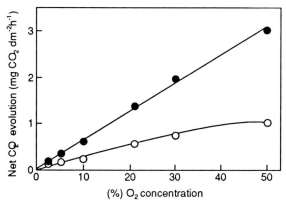

Figure 3-24. O_2 response of CO_2 evolution into CO_2-free gas by intact plants of wild-type and mutant *Arabidopsis*. Wild-type (●); *stm* mutant (○). (Somerville and Ogren, *Plant Physiol.* 67: 666-671, 1981; reproduced by permission of the ASPP.)

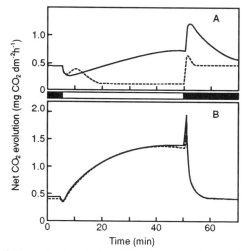

Figure 3-25. CO_2 evolution into a CO_2-free gas stream by leaves of mutant (A) and wild-type (B) *Arabidopsis* in the presence and absence of NH_3. Response of untreated leaves (-); response of leaves placed on medium containing 10 mM $(NH_4)_2SO_4$ (---). The broken bar between graphs A and B represents light and dark conditions. (Somerville and Ogren, *Plant Physiol.* 67: 666-671, 1981; reproduced by permission of the ASPP.)

but did not show photorespiration when placed on 10 mM NH_3 (Fig. 3-25A).

3-6-2.3   The Photorespiratory NH_3 Cycle

During the mitochondrial oxidation of glycine, NH_3 is released at a rate that greatly exceeds the rate of primary NO_3^- reduction. This NH_3 must be refixed and recycled to the peroxisome as glutamate before another glyoxylate molecule can be aminated. In principle, NH_3 refixation could occur by glutamate dehydrogenase (GDH), or by sequential action of glutamine synthetase (GS) and glutamate synthase (GOGAT) (see Fig. 3-16). The importance of the latter pathway was confirmed by the isolation of a class of mutants deficient in GOGAT activity. They could not refix NH_3 and the pool of mobile NH_3 (in glutamate, aspartate, and alanine) was rapidly depleted. Photosynthesis was dramatically inhibited by the combined effects of NH_3 toxicity and the shortage of amino donors for glyoxylate transamination (Somerville and Ogren, 1982).

The CO_2 released in photorespiration is believed to result from a mitochondrial reaction in which glycine is converted to equimolar amounts of CO_2, NH_3 and the C-1 group of $N5$, $N10$-methylene-tetrahydrofolate (see Section 3-6.2.2). The NH_3 could be refixed directly into glutamate by mitochondrial GDH, which seems unlikely because of the high K_m value for NH_3 exhibited by GDH in vitro. Alternatively, it has been suggested that the NH_3 could be reassimilated by the sequential action of cytoplasmic GS and GOGAT (see Fig. 3-16). Direct evidence for this was provided from experiments with mutants lacking leaf GOGAT activity (Table 3-23; Somerville and Ogren, 1980). A single nuclear mutation seemed to be responsible for the enzyme deficiency and for the conditional lethal phenotype (*gluS*).

A number of parameters were measured during the illumination of wild-type and mutant plants having to do with photosynthesis and photorespiration (Fig. 3-26; Somerville and Ogren, 1980). The observed inhibition of photosynthesis in the mutant may have been due to NH_3 accumulation. During just 25 minutes of illumination under photorespiratory conditions, one of the mutants tested (CS113) increased NH_3 20-fold to 3 mM, a concentration sufficient to uncouple photophosphorylation. The NH_3 accumulation was attributable to the virtual disappearance of free glutamate in the leaf and, thus, the lack of an acceptor for reassimilation (Fig. 3-26). The reduction in glutamate corresponded roughly to the increase in glutamine as NH_3 was initially refixed by GS. The glutamine pool in the mutant then stabilized because further synthesis was prevented by the lack of glutamate, and the GOGAT deficiency prevented conversion of glutamine to glutamate. The observation that NH_3 was not rapidly refixed into glutamate verified the suggested low in vivo activity of GDH in an assimilatory

Table 3-23. Glutamate synthase activity in leaf extracts of wild-type and mutant lines of *Arabidopsis*.

[a] Enzyme activity is expressed as µmoles of product/hour/mg protein.

[b] Equal volumes of wild-type and mutant (CS113) extract were mixed immediately before assay.

Strain	Glutamate synthase activity[a]	Rubisco activity[a]
Wild type	5.3	30.4
CS30	0.2	29.7
CS103	0.2	29.7
CS113	0.1	30.7
F_1(WTxCS113)	2.8	31.1
F_1(CS103xCS113)	0.3	31.5
Mixture[b]	0.3	31.5

(Somerville and Ogren, reprinted by permission from *Nature* vol. 286 pp.257-259. Copyright © 1980 Macmillan Magazines Ltd.)

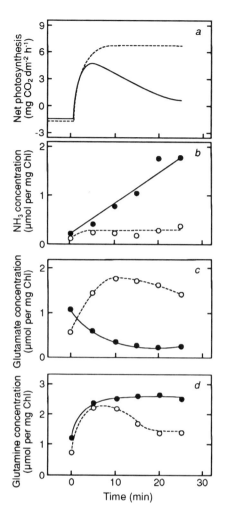

Figure 3-26. Changes in photosynthesis rate (a) and levels of NH_3 (b); glutamate (c); and glutamine (d), during illumination of wild-type and mutant *Arabidopsis* (CS113, see Table 3-23). The response of the *gluS* mutant (CS113) is indicated by the solid lines and symbols and that of the wild type by broken lines and open symbols. (Somerville and Ogren, reprinted by permission from *Nature* vol. 286 pp. 257-259. Copyright © 1980 Macmillan Magazines Ltd.)

direction and confirmed the importance of the GS/GOGAT pathway in the reassimilation of photorespiratory NH_3. In fact, the sole function of leaf GOGAT seems to be the reassimilation of photorespiratory NH_3, a point that could not be ascertained only with the use of mutants such as these (Somerville and Ogren, 1980). Other mutants were required, such as those discussed later.

3-6.2.4   Confirmation That the GS/GOGAT Pathway is Involved in Photorespiration-Chloroplast Dicarboxylate Transport

The involvement of this pathway in photorespiration was confirmed by the recovery of a mutant with greatly reduced ability to transport dicarboxylic acids across the chloroplast envelope. This is functionally equivalent to not having GOGAT, because 2-oxoglutarate and glutamate are unable to cycle between the chloroplast and the peroxisome at a sufficient rate (see Fig. 3-16).

The constituent reactions of the photorespiratory pathway of higher plants occur in three organelles, the chloroplast, the mitochondrion, and the peroxisome. Thus, transport processes must intervene at several steps of the pathway. To the extent that the transport of substrates, products, or co-factors is limiting, these transport processes may be expected to exert a strong regulatory influence on photorespiratory metabolism. Furthermore, tightly integrated with the photorespiratory carbon cycle is the photorespiratory nitrogen cycle in which ammonia released during glycine deamination is refixed by the sequential action of GS and GOGAT, as was just discussed. The resultant glutamate supports both photorespiratory ammonia refixation and glyoxylate amination. In the absence of adequate glutamate pools, glyoxylate is oxidized nonenzymatically to CO_2 at rates that significantly reduce net CO_2 assimilation, and ammonia accumulates to toxic levels. Because glutamate is synthesized in the chloroplast and consumed in the peroxisome, the transfer of glutamate and 2-oxoglutarate between these two organelles is a necessary component of photorespiratory nitrogen metabolism (Somerville and Ogren, 1983).

Studies with isolated chloroplasts have revealed the presence of a transporter, designated the dicarboxylate transporter, that catalyzes the counterexchange of several dicarboxylic acids across the chloroplast inner membrane. These compounds include malate, 2-oxoglutarate, aspartate, glutamate, and, in some cases, glutamine. Therefore, the chloroplast dicarboxylate transporter is implicated as an important component of the photorespiratory pathway. The mutant line CS156 of *A. thaliana* carries a recessive nuclear mutation responsible for the specific loss of dicarboxylate transport activity (the *dct* mutant) (Fig. 3-27). Similar uptake patterns were observed in this mutant in the case of glutamate, aspartate, and malate, but not for glutamine or inorganic phosphate where there was no difference between the wild type and the mutant (Fig. 3-27). Thus, in contrast to some other observations with other plant species, results in these experiments seemed to indicate a separate transporter for glutamine.

Physiological studies of pools of amino acids and ammonia strongly indicated that the level of glutamate in the *dct* mutant was only about half that in the wild type. Yet in vitro assays revealed the presence of sufficient levels of the enzymes of glutamate synthesis to support

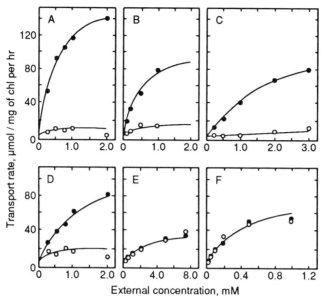

Figure 3-27. The uptake of several compounds into chloroplasts from wild-type (●) and *dct* mutant plants of *Arabidopsis* . The compounds assayed for uptake were 2-oxoglutarate (A), aspartate (B), glutamate (C), malate (D), glutamine (E), and inorganic phosphate (F). (Somerville and Ogren, *Proc. Natl. Acad. Sci. USA 80*: 1290-1294, 1983.)

normal pools of the amino acid. To examine this inconsistency, GOGAT was measured in isolated intact chloroplasts as 2-oxoglutarate + glutamine-dependent O_2 evolution (Table 3-24; Somerville and Ogren, 1983).

At low external concentrations of 2-oxoglutarate, no 2-oxoglutarate + glutamine-dependent O_2 evolution by mutant chloroplasts was detected. When the 2-oxoglutarate concentration was raised to 100 mM, however, chloroplasts from both the mutant and wild type exhibited

Table 3-24. Glutamine + 2-oxoglutarate-dependent O_2 evolution by isolated chloroplasts.

The percentage chloroplast intactness was 88 percent for wild type and 76 percent for CS156.

Substrate, mM		O_2 evolution μmol/mg of Chl per hr	
Glutamine	2-Oxoglutarate	Wild-type strain	CS156 mutant
5	3	4.77	0
5	100	4.00	10.60

(Somerville and Ogren, *Proc. Natl. Acad. Sci. USA 80*: 1290-1294, 1983.)

GOGAT activity. On the basis of these experiments (Table 3-24), it was considered probable that chloroplasts from the mutant were substantially less permeable to 2-oxoglutarate than the wild-type chloroplasts, a conclusion confirmed by the results reported in Figure 3-27.

The lethality of the *dct* mutant in atmospheres promoting photorespiration could be explained by the in vivo loss of GOGAT function. In turn, this restricted glyoxylate amination and photorespiratory ammonia refixation by limiting the supply of glutamate. If glyoxylate is not transaminated, it does not accumulate but undergoes nonenzymatic oxidation, resulting in a greatly enhanced loss of recently fixed carbon. As a result, Calvin cycle intermediates are depleted, leading to a decline in net CO_2 fixation.

One symptom of the defect in this mutant is a major reduction in the amount of a prominent polypeptide of the chloroplast envelope suggesting that this polypeptide may be a functional component of the dicarboxylate transporter. An abundant 42-kD protein in the envelopes from wild type is low or absent in envelopes from the mutant (Somerville and Somerville, 1985).

3-6.2.5   *Can the Oxygenase Activity of Rubisco be Reduced?*

Quantitative measurements of photorespiration are difficult to achieve because photosynthesis is diametric. The CO_2 evolved in the former may be partly consumed in the latter. Thus, true values can be obtained for neither. In the presence of adequate amino group availability, measurement of the accumulation of glycine in a mutant lacking serine *trans*-hydroxymethylase (*stm* mutant) should provide an accurate estimate of photorespiration and its ratio to true photosynthesis.

Glycine rapidly accumulated in leaves of the *stm* mutant exposed to strong photorespiratory conditions but not at low O_2. A linear relationship was seen between the moles of glycine accumulated to moles of CO_2 fixed, a result consistent with the assumption that the carboxylase/oxygenase activities of rubisco were the sole determinants of this ratio (Fig. 3-28; Somerville and Somerville, 1983). If 1 mole of CO_2 is released for every 2 moles of glycine that traverse the photorespiratory pathway, these results (Fig. 3-28) indicate that in standard atmospheric conditions photorespiration is about 27 percent of the rate of true photosynthesis or about 36 percent of the apparent rate, assuming no internal refixation of photorespiratory CO_2.

3-6.2.6   *Conclusions and Discussion of Other Photorespiratory Mutants*

Results, such as those in Figure 3-28, emphasize the extent to which photorespiration modifies photosynthetic productivity and why efforts

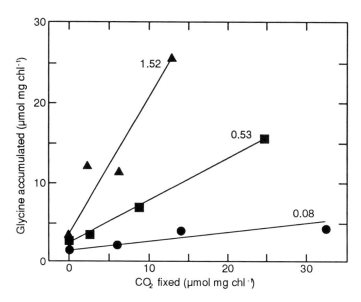

Figure 3-28. The effect of different O_2 concentrations on the relationship between glycine accumulation and photosynthetic CO_2 fixation in a *stm* mutant of *Arabidopsis*. The medium supporting the leaves contained 30 mM each of serine and NH_4Cl. The gas phase contained: 323 µl CO_2/L, 2 percent O_2, balance N_2 (●); 309 µl CO_2/L, 21 percent O_2, balance N_2 (■); 349 µl CO_2/L, 50 percent O_2, balance N_2 (▲). The numbers in the figure are the slopes of the associated lines. (Somerville and Somerville, *J. Exp. Bot. 34:* 415-424, 1983; reproduced by permission of Oxford University Press.)

are being directed toward reducing or eliminating it from agronomically important plants. Yet, although all photosynthetic organisms possess a bifunctional rubisco, not all species exhibit photorespiration. Nature has effectively controlled photorespiration in some species, but always by CO_2 concentrating mechanisms such as occur in C-4 photosynthesis and in the unicellular algae and cyanobacteria. Why has the control of photorespiration not been achieved by modification of rubisco? No satisfactory answer to this question has yet been provided. In fact, as one progresses up the evolutionary ladder from photosynthetic bacteria to angiosperms, the ratio of carboxylase to oxygenase activity increases. Thus, as variation does exist in nature it is possible that additional advantageous changes can be made in the enzyme by mutagenesis and the application of appropriate selection pressure. Indeed, in the green alga *Chlamydomonas reinhardtii*, it has been found possible to modify the CO_2/O_2 specificity of rubisco by mutation (Chen and Spreitzer, 1989). In the case of higher plants, however, results of selection for a modified photorespiration capability has led only to changes in other aspects of CO_2/O_2 utilization (Somerville and Ogren, 1982). One such case is discussed below.

Results in Figure 3-28 indicated that at high O_2 levels, the ratio of net photorespiration/net photosynthesis was high. Zelitch (1989, 1990a,b) exploited this fact to isolate tobacco plants with O_2-resistant photosynthesis. A population of dihaploid tobacco plantlets were grown in a 42 percent O_2/low CO_2 atmosphere and screened for normal green color among the population that generally turned yellow under these atmospheric conditions. Several of the plants with O_2 resistance had superior net photosynthesis under certain conditions and a lower stoichiometry of photorespiratory CO_2 formation. These plants also averaged about 40 percent greater catalase activity than the wild type.

The photorespiratory metabolites glyoxylate and hydroxypyruvate are rapidly decarboxylated by hydrogen peroxide (H_2O_2) which is generated in the pathway by glycolate oxidase (see Fig. 3-16). Others found a barley mutant grossly deficient in catalase to be unable to survive under photorespiratory conditions (see Zelitch, 1989). When photorespiration increases greatly, as at high O_2 levels for example, the excess H_2O_2 produced might successfully attack the vulnerable keto acids of the pathway and generate CO_2. A plant could be, therefore, O_2 resistant because it had a greater catalase activity and showed superior net photosynthesis under certain conditions because less CO_2 is produced internally. This seems to be the case in these mutants of tobacco (Zelitch, 1989,1990a,b). Another attempt to isolate O_2-insensitive plants has led to the identification of a mutant of *Flaveria linearis* with a reduced cytosolic fructose-1,6-bisphosphatase, not a modified rubisco (Sharkey et al., 1988).

Because of the general interest in and importance of photorespiration, the literature on the mutants generated in this pathway is already voluminous and growing rapidly. Three of the most recent publications that can be used to trace earlier work on the physiology of this class of mutants are in barley (Murray et al., 1989), in tobacco (McHale et al., 1989), and in *Arabidopsis* (Grumbles, 1989).

References

Archer, E.K. and Bonnett, H.T. 1987. Characterization of a virescent chloroplast mutant of tobacco. *Plant Physiol. 83*: 920-925.Archer, E.K., Hakansson, G., and Bonnett, H.T. 1987. The phenotype of a virescent chloroplast mutation in tobacco is associated with the absence of a 37.5-kD polypeptide. *Plant Physiol. 83*: 926-932.

Bellemare, G., Bartlett, S.G., and Chua, N-H. 1982. Biosynthesis of chlorophyll a/b-binding polypeptides in wild type and the *chlorina f2* mutant of barley. *J. Biol. Chem. 257*: 7762-7767.

Benedict, C.R. and Kohel, R.J. 1969. The synthesis of ribulose-1,5-diphosphate carboxylase and chlorophyll in virescent cotton leaves. *Plant Physiol. 44*: 621-622.

Benedict, C.R. and Kohel, R.J. 1970. Photosynthetic rate of a virescent cotton mutant lacking chloroplast grana. *Plant Physiol. 45*: 519-521.

Benedict, C.R. and Ketring, D.L. 1972. Nuclear gene affecting greening in virescent peanut leaves. *Plant Physiol. 49*: 972-976.

Benedict, C.R., McCree, K.J., and Kohel, R.J. 1972. High photosynthetic rate of a chlorophyll mutant of cotton. *Plant Physiol. 49*: 968-971.

Bennett, J., Williams, R., and Jones, E. 1984. Chlorophyll-protein complexes of higher plants: Protein phosphorylation and preparation of monoclonal antibodies. In *Advances in Photosynthetic Research*, vol. 3, Sybesma, C. (ed.), Nijhoff/Junk, The Hague, Netherlands, pp. 99-106.

Browse, J., McCourt, P., and Somerville, C.R. 1985. A mutant of *Arabidopsis* lacking a chloroplast-specific lipid. *Science 227*: 763-765.

Browse, J., McCourt, P., and Somerville, C. 1986. A mutant of *Arabidopsis* deficient in C18:3 and C16:3 leaf lipids. *Plant Physiol. 81*: 859-864.

Browse, J., Kunst, L., Anderson, S., Hugly, S., and Somerville, C. 1989. A mutant of *Arabidopsis* deficient in the chloroplast 16:1/18:1 desaturase. *Plant Physiol. 90*: 522-529.

Burke, J.J., Steinback, K.E., and Arntzen, C.J. 1979. Analysis of the light-harvesting pigment-protein complex of wild type and a chlorophyll b-less mutant of barley. *Plant Physiol. 63*: 237-243.

Campbell, W.J. and Ogren, W.L. 1990. Electron transport through photosystem 1 stimulates light activation of ribulose bisphosphate carboxylase/oxygenase (rubisco) by rubisco activase. *Plant Physiol. 94*: 479-484.

Chen, Z. and Spreitzer, R.J. 1989. Chloroplast intragenic suppression enhances the low CO_2/O_2 specificity of mutant ribulose-bisphosphate carboxylase/oxygenase. *J. Biol. Chem. 264*: 3051-3053.

Duranton, J. and Brown, J. 1987. Evidence for a light-harvesting chlorophyll a-protein complex in a chlorophyll b-less barley mutant. *Photosyn. Res. 11*: 141-151.

Edwards, G.E. and Jenkins, C.L.D. 1988. C4 photosynthesis: Activities of photosynthetic enzymes in a virescent mutant of maize having a low-temperature-induced chloroplast ribosome deficiency. *Aust. J. Plant Physiol. 15*: 385-395.

Ellis, R.J. 1979. The most abundant protein in the world. *Trends in Biochemical Science 4*: 241-244.

Ghirardi, M.L., McCauley, S.W., and Melis, A. 1986. Photochemical apparatus organization in the thylakoid membrane of *Hordeum vulgare* wild type and chlorophyll b-less *chlorina f2* mutant. *Biochim. Biophys. Acta 851*: 331-339.

Goodchild, D.J., Highkin, H.R., and Boardman, N.K. 1966. The fine structure of chloroplasts in a barley mutant lacking chlorophyll b. *Exptl. Cell Res. 43*: 684-688.

Grumbles, R.M. 1989. Effect of glutamate synthase deficiency on CO_2 exchange, ATP level, Calvin cycle enzymes and ribulose 1,5-bisphosphate levels in *Arabidopsis* leaves. *J. Plant Physiol. 134*: 691-696.

Henriques, F. and Park, R.B. 1975. Further chemical and morphological characterization of chloroplast membranes from a chlorophyll b-less mutant of *Hordeum vulgare. Plant Physiol. 55:* 763-767.

Highkin, H.R. 1950. Chlorophyll studies on barley mutants. *Plant Physiol. 25:* 294-306.

Highkin, H.R. and Frenkel, A.W. 1962. Studies of growth and metabolism of a barley mutant lacking chlorophyll b. *Plant Physiol. 37:* 814-820.

Kobayashi, H., Bogorad, L., and Miles, C.D. 1987. Nuclear gene-regulated expression of chloroplast genes for coupling factor one in maize. *Plant Physiol. 85:* 757-767.

Kobza, J. and Seemann, J.R. 1988. Mechanisms for light-dependent regulation of ribulose-1,5-bisphosphate carboxylase activity and photosynthesis in intact leaves. *Proc. Natl. Acad. Sci. USA. 85:* 3815-3819.

Kunst, L., Browse, J., and Somerville, C. 1988. Altered regulation of lipid biosynthesis in a mutant of *Arabidopsis* deficient in chloroplast glycerol-3-phosphate acyltransferase activity. *Proc. Natl. Acad. Sci. USA. 85:* 4143-4147.

Kunst, L., Browse, J., and Somerville, C. 1989a. A mutant of *Arabidopsis* deficient in desaturation of palmitic acid in leaf lipids. *Plant Physiol. 90:* 943-947.

Kunst, L., Browse, J., and Somerville, C. 1989b. Altered chloroplast structure and function in a mutant of *Arabidopsis* deficient in plastid glycerol-3-phosphate acyltransferase activity. *Plant Physiol. 90:* 846-853.

Leto, K.J. and Miles, D. 1980. Characterization of three photosystem II mutants in *Zea mays* L. lacking a 32,000-dalton lamellar polypeptide. *Plant Physiol. 66:* 18-24.

Markwell, J.P., Danko, S.J., Bauwe, H., Osterman, J., Gorz, H.J. and Haskins, F.A. 1986. A temperature-sensitive chlorophyll b-deficient mutant of sweetclover (*Melilotus alba*). *Plant Physiol. 81:* 329-334.

McCourt, P., Browse, J., Watson, J., Arntzen, C.J., and Somerville, C.R. 1985. Analysis of photosynthetic antenna function in a mutant of *Arabidopsis thaliana* (L.) lacking *trans*-hexadecenoic acid. *Plant Physiol. 78:* 863-858.

McCourt, P., Kunst, L., Browse, J., and Somerville, C.R. 1987. The effects of reduced amounts of lipid unsaturation on chloroplast ultrastructure and photosynthesis in a mutant of *Arabidopsis. Plant Physiol. 84:* 353-360.

McHale, N.A., Havir, E.A., and Zelitch, I. 1989. Photorespiratory toxicity in autotrophic cell cultures of a mutant of *Nicotiana sylvestris* lacking serine:glyoxylate aminotransferase activity. *Planta 179:* 67-72.

Metz, J.G. and Miles, D. 1982. Use of a nuclear mutant of maize to identify components of photosystem II. *Biochim. Biophys. Acta 681:* 95-102.

Metz, J.G., Miles, D., and Rutherford, A.W. 1983. Characterization of nuclear mutants of maize which lack the cytochrome f/b-563 complex. *Plant Physiol. 73:* 452-459.

Miles, C.D. and Daniel, D.J. 1973. A rapid screening technique for photosynthetic mutants of higher plants. *Plant Sci. Lett. 1:* 237-240.

Miles, C.D. and Daniel, D.J. 1974. Chloroplast reactions of photosynthetic mutants in *Zea mays*. *Plant Physiol. 53*: 589-595.

Murray, A.J.S, Blackwell, R.D., and Lea, P.J. 1989. Metabolism of hydroxypyruvate in a mutant of barley lacking NADH-dependent hydroxypyruvate reductase, an important photorespiratory enzyme activity. *Plant Physiol. 91*: 395-400.

Ogren, W.L. 1984. Photorespiration: pathways, regulation and modification. *Ann. Rev. Plant Physiol. 35*: 415-442.

Ogren, W.L., Salvucci, M.E., and Portis, A.R. 1986. The regulation of rubisco activity. *Phil. Trans. R. Soc. Lond. B. 313*: 337-346.

Polacco, M.L., Chang, M.T., and Neuffer, M.G. 1985. Nuclear, virescent mutants of *Zea mays* L. with high levels of chlorophyll (*a*/*b*) light-harvesting complex during thylakoid assembly. *Plant Physiol. 77*: 795-800.

Polacco, M., Vann, C., Rosenkrans, L., and Harding, S. 1987. Nuclear genes that alter assembly of the chlorophyll a/b light-harvesting complex in *Zea mays*. *Dev. Genet. 8*: 389-403.

Salisbury, F.B. and Ross, C.W. 1985. *Plant Physiology*, 3rd ed., Wadsworth, Belmont, CA.

Salvucci, M.E. 1989. Regulation of rubisco activity in vivo. *Physiol. Plant. 77*; 164-171.

Salvucci, M.E., Portis, A.R., and Ogren, W.L. 1985. A soluble chloroplast protein catalyzes ribulosebisphosphate carboxylase/oxygenase activation in vivo. *Photosyn. Res. 7*: 193-201.

Salvucci, M.E., Portis, A.R., and Ogren, W.L. 1986. Light and CO_2 response of ribulose-1,5-bisphosphate carboxylase/oxygenase activation in *Arabidopsis* leaves. *Plant Physiol. 80*: 655-659.

Sharkey, T.D., Kobza, J, Seemann, J.R., and Brown, R.H. 1988. Reduced cytosolic fructose-1,6-bisphosphatase activity leads to loss of O_2 sensitivity in a *Flaveria linearis* mutant. *Plant Physiol. 86*: 667-671.

Somerville, C.R. 1986. Analysis of photosynthesis with mutants of higher plants and algae. *Ann. Rev. Plant Physiol. 37*: 467-507.

Somerville, C.R. and Ogren, W.L. 1979. A phosphoglycolate phosphatase-deficient mutant of *Arabidopsis*. *Nature 280*: 833-836.

Somerville, C.R. and Ogren, W.L. 1980. Inhibition of photosynthesis in *Arabidopsis* mutants lacking leaf glutamate synthase activity. *Nature 286*: 257-259.

Somerville, C.R. and Ogren, W.L. 1981. Photorespiration-deficient mutants of *Arabidopsis thaliana* lacking mitochondrial serine transhydroxymethylase activity. *Plant Physiol. 67*: 666-671.

Somerville, C.R. and Ogren, W.L. 1982. Genetic modification of photorespiration. *TIBS* (Trends in Biochemical Science) *7*: 171-174.

Somerville, C.R., Portis, A.R. Jr., and Ogren, W.L. 1982. A mutant of *Arabidopsis thaliana* which lacks activation of RuBP carboxylase in vivo. *Plant Physiol. 70*: 381-387.

Somerville, C.R. and Ogren, W.L. 1983. An *Arabidopsis thaliana* defective in chloroplast decarboxylate transport. *Proc. Natl. Acad. Sci. USA. 80*: 1290-1294.

Somerville, S.C. and Somerville, C.R. 1983. Effect of oxygen and carbon dioxide on photorespiratory flux determined from glycine accumulation in a mutant of *Arabidopsis thaliana*. *J. Exp. Bot. 34*: 415-424.

Somerville, S.C. and Somerville, C.R. 1985. A mutant of *Arabidopsis* deficient in chloroplast dicarboxylate transport is missing an envelope protein. *Plant Sci. Lett. 37*: 217-220.

Taylor, W.C., Barkan, A., and Martienssen, R.A. 1987. Use of nuclear mutants in the analysis of chloroplast development. *Dev. Genet. 8*: 305-320.

Thornber, J.P. and Highkin, H.R. 1974. Composition of the photosynthetic apparatus of normal barley leaves and mutant lacking chlorophyll b. *Eur. J. Biochem. 41*: 109-116.

Tomarchio, L., Triolo, L., and DiMarco, G. 1983. Photosynthesis, ribulose-1,5-bisphosphate of virescent and normal green wheat leaves. *Plant Physiol. 73*: 192-194.

Zelitch, I. 1989. Selection and characterization of tobacco plants with novel O_2-resistant photosynthesis. *Plant Physiol. 90*: 1457-1464.

Zelitch, I. 1990a. Further studies on O_2-resistant photosynthesis and photorespiration in a tobacco mutant with enhanced catalase activity. *Plant Physiol. 92*: 352-357.

Zelitch, I. 1990b. Physiological investigations of a tobacco mutant with O_2-resistant photosynthesis and enhanced catalase activity. *Plant Physiol.93*: 1521-1524.

II.

Mutants Affecting Plant Development

Development is applied in its broadest sense to the whole series of changes that an organism goes through during its life cycle, but it may be applied equally to individual organs, to tissues, or even to cells. Development is most clearly manifest in changes in the form of an organism, such as when it advances from a vegetative to a flowering condition. Similarly, we may speak of the development of a leaf, from a simple primordium to a complex mature organ.

The problems of development can be studied in a number of ways, but basically there are two major approaches. (1) the morphological and anatomical and (2) the physiological and biochemical. Developmental morphology and anatomy were formerly concerned with describing the visible changes occurring during development using experimental techniques. Development, however, cannot be fully understood without a study of the manifold biochemical and physiological processes underlying and determining the morphological changes.

When we come to consider the biochemistry and physiolgoy of development we find a further dichotomy of approach. On the one hand, a considerable body of knowledge has been acquired about the role of hormones and other metabolites as "internal" factors controlling growth and differentiation; on the other hand, the profound importance of environmental factors, such as day length and temperature, in the regulation of some of the major phases in the plant life cycle has been clearly demonstrated, although there is considerable evidence that a number of these environmental influences are mediated through effects on the levels and distribution of hormones within the plant (Wareing and Phillips, 1981).

Although the general framework of developmental processes has been established by these morphological, anatomical, physiological, and biochemical studies it is clear that we require other means of investigating these problems if we are to expose the underlying regulatory mechanisms. There is growing conviction among plant scientists that the genetic approach, coupled with the techniques of biochemistry and molecular biology, offers unparalleled opportunities for unraveling the molecular basis of plant development (Thomas and Grierson, 1987).

Research on mutants in microorganisms and animals has provided a tremendous boost to our understanding of the regulation of metabolism and development and it is reasonable to expect that the study of plant mutants will be equally rewarding. Indeed, it is worth recalling that the rules of inheritance were established by studies of the garden pea and transposable elements were detected first in maize. These days, experimental approaches to the developmental biology of plants do, indeed, range over the fields of morphology, physiology, biochemistry, and molecular biology. A major aim of this segment of the book is to give a taste of this diversity and to illustrate the value of the study of mutants as they affect the developmental processes in the major stages of the life cycle of plants.

Thus, in Chapter 4 some examples are discussed in which the vegetative growth of the plant is changed by mutations that affect either the internal environment of the plant by altering the levels of hormones or by genetic changes that affect the response of the plant to its surrounding environment. In Chapter 5, examples are given of the influence of mutation on some events associated with flowering and fruiting, and in Chapter 6, the enormous field of seed biochemical genetics is addressed, if only partially, through the discussion of several examples of how seed development can be modified by mutation.

4. Mutants Affecting Vegetative Growth and Development

Plant hormones play a central role in normal plant development. Despite their importance, very little is known about either the sites at which they act or the mechanisms by which they exert their effects. The characterization of mutants blocked in some step of hormone action is a promising experimental approach to the complexities of plant growth regulation. Recent years have seen much interest in this approach (see Reid, 1990, for review), resulting in the generation of new mutants with altered hormone responses. Several of the case studies in this Chapter deal with examples of the expanding list of putative hormone response mutants and provide information on the emerging ideas about their possible nature (Scott, 1990).

One group of plant growth substances, the auxins, are thought to play an essential role as a regulatory molecule in processes as diverse as cell expansion during shoot elongation and the differentiation of vascular elements. An important, unanswered question is whether or not auxin mediates these various developmental processes by one primary site of action or through several distinct mechanisms. Case Study 4-1 deals with mutants that are resistant to the exogenous application of synthetic auxins and that have been used in attempts to understand the mode of action of this class of plant hormones (Estelle and Somerville, 1987; Guern, 1987).

During the last decade an enormous amount has been written on the gibberellins (GAs). Considerable advances have been made regarding their biosynthesis and metabolism. Information is still lacking, however, on the developmental processes controlled by endogenous GAs, their site of action, and the mechanisms by which they elicit a physiological response. Indeed, it is still debated if endogenous GA levels are responsible for controlling developmental processes and which of the many identified GAs are biologically active. The isolation of mutants either deficient in GA synthesis or insensitive to added GA has allowed some investigation of the mechanisms by which this class of hormones act (Reid, 1986). Examples are discussed in Case Study 4-2 (GA-deficient mutants) and Case Study 4-3 (GA-insensitive mutants).

Very few mutants have been isolated that affect cytokinins or ethylene. In fact, it was not possible to include here individual case studies on these growth substances. However, in Case Study 4-4, some examples are discussed of the isolation of a few mutants affecting cytokinins or ethylene that provide encouragement that others will, in time, be found and will be used in physiological investigations along with those of the other classes of growth substances.

The gravitropic bending of a plant organ is the culmination of a highly coordinated sequence of events in which hormones are involved. These can be separated into gravity perception, transduction of the stimulus, and the growth response. It is the relationship between these events that dictates whether an organ will be positively or negatively gravitropic or grow at an angle to the gravity vector. If the site of action of a mutation affects any step in the program, then the chances of the gravitropic response being impaired in some manner is high. Such a modification in the sequence could ultimately lead to a change in gravitropic orientation or sensitivity, or even the inability to respond to a gravitropic stimulus (Roberts, 1987). By far the largest group of gravitropic mutants documented is composed of plants whose ability to respond to gravity has been partially lost. Several examples of such mutations are described in Case Study 4-5.

Plant growth and development can also be controlled by means other than through plant hormones. Plants use specific regulatory photoreceptors to perceive light signals and appropriately adjust their developmental pattern (sometimes by modified responses to hormones). Two major plant regulatory photoreceptors have been identified: the bluelight receptor and phytochrome. Currently, little is known about either the blue light receptor, thought to be a flavoprotein, or the molecular events of blue-light-induced responses. The phytochrome system is better understood.

The transduction chain in plants from photoperception to the ultimate observed physiological response involves many steps, most of which are not light regulated. Until recently, photomorphogenesis has been studied mainly by applying light treatments and observing their ultimate effects. With pigments such as phytochrome where photoreversible inductive effects require short periods of irradiation, this has provided an elegant experimental approach. Complications arise, however, because of the coaction of several photoreceptive systems regulating the same processes or because of the multiple effects induced by one photoreceptor. For the best characterized photoreceptor, phytochrome, the possible presence of two phytochrome pools in higher plants provides an additional complication (Vanderhoef and Kosuge, 1984). The isolation of mutants that exhibit specific phenotypic aberrations offers a potentially powerful means of elucidating the steps of sensory transduction chains. Case Study 4-6 deals with mutants in which phytochrome-mediated responses have been modified.

Case Study 4-1:   Auxin Mutants

Because the concentration of phytohormones in plant tissues is very low, conventional approaches to metabolic studies and mode of action have been fraught with technical difficulties. A genetic approach, however, has been used effectively to study the biosynthesis of gibberellins and abscisic acid in maize (see later Case Studies). A comparable genetic analysis of auxin biosynthesis and response is complicated because a dramatic disruption of these processes by mutation is likely to result in lethality early in the development of the plant. No completely auxin-deficient mutants have been identified in any species and, since auxin is required for growth of tissue in culture, auxin would appear to be essential for viability (Blonstein et al., 1988). In principle, it should be possible to isolate conditionally auxin-deficient mutants. The identification of a conditional mutation in the auxin biosynthetic pathway, however, is problematic as it is not clear that exogenous application of auxin will rescue an auxin-deficient plant. The role of auxin is thought to be regulatory and precise spatial patterns of concentration (i.e., gradients) are likely to be critical. Similarly, the morphological phenotype of a mutant defective in auxin response is unknown and may not be specific.

As an alternative approach, mutants resistant to the exogenous application of synthetic auxins have been isolated in several tissue culture systems and in *Arabidopsis* seedlings. Among other possibilities such mutants might be resistant because the level of endogenous hormone is different from that in the wild type or because of changes at the primary site of auxin action. Following is an account of mutants isolated either for their resistance to 2,4-dichlorophenoxy acetic acid (2,4-D) (*Arabidopsis*) or to naphthalene acetic acid (NAA) (tobacco cells) (Estelle and Somerville, 1987), or that arose spontaneously (tomato) (see Zobel, 1973).

4-1.1   The Isolation and Characterization of Auxin Mutants

4-1.1.1   Arabidopsis

When wild-type *Arabidopsis* seeds were placed on agar-solidified minimal medium containing 5μM 2,4-D, the seedlings emerged from the seedcoat, accumulated chlorophyll but did not grow. About 150,000 M_2 seeds were screened in this way. Thirteen plants showing significant growth (ME-1 to ME-13) were recovered (Estelle and Somerville, 1987).

The physical appearance of 12 of the resistant plants was dramatically different from wild-type at all stages of development. Mutant rosettes were small compared with wild-type and had small crinkled leaves with shortened petioles. The roots were thinner than wild-type and not as highly branched. Wild-type plants typically produced between one and five primary inflorescences, whereas ME-3 plants, for example, produced up to 25 or 30 thin-stemmed primary inflorescences. This growth behavior resulted in the formation of a short bush instead of the vertically oriented wild-type plant and suggested that the mutation caused a reduction in apical dominance.

The flowers on the mutants were smaller than wild-type and 8 or 12 were self-sterile. In ME-5, for example, flowers produced about 400 grains per flower, whereas wild-type flowers had nearly 4,000 grains per flower. In addition, stamens were substantially shorter than the pistil, therefore, the pollen was released some distance below the stigma leading to the lack of self-fertilization. All of the mutants were recessive as demonstrated by the examples given in Table 4-1 (Estelle and Somerville, 1987). All mutants fell into the same complementation group, a gene designated as *Axr-1* for auxin resistance.

In one series of experiments, when wild-type *Arabidopsis* plants were grown on minimal medium supplemented with sufficiently high concentrations of either 2,4-D or auxin (IAA), root growth was inhibited. The mutants ME-3 and ME-6 were both less sensitive. The concentration of 2,4-D required for 50 percent inhibition of root growth was 0.037μM in the wild type and between 1 and 2μM in the case of the mutants (Fig.

Table 4-1. Genetic segregation of 2,4-D resistance in mutant lines of *Arabidopsis*, ME-3 and ME-6.
[a] χ^2 calculated based on an expected ratio of 3 sensitive to 1 resistant
[b] $P > 0.5$

Cross	Number of plants		
	Resistant	Sensitive	χ^{2a}
ME-6 x ME-6	153	0	
ME-6 x wild-type F_1	0	52	
ME-6 x wild-type F_2	37	83	2.15[b]
ME-3 x ME-3	122	0	
ME-3 x wild-type F_1	0	41	
ME-3 x wild-type F_2	43	131	0.008[b]

(Estelle and Somerville, *Mol. Gen. Genet. 206*: 200-206, 1987.)

4-1A,B). The sensitivity to IAA was also altered in the two mutants relative to the wild-type, although the difference was not as great as for 2,4-D (Fig. 4-1B).

The nature of the *Axr-1* gene product or effect has not been identified but, in this study, it was not found to be due either to a change in the uptake or the metabolism of 2,4-D. The evidence suggested that this gene encodes an essential function associated with auxin action, and that mutations in *Axr-1* that produce resistance are accompanied by a

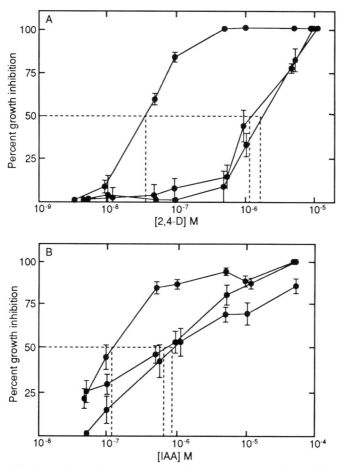

Figure 4-1. The effect of exogenous auxin on root growth of wild type and mutant seedlings. Inhibition of root growth in wild-type (●), ME-3 (▲), and ME-6 (■) plants is expressed relative to root growth on nonsupplemented medium. Dashed lines show the level of compound producing a 50 percent inhibition of root growth. (**A**) The levels of 2,4-D producing 50 percent inhibition were: wild type, 0.037 μM; ME-3, 0.68μM; ME-6, 0.84 μM. (**B**) The levels of IAA producing 50 percent inhibition were: wild-type, 0.12 μM; ME-3, 0.68 μM; ME-6, 0.84 μM. (Estelle and Somerville, *Mol. Gen. Genet.* 206: 200-206, 1987.)

reduction in function of the gene product and a disruption of normal development. Estelle and Somerville (1987) suggested the possibility that the *Axr-1* gene codes for an auxin receptor and that resistance is due to an alteration that has a greater effect on the affinity of this receptor for 2,4-D than for IAA.

4-1.1.2   Tobacco

Many types of protoplasts require an auxin and a cytokinin for the induction of cell division. The growth regulator requirements of the descendant cells grown at low densities are modified when compared with those of the initial protoplast cultures, particularly with the supply of auxins. Cells grown at low density require an auxin, such as NAA, to proliferate, but low NAA concentrations (0.05 to 1.0 μM) are required for optimal stimulation of growth, whereas standard NAA concentrations promoting protoplast division (5 to 20 μM) are strongly toxic to growth at low cell density. This discrepancy has been explained in terms of particularly intense metabolism of NAA by protoplasts. Why high auxin concentrations were toxic to protoplast-derived cells remained to be understood. This question stimulated the search for NAA-resistant clones in the hope that by studying their metabolism an understanding of the basis of NAA toxicity might be possible (Muller et al., 1985).

NAA-resistant clones were selected from mutagenized populations of haploid mesophyll protoplasts. The resistance of the selected clones was confirmed by measuring the toxicity of NAA to cells derived from protoplasts obtained from the leaves of plants derived from these clones (called p-cells). This lengthy procedure was employed primarily so that resistance could be measured under conditions in which the physiology of developing cells was as close as possible to that of cells initially exposed to selection. From among the 86 independently derived calli rescued, two clones were found (#35 and 36) that were more resistant to NAA than the wild type and that did not, as a consequence, develop roots. This inability to root in vitro was observed reproducibly for cuttings derived from these plantlets and propagated over a 2-year period. NAA resistance was tested under low density growth conditions on p-cells derived from rooting and nonrooting plantlets obtained in different crosses. Dose-response curves were compared to those obtained for the diploid wild type (Fig. 4-2; Muller et al., 1985).

Sensitive wild-type cells were systematically killed when NAA concentrations were higher than 5 μM (Fig. 4-2). A reproducible four- to fivefold increased tolerance to high NAA concentrations was observed for p-cells derived from nonrooting clones. This increased tolerance was associated with decreased growth stimulation by low NAA concentrations, dose-response curves being shifted toward higher concentrations. Testing of 17 clones derived from mutant 35 and 24

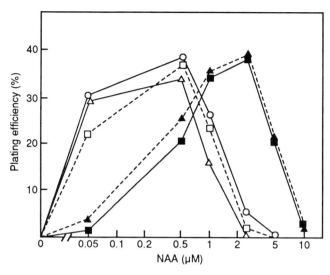

Figure 4-2. Stimulation by NAA of the growth of wild-type and mutant cells of *Nicotiana tabacum.* Protoplasts were isolated from rooting and grafted, nonrooting clones obtained among the progeny of mutant clones 35 and 36. Protoplast-derived cells were then plated in medium without NAA at low density and incubated in the presence of variable NAA concentrations. Average values of plating efficiencies corresponding to four nonrooting clones (■) and eight rooting clones (□) derived from mutant 35, and from nine nonrooting clones (▲) and nine rooting clones (△) derived from mutant 36 were calculated and compared with the corresponding plating efficiencies of wild-type-derived cells (o). Plating efficiencies of cells derived from rooting plantlets were systematically nil in the presence of 5 μM NAA. (Muller et al., *Mol. Gen. Genet. 199:* 194-200, 1985.)

from mutant 36 has confirmed the linkage of NAA resistance to the nonrooting phenotype, without exception (Table 4-2). As can be seen in Table 4-2, the nonrooting (designated as *Rac*) factor was dominant over rooting. Thus, heterozygotes (*Rac*/+) had a nonrooting phenotype. This would allow a genetic study to be undertaken of root morphogenesis and its control by growth regulators. From a genetic point of view the *Rac* phenotype is a very convenient marker and could be screened easily at early stages of seedling development.

Auxin tolerance of mutant cells and tissues was not correlated with an increased rate of conjugation of auxins in these tissues or to a perturbation of auxin transport. An altered rate of biosynthesis of auxin was probably also not the basis for resistance of the mutant to NAA and related molecules. A direct measurement of auxin concentrations in wild-type and mutant plants, however, was not performed (Caboche et al., 1987). The only difference of significance between the wild-type and mutant plants in this study was found to be in the rate at which auxin was transported through tissues. [³H]IAA

Table 4-2. Segregation of NAA resistance tested at the cellular level and of the nonrooting phenotype in the progenies of clones 35 and 36 of *Nicotiana tabacum*. Seventeen progeny plants of clone 35 and 24 progeny plants of clone 36 were individually tested for NAA resistance on protoplast-derived cells grown at low densities. Cells were considered sensitive when no colony was able to develop in the presence of 5 µM NAA in the culture medium.

Resistance to NAA	Progeny obtained by cross-pollination with the wild type or by self-pollination of			
	Clone 35		Clone 36	
	Rooting	Non-rooting	Rooting	Non-rooting
+	0	9	0	15
-	8	0	9	0

(Muller et al., *Mol. Gen. Genet. 199:* 194-200, 1985.)

transport was found to be 0.68 cm/hour for mutant plants as opposed to 0.38 cm/hour for wild-type plants. The difference in morphologies of mutant plants lacking roots and wild-type plants could have accounted for this discrepancy. No real correlation to any factor was found (Caboche et al., 1987).

4-1.1.3   Tomato

Experiments with *Arabidopsis* and tobacco failed to identify the specific lesions in the hormone mutants isolated. In the case of the *diageotropica* *(dgt)* mutant of tomato, however, recent studies have provided convincing evidence for the tissue specific alteration of auxin receptor polypeptides (Hicks et al., 1989).

The *dgt* mutant of tomato is a recessive mutant of the parental variety, VFN8, and arose spontaneously at a single locus. Tomato plants homozygous for the mutation have diagravitropic (unsupported horizontal) shoot growth, abnormal vascular tissue, altered leaf morphology, and no lateral root branching. Although the endogenous levels of auxin are the same in *dgt* and VFN8 shoot apices, the mutants are insensitive to exogenously applied auxin in ethylene production and stem elongation. These morphological abnormalities suggest that the *dgt* lesion is associated with a primary site of auxin perception or action.

To label and identify potential auxin receptors, Hicks et al. (1989) used a radioactively labeled, photoaffinity auxin analog, [³H]5-N$_3$-IAA (azido-IAA), which is an active auxin in several different bioassays. Its uptake and transport characteristics in stems are also similar to those of auxin. Thus, azido-IAA would be expected to bind to auxin receptors with an affinity similar to that of IAA. In fact, both the roots and stems of VFN8 had a polypeptide doublet of 40- and 42-kD that was labelled to high specific activity with azido-IAA. These labeled polypeptides were not detected in membranes from *dgt* stems although they were present in *dgt* roots at an intensity equal to that in VFN8 roots (Hicks et al., 1989). Thus, indications were that the 40- and 42-kD auxin-binding polypeptides were greatly diminished or had much reduced auxin-binding capacity in mutant stems. In either case, the alteration was developmentally regulated. These results were consistent with the hypothesis that *dgt* plants have an altered auxin receptor and identification of the two polypeptides that appear to be affected by the *dgt* lesion may make it possible to dissect the mechanism of auxin action. Thus, this mutant currently holds great promise in the search for genetically altered material and aid in the understanding of hormone action in plants. To date, however, the nature of the *dgt* lesion, its relationship to the 40- and 42-kD azido-IAA-binding polypeptides, and their relationship to putative auxin-specific receptor proteins have not been ascertained.

4-1.2   The Use of Auxin Mutants in Physiological Studies

4-1.2.1   *Effect of Auxin on Transmembrane Potential in Tobacco*

There is considerable evidence for interactions between auxin and membranes. These interactions include ultrastructural changes, changes in physical properties and in functional properties such as ionic fluxes. In the last case, the passive permeability to cations has been found to be enhanced in artificial lipid vesicles by auxin. An H^+ excretion, probably linked to the plasmalemma proton pump, was stimulated by auxin in several tissues, including pea stem segments or maize coleoptiles. This effect is usually considered to account for measured changes in the transmembrane potential difference (Em) after auxin treatment of pea internode segments or coleoptiles (Ephritikhine et al., 1987).

Causal links between these effects of auxin on membrane properties and the biological activity of auxin are difficult to establish. A novel opportunity to study the basis of the biological activity of auxins at the cellular level has been provided recently by the selection of a NAA-tolerant mutant from tobacco mesophyll protoplasts (Section 4-1.1.2). Through the use of this mutant, Ephritikhine and co-workers (1987) tested the hypothesis that the wild-type and mutant protoplasts differ in their membrane properties.

In culture medium without any auxin, the transmembrane Em values were generally positive for both wild-type and mutant protoplasts (Fig. 4-3). For the wild type, Em was lower when the medium was supplied with NAA; a maximum effect (Em = -11.5 ± 0.4 mV) was

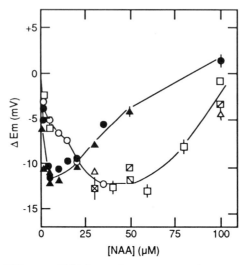

Figure 4-3. Effects of NAA on the transmembrane potential difference (Em) of wild type (●,▲) and mutant (○,△) protoplasts of tobacco mesophyll. (Ephritikhine et al., *Plant Physiol. 83:* 801-804, 1987; reproduced by permission of the ASPP.)

reached with 5 μM NAA. Increasing NAA concentrations above this level reduced the hyperpolarization until it was decreased by 50 percent in the presence of 35 μM NAA and nullified for 100 μM. The mutant displayed an electrical response of the same intensity. No difference in the timing, the stability, and the reproducibility of the response was noted between wild-type and mutant protoplasts. For the mutant, however, a marked shift in the dose-response curve was observed for the effective concentrations of NAA, as the maximal Em decrease (Em = -12.6±0.6 mV) was obtained at concentrations of about 40 μM; at 85 μM NAA the hyperpolarization was reduced by 50 percent and totally inhibited at 100 μM (Fig. 4-3; Ephritikhine et al., 1987).

It has been widely suggested that auxin-induced hyperpolarization involves the stimulation of the proton-pumping ATPase, and it would be tempting to speculate that wild-type and mutant protoplasts differ either in the intrinsic properties of this electrogenic system or in its responsiveness to auxin. Other explanations are possible, however, such as modifications of plasmalemma ionic conductances that could also account for the Em variations measured. The fact remains that the wild-type and mutant protoplasts exhibited different sensitivities to auxins as revealed by the dose-respone curves of Em modifications. Thus, these protoplasts isolated from the two genotypes appear to be good material to study a membrane response to auxins related to their biological activity. The rapidity of the membrane response (less than 2 minutes) with regard to the delay of the growth response (3 to 4 weeks) suggests that the electrical response could be one of the first events involved in the growth response of protoplasts and protoplast-derived cells to auxins. How all this might be linked to rooting and nonrooting has not yet been addressed in these studies (Ephritikhine et al., 1987).

4-1.2.2   *Ethylene Responses in Tomato*

The *dgt* mutant of tomato is characterized by diagravitropic growth of the shoot and roots, dark green hyponastic leaf segments, thin rigid stems, and roots that lack lateral root primordia (Zobel, 1973). Mutant plants gave no response to 10 µM IAA, whereas strong epinasty was observed in isogenic controls. On the basis of these preliminary results, mutant and control plants were exposed to IAA at 10 µM and 100 µM. Because IAA-induced epinasty was known to be caused by ethylene production, other plants were sprayed with 10 and 100 µM Ethephon (Table 4-3; Zobel, 1973).   Both IAA concentrations produced near maximum epinasty in the normal, whereas only the high concentration produced epinasty in *dgt* (Table 4-3). Both Ethephon applications, on the other hand, produced near maximum epinasty in *dgt* and normal plants. Furthermore, ethylene at even the lowest concentration tested (0.005 µl/L) brought about morphological normalization of mutant plants (Zobel, 1973).

The most definite conclusion to be drawn from the results of Zobel (1973) was that the *dgt* mutant required extremely low concentrations of ethylene for normal morphological development. In its sensitivity to ethylene the mutant provided a useful tool for further characterization of the control of plant development which auxin and ethylene mediate, an example of which is the regulation of petiole epinasty.

Table 4-3. Responses of nonflowering *dgt* (dgt) and normal (+) tomato plants to IAA and Ethephon.
[a] Increase in branching although not normal.
[b] Strong epinasty and normal gravitropic response.
[c] Extreme epinasty and normal gravitropic response.

Treatment	Concn	dgt		Normal	
		Roots	Shoots	Roots	Shoots
IAA spray	10^{-4}M	+[a]	+[b]	+	+[c]
IAA spray	10^{-5}M	dgt	dgt	+	+[b]
Ethrel spray	10^{-4}M	+	+[c]	+	+[c]
Ethrel spray	10^{-5}M	+	+[b]	+	+[b]

(Zobel, *Plant Physiol.* 52: 385-389, 1973; reproduced by permission of the ASPP.)

Epinasty can be defined as growth in a direction away from the plant axis as a result of differential growth in a plageotropic organ (Ursin and Bradford, 1989). Epinastic movement of petioles is the result of greater expansion of adaxial cells as compared with abaxial cells in specific regions of the petiole. In the tomato, both auxin and ethylene will induce petiole epinasty, with auxin action being dependent on induction of ethylene synthesis.

Contrasting theories for the mechanisms of ethylene and auxin action in the induction of epinasty have been proposed. In one hypothesis, ethylene alters lateral auxin transport, resulting in a transverse auxin gradient within the petiole. Higher auxin levels in the adaxial portion of the petiole result in greater adaxial growth and, hence, bending. An alternative hypothesis is that the adaxial petiole cells responsible for epinastic growth are so-called type III target cells that elongate in response to ethylene but that growth is dependent on the presence of auxin.

Separation of the roles of auxin and ethylene in petiole epinasty has been complicated by the intimate relationship between auxin effects on ethylene synthesis and possible ethylene effects on auxin transport or synthesis. In the *dgt* mutant of tomato, however, tissues show only a slight stimulation of ethylene production in response to high auxin concentrations. Thus, the effects of auxin and ethylene on epinastic growth can be investigated independently without the use of inhibitors. Also, because growth itself is highly insensitive to auxin in *dgt*, it should be possible to determine whether auxin action is required for the response to ethylene.

Ursin and Bradford (1989) compared the epinastic growth responses of wild-type (VFN8) and *dgt* petioles to a range of IAA concentrations. The response of *dgt* to auxin was only significant at 10 µM IAA where the response was still approximately half of that observed in VFN8 (Fig. 4-4). This slight promotion of epinasty by high IAA levels in *dgt* was completely inhibited by pretreatment with aminoethoxyvinylglycine (an inhibitor of ethylene synthesis), indicating that it was dependent on ethylene synthesis. Exogenous ethylene (1.3 µl/L) promoted an epinastic response in auxin-depleted VFN8 petioles, which was approximately one-half as large as the response promoted by 10 µM IAA alone. The effect of added ethylene on auxin-treated petioles was synergistic at 0.1 µM IAA and additive at higher auxin concentrations (Fig. 4-4). The response of *dgt* to ethylene in the absence of exogenous auxin was more than double that of VFN8.

The relative insensitivity of petioles of *dgt* to high auxin concentrations was attributed by Ursin and Bradford (1989), from their results (Fig. 4-4), to the poor ability of auxin to stimulate ethylene synthesis in the mutant. Furthermore, they concluded that the ability of *dgt* petioles to become epinastic in response to ethylene, despite their insensitivity to auxin, demonstrated that epinasty in the tomato is a direct response to ethylene quite independently of auxin and that this

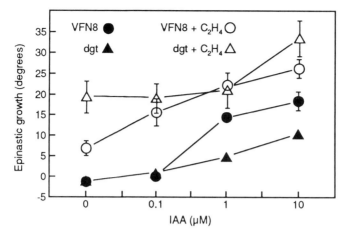

Figure 4-4. Epinastic growth responses of VFN8 and *dgt* petiole explants of tomato to IAA with and without 1.3 μl/L added ethylene. (Ursin and Bradford, *Plant Physiol. 90:* 1341-1346, 1989; reproduced by permission of the ASPP.)

separation of the effects of auxin and ethylene on cell expansion should make it possible to use this mutant to investigate also, independently, the influence of these two growth substances on such events as the extensibility and orientation of cell wall microfibrils and their control (Daniel et al., 1989; Ursin and Bradford, 1989).

Case Study 4-2:   The Gibberellin-deficient Mutants

During the last decade, an enormous amount has been written on the gibberellins. Considerable advances have been made regarding their biosynthesis and metabolism. Information is still lacking, however, on the developmental processes controlled by endogenous GAs, the site of action of the GAs and the mechanisms by which GAs elicit a physiological response. Indeed, it is still debated if endogenous GA levels are responsible for controlling developmental processes and which of the many identified GAs are biologically active. One way of determining which of the GAs are involved in development would be through the use of mutants that have reduced amounts of the growth substances.

The selection of mutants influencing GA metabolism or sensitivity has usually been based on phenotypic changes in characteristics in response to GAs applied to plants. These characteristics include internode length, apical senescence, sex expression, and seed dormancy. Selection based on changes in morphological characters can provide only circumstantial evidence that a mutation is involved directly with

GA metabolism or sensitivity. For example, a GA metabolism mutant may be indicated if an accurate phenocopy of the wild type is produced after treatment of the mutant type with an active GA such as GA_3 or GA_1. Even where good application data are available, however, direct proof that GA metabolism exists is difficult to obtain because of the extremely low levels of GAs in vegetative tissue. Proof that a GA metabolism mutant exists requires endogenous GAs, but a demonstration of the differences in metabolism between the mutant and the wild-type is often difficult to obtain (Reid, 1986).

GA metabolism mutants may be subdivided into GA synthesis mutants and GA breakdown or utilization mutants. Altered GA levels are indicative of both groups. To date, however, no mutants falling into the breakdown or utilization categories have been unequivocally shown to exist. Examples are described of mutants that are deficient in the synthesis of GAs and how these variants have been useful in some physiological studies.

4-2.1   Mutants in the Pathway Leading to GA_1 Biosynthesis

An important impetus in determining the action of GA synthesis mutants has been the elucidation of GA biosynthetic pathways in certain higher plants. The pathway from mevalonic acid (MVA) to GA_{12}-aldehyde appears common to all the plants examined; after GA_{12}-aldehyde several pathways differing largely in the pattern and timing of hydroxylation occur. Among the latter, the early 13-hydroxylation pathway, which is similar in maize, pea, and rice, appears to be the most significant in the GA mutants included in this Case Study (Fig. 4-5; Reid, 1986).

Available data suggest that GA_1 may be the only native GA active per se in the control of shoot elongation in maize (however, see the recent report by Fujioka et al., 1990). Conceptually, the idea of a single active GA would be significant as it would simplify studies on the primary action of endogenous GAs in the control of elongation growth. GA_1 is a member of the early 13-hydroxylation pathway that originates as a branch from GA_{12}-aldehyde (Fig 4-5). A number of mutants have been isolated from maize, pea, and rice that control steps in the biosynthetic pathway leading to GA_1. Each mutant appears to control a specific step in the early 13-hydroxylation pathway leading to GA_1. Here, it is demonstrated how such mutants were used to uncover some steps in GA biosynthesis (Phinney, 1984).

It is difficult to determine if more than one member of a biosynthetic pathway is active. The problem applies not only to intermediates in the GA biosynthetic pathways, but also to intermediates leading to the biosynthesis of the other plant hormones. In microorganisms, the task has been simplified by the use of single gene heterotrophic mutants. These mutants fail to grow in the absence of a specific compound and

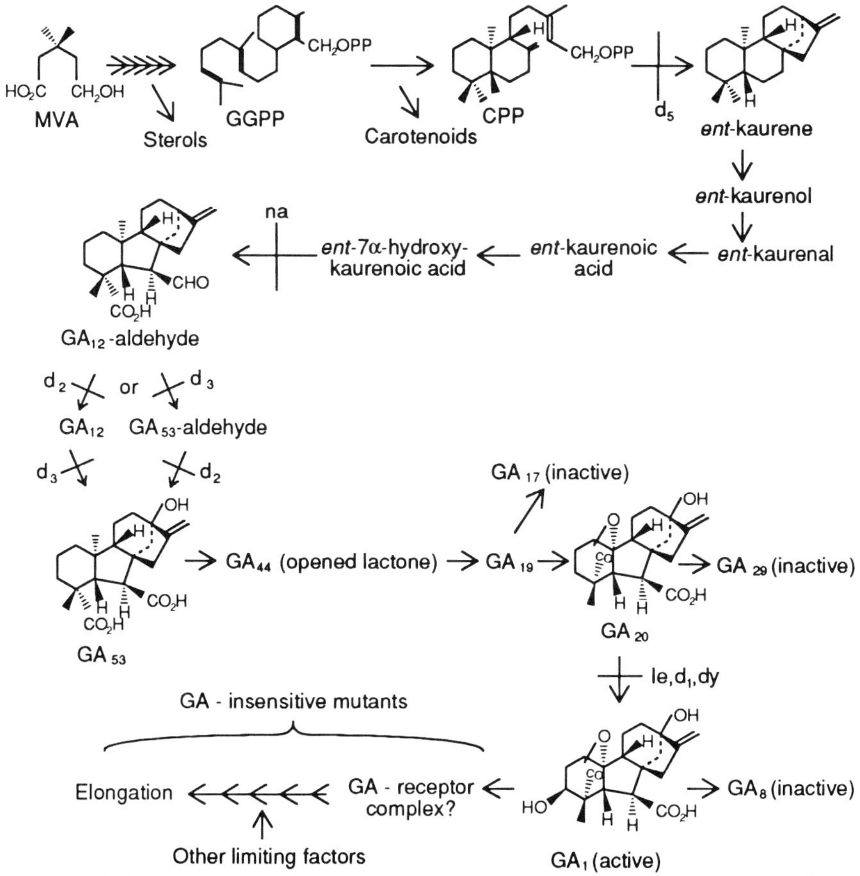

Figure 4-5. The proposed sites of action of GA mutants in maize (d_p, d_2, d_3 and d_5), peas (*le, na*), and rice (*dy*) in the early 13-hydroxylation gibberellin biosynthetic pathway from mevalonic acid (MVA) to the biologically active product GA_1, and the possible sites of action of GA insensitive mutants. GGPP = geranylgeranylpyrophosphate; CPP = copalylpyrophosphate. (Reid, in *A Genetic Approach to Plant Biochemistry*, A.D. Blonstein, P.J. King (eds.), Springer-Verlag, New York, pp. 1-34, 1986.)

resume growth with the exogenous application of that compound (for example, the auxotrophic mutants discussed in an earlier section). The use of such mutants can greatly simplify the elucidation of a metabolic pathway, especially when the genes block steps in a pathway without any major branch points. In this type of system, intermediates are metabolized stepwise to a product and it is only the product that is active. The experimental protocol is to obtain a series of nonallelic mutants that grow only in the presence of a specific compound. A series of presumptive biosynthetic intermediates to this compound are then obtained and tested for their ability to support normal growth. The pattern of growth response for each mutant (i.e., positive or negative) will determine the order of intermediates in the pathway and locate the specific steps blocked by each mutant gene (Phinney, 1984).

In *Zea mays* the four dwarfing genes, d_1 *(dwarf-1)*, d_2, d_3, and d_5, are simple nonallelic recessives. They grow normally with continued application of exogenous GA. Other plant growth regulators have no effect. The mutant phenotypes are expressed from germination to maturity and in the dark as well as in the light. At maturity, mutants are 20 to 25 percent the height of normals. Internodes are short as are leaves and inflorescences. The shortened internodes of the dwarf shoots are the result of a reduction in both cell number and cell length. Fruits and embryos of dwarfs are normal in size. The genes are apparently not expressed until the early stages of germination, at which time the mesocotyl and coleoptile begins to elongate. The four maize dwarfs are GA mutants because (1) they respond by normal growth to microgram amounts of exogenous GA, and (2) they either lack endogenous GA-like substances or these substances are present but qualitatively different from those in normals (Phinney, 1984). The nature of the lesions in some of these mutants is outlined in the following discussion.

4-2.1.1   The dwarf-5 *Mutant*

The *dwarf-5* (d_5) gene apparently controls the cyclization of copalyl pyrophosphate (CPP) to *ent*-kaurene (Fig. 4-5). This was deduced from the fact that d_5 mutants elongate in response to GA_1, GA_{20} and to *ent*-kaurene (Table 4-4).

Table 4-4. Relative activies of endogenous maize GAs and their precursors, following their application to seedlings of *dwarf-1, dwarf-2, dwarf-3* and *dwarf-5* mutants of *Zea mays*.

The activity for GA_1 on each mutant was set at 100 percent. Activities of all other GAs and precursors are given as a percentage of the activity for GA_1 (length of leaf sheath 1+2 in mm).

Mutant	*ent*-kaurene	GA_{12}-aldehyde	GA_{53}-aldehyde	GA_{53}	GA_{20}	GA_1
dwarf-5	1	5	10	10	100	100
dwarf-3	0	0	10	10	100	100
dwarf-2	0	0	0	10	100	100
dwarf-1	0	0	0	1	1	100

(Phinney, in *The Biosynthesis and Metabolism of Plant Hormones*, A. Crozier, J.R. Hillman (eds.), Cambridge University Press, London, pp. 17-41, 1984.)

To locate the position of the block in the GA-biosynthetic pathway, the in vitro biosynthesis of *ent*-kaurene in d_5 seedlings was compared with that in normals. *Ent*-kaurene is formed from MVA by intermediates including geranylgeranylpyrophosphate (GGPP) (Fig. 4-5). Cell-free extracts were prepared from 6-day-old etiolated normal and d_5 *Zea mays* seedlings using shoots only, that is, the coleoptiles and their enclosed young leaves (Table 4-5; Hedden and Phinney, 1979). The cyclization step leading to *ent*-kaurene is apparently controlled by the normal allele of the d_5 gene, as the mutant gene resulted in a marked reduction in the synthesis of *ent*-kaurene. The reduction was observed whether MVA, GGPP, or CPP was used as substrate, which indicated that the step between CPP and *ent*-kaurene was altered by the mutant gene. Thus, the reduced rate of *ent*-kaurene synthesis was responsible for the apparent absence of endogenous GAs in the d_5 plants (Hedden and Phinney, 1979).

4-2.1.2   The dwarf-1 *mutant*

The *dwarf-1* (d_1) gene appears to control the 3--hydroxylation of GA_{20} to GA_1. In the d_1 assay, the maize GAs showed a clear break in activity between GA_1 and GA_{20}. GA_{20} had less than 1 percent the activity of GA_1 and precursors before GA_{20} were biologically inactive (Tab'~ 4-4; Fig. 4-6; Phinney, 1984). These observations indicated that the d_1 lesion blocks the conversion of GA_{20} to GA_1 (Fig. 4-5).

Feeding studies with $[^{13}C,^3H]$-GA_{20} clearly demonstrated that the d_1 gene controls the 3--hydroxylation of GA_{20} to GA_1. In these feeds GA_1 was identified as a metabolite in normal and d_5 seedlings, but not in d_1

Table 4-5. Incorporation of radioactivity into *ent*-kaurene and *ent*-isokaurene from mevalonic acid (MVA)-[2-^{14}C], geranylgeranyl-pyrophosphate (GGPP)-[^{14}C] and copalylpyrophosphate (CPP)-[^{3}H] incubated in cell-free extracts of normal and *dwarf-5* shoots of maize. Results are expressed in cpm/g fresh weight.

Substrate	Normal			Dwarf-5		
	MVA	GGPP	CPP	MVA	GGPP	CPP
ent-Kaurene	673	103	766	16	10	153
ent-Isokaurene	84	26	81	127	37	608
ent-Kaurene/ *ent*-Isokaurene	8.0	4.0	9.5	0.13	0.27	0.25

(Hedden and Phinney, *Phytochem. 18:* 1475-1479; reprinted by permission of Pergamon Press PLC, Copyright 1984.)

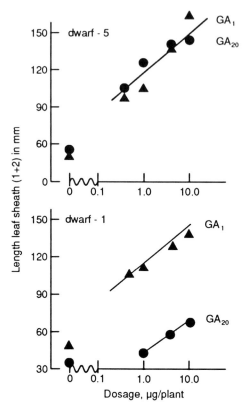

Figure 4-6. Typical three-point bioassays used to calculate the relative activities (shown in Table 4-4). (Phinney, in *The Biosynthesis and Metabolism of Plant Hormones,* A. Crozier, J.R. Hillman (eds.), Cambridge University Press, London, pp. 17-41, 1984.)

seedlings (Table 4-6). The absence of $[^{13}C,^{3}H]GA_1$ in feeds of $[^{13}C,^{3}H]GA_{20}$ to d_1 together with the formation of $[^{13}C,^{3}H]GA_1$ in feeds to normal and d_5 provided direct evidence that the d_1 mutation blocks the step from GA_{20} to GA_1. One interpretation, therefore, is that the d_1 gene codes for an altered 3--hydroxylase that inefficiently metabolizes GA_{20} to GA_1 (Spray et al., 1984).

Table 4-6. Summary of metabolites isolated from feeds of $[^{13}C, ^{3}H]GA_{20}$ to seedlings of normal, d_5 and d_1 maize.

$[^{13}C,^{3}H]GA$	Normal	*Dwarf-5*	*Dwarf-1*
GA_{20}	Present	Present	Present
GA_1	Present	Present	Absent

(Spray et al., *Planta 160:* 464-468, 1984.)

Additional evidence for the position of the d_1 block came from studies with grafts between d_1 and d_5 seedlings. In these grafts, the d_5 member showed appreciable shoot elongation over the controls, which was to be expected if d_1 accumulates intermediates subsequent to the d_5 block, which are translocatable (Katsumi et al., 1983).

4-2.1.3   The dwarf-2 Mutant

In the *dwarf-2* (d_2) assay, GA_1, GA_{20} and GA_{53} were active at levels similar to those observed in the d_5 test system. A break in activity occurred between GA_{53} and GA_{53}-aldehyde so that GA_{53}-aldehyde, GA_{12}-aldehyde, and *ent*-kaurene were inactive (Table 4-4). Thus, the d_2 mutant appears to block the C-7 oxidation steps (Fig. 4-5):

GA_{53}-aldehyde ----> GA_{53} and/or GA_{12}-aldehyde ------> GA_{12}

4-2.1.4   The dwarf-3 Mutant

The *dwarf-3* (d_3) mutant probably blocks the C-13 oxidation of GA_{12}-aldehyde to GA_{53}-aldehyde and/or GA_{12} to GA_{53}, a conclusion based exclusively on bioassay data. In the d_3 assay, a break in activity was observed between GA_{53}-aldehyde and GA_{12}-aldehyde. GA_1, GA_{20}, GA_{53}, and GA_{53}-aldehyde were active at levels similar to those observed in the d_5 assay. Both GA_{12}-aldehyde and *ent*-kaurene were inactive (Table 4-4; Phinney, 1984).

In summary, the available data suggest that the d_1, d_2, d_3 and d_5 genes of maize control the following steps in the biosynthetic pathway leading to the biologically active GA_1 (Fig. 4-5):

1. d_5 gene: CPP ---> *ent*-kaurene (activity of *ent*-kaurene synthetase)

2. d_3 gene: GA_{12}-aldehyde ---> GA_{53}-aldehyde and/or GA_{12} ---> GA_{53} (C-13 hydroxylase)

3. d_2 gene: GA_{12} ---> GA_{53} and/or GA_{12}-aldehyde ---> GA_{53}-aldehyde (C-7 oxidase)

4. d_1 gene: GA_{20} ---> GA_1 (C-3--hydroxylase)

These results with mutants of maize are paralleled by others with similar mutants of rice and pea. Together, these examples support a cause and effect relationship between the presence of endogenous GA_1 and elongation growth in higher plants. They clearly show that the level of GA_1 is a limiting factor in the control of shoot elongation. Here are two examples:

1. The three genes, d_5 (maize), as we have already seen, *na* (pea) and *dx* (rice) each control an early step in the GA biosynthetic pathway leading to GA_1. As a result, endogenous GA-like substances are absent in these mutants. The three mutants resume normal growth in the presence of exogenous GA. Two of the mutants, d_5 and *na*, elongate when grafted to a normal (tall), presumably as a result of the movement of endogenous GA_1 from the normal to the mutant tissue.

2. The three genes, d_1 (maize), *le* (pea) and *dy* (rice) probably control the 3--hydroxylation of GA_{20} to GA_1. The available evidence suggests that the mutants lack endogenous GA_1, or GA_1 is present at very low levels compared with normals. These three mutants also accumulate GA-like substances that are presumably intermediates before the genetic blocks. Two of the mutants, d_1 and *le*, have been shown to elongate when grafted to normals. This elongation is probably due to the movement of endogenous GA_1 from the normals to the mutants (Phinney, 1984).

Maize provides the most compelling evidence that GA_1 may be the only endogenous GA that is active in the control of shoot elongation. Because the d_1 gene blocks the late step, GA_{20} to GA_1, this mutant can be used to evaluate the activity of all members of the biosynthetic pathway before the block. In the d_1 assay *ent*-kaurene, *ent*-kaurenol, *ent*-kaurenoic acid, and the aldehydes of GA_{12} and GA_{53} are biologically inactive; GA_{53} and GA_{20} are less than 1 percent as active as GA_1. These precursors of GA_1 seem to play a minor role, if any, in the control of shoot elongation in maize. It is more likely that the observed bioactivities are the result of a "leaky" mutant that enables small amounts of GA_{53} and GA_{20} to be metabolized to GA_1. Leaky mutants would also be a likely explanation for the limited shoot growth associated with the d_1, d_2, d_3, and d_5 mutants of maize.

In summary, GA_1 may be the critical GA controlling shoot elongation growth in maize, pea, and rice. Very recent evidence concerning the biosynthesis of GA_3 from GA_{20} via GA_5 in maize, however, suggests that such a conclusion should be regarded with some caution (Fujioka et al., 1990). Whether GA_1 control of shoot elongation is a common phenomenon in all higher plants is an hypothesis that has not yet been tested, as the widespread occurrence of this GA in shoots has yet to be documented (Phinney, 1984). Recent evidence from the use of tall and dwarf wheat genotypes, however, adds further weight to the argument that GA_1 does indeed play a central role in elongation growth (Pinthus et al., 1989).

4-2.2   The Use of GA-deficient Mutants in Physiological Studies

4-2.2.1   GA-induced Changes in mRNAs and Polypeptides in Maize and Pea

The result of GA application to dwarf plants is enhanced cell division and cell elongation. Basic molecular questions relevant to GA-induced growth, however, remain unanswered.

To determine the feasibility of cloning genes whose expression is dependent on the presence of GA, dwarf varieties of corn (d_5) and pea (Progress #9, which has endogenous GAs but not the biologically active form GA_1, which is present in only trace amounts) were chosen. The method utilized in vivo protein labeling with [^{35}S]methionine coupled with two-dimensional gel electrophoresis to obtain a general picture of GA-induced polypeptide changes. In addition, cell-free translation of purified mRNA followed by two-dimensional gel electrophoresis was used to assess the spectrum of changes in the mRNA pool at discrete times after GA application (Chory et al., 1987).

Sections from either etiolated maize or green pea seedlings were incubated in the presence of [^{35}S]methionine for 3 hours with or without GA. Labelled proteins from soluble and particulate fractions were analyzed by two-dimensional gel electrophoresis and specific patterns of protein synthesis were observed upon treatment with GA. Polyadenylated mRNAs from etiolated or green maize shoots and green pea epicotyls treated or not with GA (a 0.5- to 16-hour time course) were assayed by translation in a rabbit reticulocyte extract and the products separated by two-dimensional gel electrophoresis. Both increases and decreases in the levels of specific polypeptides were seen for pea and corn, and these changes were observed within 30 minutes of treatment with GA (Table 4-7).

These data (Table 4-7) indicate that GA induces changes in the expression of a subset of gene products within elongating dwarfs. This may be due to changes in transcription rate, mRNA stability, or increased efficiency of translation of certain mRNAs. Results taken together indicate that GA_3 increases the translational activity in the rabbit reticulocyte system of 6 to 15 poly(A)RNAs within 30 minutes of GA_3 application. This is true regardless of the experimental system chosen.

Table 4-7. Summary of polypeptide changes in maize as a result of GA_3 treatment.   See text for explanation of treatments.
[a] Tabulated are molecular weights relative to known protein standards.
[b] Shown are the values for the 0.5-hour time point. The changes are the same for 2, 4, 8, and 16 hours.
[c] Shown are the values for the 2-hours time point. The changes are the same for 4, 8, 24, and 48 hours.
[d] Refers to changes seen in polypeptides both during in vivo and in vitro labeling.
[e] Refers to changes seen in polypeptides in both green and etiolated tissue.

Synthesis Repressed		Synthesis Enhanced		
			In vitro	
In vivo	In vitro[b]	In vivo	Etiolated tissue[b]	Green[c]
26.5[a]	23.5[d]	94.4[d]	93.3[d]	36.6
15.8	22.0	92.9[d]	93.3[d]	35.3
12.3[d]	20.3[d]	68.0	42	28.3[c]
11.8	13.5	57.2	31	26.1
9.7		57.1	28.5[c]	25.1
9.1		52	24.7[d]	10.1
		25.1[d]	23.9	
		22.6	17.7	
		18	14.5	
		11.5	14.3	
		9.3	14	
		8.7	13.3	
		7.3	12.2	
			11.3	

(Chory et al., Plant *Physiol. 83:* 15-23, 1987; reproduced by permission of the ASPP.)

The identity of the polypeptides that are modulated as a function of GA treatment remains unknown (Chory et al., 1987). Thus, it seems likely that increased and decreased transcription of certain genes will be one of the mechanisms involved in GA action during stem elongation. The screening of complementary DNA (cDNA) libraries for genes that are differentially regulated after GA application is the natural outcome of these investigations using mutants such as d_5 and Progress #9.

4-2.2.2   *Phytochrome Control of GA Synthesis in Dwarf Pea*

Light-grown dwarf peas lack the ability to convert GA_{20} to GA_1 yet GA_1 and GA_8 can be found in extracts of etiolated dwarf pea shoots. From these observations it can be hypothesized that etiolated dwarf pea plants lose the ability to convert GA_{20} to GA_1 (Fig. 4-5) upon formation of the far-red form of phytochrome (P_{fr}) as a consequence of exposure to red light. Light-grown plants of the *lele* genotype are deficient in this conversion, and hence, show the dwarf phenotype. The *Le* locus affects the 3--hydroxylation of GA_{20} to GA_1, the form most active in promoting stem elongation (Campell and Bonner, 1986). This hypothesis was tested as follows.

For continuous growth recording experiments, seeds of Alaska or Progress #9 pea were grown in a dark cabinet at 25°C until the third internodes were approximately 1 cm long. When exposed to light, red light (R) treatments consisted of 4 minutes R given at 4 hour intervals; R + far red (FR) treatments were 4 minutes R and 8 minutes FR started simultaneously with the R, also applied at 4 hours intervals (Table 4-8).

GA_{20} and GA_1 were equally effective when applied to Alaska pea plants (Table 4-8). The apparent disparity in elongation rates between GA_{20}- and GA_1-treated Alaska peas was artifactual; it resulted from the use of seedlings that were growing more rapidly than usual, before treatment, for the GA_{20} experiments. GA_{20} and GA_1 increased the elongation rate, relative to untreated controls, to approximately the same extent (data not shown). Alaska pea plants elongated rapidly

Table 4-8. Summary of continuous growth measurements for Progress and Alaska peas. Mean third internode elongation rates over the 48 hours after initial GA application. Figures are mean ± SE.

Cultivar	Treatment	Mean Elongation Rate		
		Dark	R μm/h	R + FR
Progress	None	111±9	38± 4	75± 7
Progress	GA_{20}	304± 61	70± 5	172±22
Progress	GA_1	388± 51	207±56	
Alaska	None	315± 43	142±18	
Alaska	GA_{20}	819±103	493±84	
Alaska	GA_1	489± 48	253±55	

(Campell and Bonner, *Plant Physiol. 82:* 909-915, 1986; reproduced by permission of the ASPP.)

after GA_{20} application, whether dark grown or R treated. In contrast, the response of Progress peas to GA_{20} application was altered by light treatment. The growth rate of dark-grown plants was greatly increased by GA_{20}. GA_{20} only slightly increased the elongation rate of R-treated plants. GA_{20} increased the growth rate of R + FR-treated plants to a larger extent, although the increase was not as great as in dark-grown plants. R greatly reduced the growth rate of GA_{20}-treated Progress peas relative to GA_1-treated plants, but the relative effectiveness of the two GAs in Alaska plants was essentially unchanged by R irradiation (Table 4-8).

These results (Table 4-8) suggested that the presence of P_{fr} was required for the full phenotypic manifestation of the *lele* genotype. It was concluded that the deficiency in 3--hydroxylation of GA_{20} to GA_1 in genotype *lele* was due to a P_{fr}-induced blockage in the expression of that activity. The simplest interpretation of the results is that one effect of P_{fr} in inhibiting stem elongation in deetiolated dwarf pea seedlings is to confer a phytochrome-mediated photosensitivity in the mutant on the expression of *Le* or on the stability or integrity of the 3--hydroxylase. More recent investigations, however, suggest that this short-term response to R is not mediated by either a reduction in the level of GA or a reduction in the level or affinity of a GA receptor (Behringer et al., 1990a). Of course, P_{fr} also has more generalized effects on stem elongation independent of its interaction with the *le* allele (Campell and Bonner, 1986). The availability of the *le* mutants, however, provided a model system through which to pinpoint where phytochrome might influence the formation of GAs.

4-2.2.3   A Mutation Conferring an Enhanced Response to GA_1

Dwarf mutants have been described in many plant species and are currently incorporated into many commercial cultivars. As pointed out in the introduction to this Case Study, they can be broadly split into two groups, those that block steps in the GA biosynthetic pathway (as illustrated by the maize mutants) and those that reduce sensitivity to the GAs (to be described later).

Much less common than the dwarf types are mutants that result in increased stem elongation. Long-stemmed mutants that influence GA sensitivity are particularly rare suggesting that they may be concerned with the direct consequences of GA reception and subsequent steps leading to elongation. Such mutations may provide information on the direct sequence of events between GA reception and stem elongation (Reid and Ross, 1988).

In peas, genes occupying nine loci have been shown to interact to determine internode length. The genes *le, lh, ls,* and *na* reduce stem elongation by blocking specific steps in the GA biosynthetic pathway, whereas genes *lk, lw*, and *lm* reduce stem elongation by reducing the GA response. The gene combination *la-cry* results in a long, thin, slender phenotype in which it appears as if all GA responses are saturated, that is, the GA receptor acts as if loaded with GA_1. In another case, compared with the parental cv. Sparkle (*Lv-le*), the internode length of the mutant NEU3 (*lv-le*) was substantially increased particularly between nodes 4 and 9 (Table 4-9; Reid and Ross, 1988). Whereas cv. Sparkle showed a pronounced increase in internode length when exposed to an extension of an 8 hours photoperiod of natural light by incandescent light at nodes 6-9, NEU3 showed no significant response. The relative difference in internode length between the lines Sparkle and NEU3 was much reduced in complete darkness (Table 4-9).

The elongated internodes of line NEU3 result from a single recessive gene termed *lv*. This gene is of physiological interest as it results in increased stem elongation. The present results suggested that the gene *lv* increases sensitivity to the active GA in peas, GA_1, as *lv* plants showed a greater response to all levels of GA_1 applied compared with *Lv* lines. Gene *lv* did not appear to be involved with GA synthesis or breakdown, as the levels of GA-like activity appeared similar in parental and mutant lines. Rather, the results suggested that *lv-le* (ie., NEU3) plants possess relatively long internodes because of an enhanced sensitivity to the small amount of GA_1 produced by the leaky *le* gene (Fig. 4-7).

These elongated mutants, combined with the GA-insensitive dwarf mutants of peas may provide the tools to examine the partial processes involved in the broad phenomenon known as GA sensitivity. It may be possible to obtain a range of mutants with changed overall sensitivity (see Reid and Ross, 1988; Behringer et al., 1990b; Croker et al., 1990) and use them to explore directly the partial processes involved at the molecular level, as illustrated in the following example.

Table 4-9. Mean stem length $\pm$ SE (cm) for pea plants of cv. Sparkle (*Lv-le*) and line NEU3 (*lv-le*) exposed to an 8-hour natural photoperiod, an 8-hour natural photoperiod plus 16-hours of weak incandescent light (24), complete darkness (0), or to the natural photoperiod extended to 18-hours by light from a mixed fluorescent-incandescent source. (Reid and Ross, *Physiol. Plant. 72*: 595-604, 1988.)

Photoperiod	Internodes	Sparkle	NEU3
8	6-9	11.42±0.60	29.20±0.45
24	6-9	19.07±0.67	27.38±1.58
18	1-3	2.73±0.08	4.59±0.19
0	1-3	16.78±0.34	20.06±0.41

(Reid and Ross, *Physiol. Plant. 72:* 595-604, 1988.)

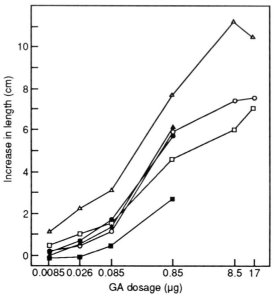

Figure 4-7. The increase in internode length between nodes 4 and 6 (cm) (compared with controls) plotted against $\log_{10}$ of the dosage of GA_1 ($\bigcirc$,$\square$,$\triangle$) or GA_{20} ($\bullet$,$\blacksquare$,$\blacktriangle$) for pea lines WL1771 (*Lv Le* $\bigcirc$,$\bullet$), Sparkle (*Lv le* $\square$,$\blacksquare$) or NEU3 (*lv le* $\triangle$,$\blacktriangle$). (Reid and Ross, *Physiol. Plant.* 72:595-604, 1988.)

4-2.2.4   *Elongation and Enzyme Induction in a* Slender *Barley Mutant*

Barley plants homozygous for the recessive *sln-1* (*slender*) allele behave as if they are continually saturated with GA. The mutant phenotype is characterized by rapid extension growth in the seedling and adult stages of growth. The mature plant is limber and unable to support itself; basal internode elongation occurs concurrently with tillering, resulting in aerial branching; root initials form on lower aerial nodes; stems and leaves are narrower than normal; heads are at least twice as long as normal, with reduced size of floral parts; the flowers are completely sterile; and no pollen is produced and no seed is set upon cross-pollination. Phenocopies of the mutant can be obtained by treating normal plants with exogenous GA. Because the mutated gene seemed to be concerned with some aspect of GA synthesis (see Croker et al., 1990) or utilization and since the influence of GA on the aleurone of barley is well understood, an examination of the affect of the mutation on aleurone enzymes seemed appropriate (Lanahan and Ho, 1988).

The molecular biology and biochemistry of the response of the barley aleurone layer to GA is understood in some detail. Upon imbibition the embryo synthesizes GA, which then diffuses throughout the germinating seed, and mediates the degradation of stored reserves of protein and carbohydrates in the endosperm by eliciting the synthesis and secretion of a number of hydrolytic enzymes from aleurone cells. Normal aleurone layers of barley are completely dependent on the GA synthesized by the embryo for the induction of hydrolase secretion. Among the most characterized of the induced hydrolases is α-amylase, the genes for which are activated by GA at the transcriptional level, although the mechanism of this activation is unknown. Another phytohormone, abscisic acid (ABA), has antagonistic effects on this GA-induction process.

Despite the wealth of genetic information on the slender mutant in pea, the biochemical basis of stem elongation is not well defined. Because the aleurone layers in barley are considerably more amenable to the study of the molecular action of GA, this mutant presented a good opportunity to determine if the aleurone layers of slender barley exhibited the GA-saturated phenotype.

Half-seeds were incubated in solutions of GA_3, ABA, or no hormones. Aliquots of the incubation media were assayed for total α-amylase, protease, and nuclease activities (Fig. 4-8; Lanahan and Ho, 1988). Secretion of nuclease, α-amylase, and protease activities was dependent on exogenous GA_3 in normal aleurone layers, whereas slender aleurone layers did not require exogenous GA_3 to secrete these hydrolytic activities.

The phenotype of slender aleurone layers could be the consequence of a number of factors. It is conceivable that slender half-seeds contain high levels of endogenous GA or are capable of synthesizing GA. Upon imbibition this GA would then induce the aleurone cells to synthesize and secrete hydrolytic enzymes. It is also possible that slender aleurone

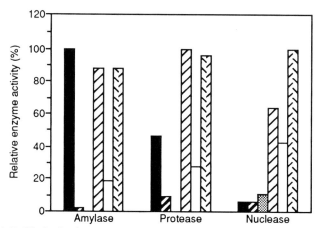

Figure 4-8. Hydrolytic enzyme activities secreted from normal and slender aleurone layers of barley. Normal and slender half-seeds were incubated in medium containing either GA_3, ABA, or no hormones (NH) for 48 hours. Incubation media were assayed for a-amylase, protease, and nuclease activity. Normal GA (■); normal NH (▨); normal ABA (▦); *slender* GA (▧); *slender* ABA (□); *slender* NH (▨). (Lanahan and Ho, *Planta 175:* 107-114, 1988.)

layers have lost their dependence on GA for induction, and will synthesize and secrete the hydrolytic enzymes constitutively. To distinguish between these possibilities the levels of GA-like compounds in slender seeds and various control tissues were determined using a bioassay for GA-like activity (Lanahan and Ho, 1988).

Gibberellin extracts from slender half-seeds, various other tissues or known amounts of GA_3 were used to induce normal Himalaya barley aleurone layers (which require the addition of GA to synthesize hydrolytic enzymes) to synthesize α-amylase. By determining the activities of α-amylase secreted into the medium by the Himalaya aleurone layers, the levels of the GA-like substances in the extract could be estimated (Fig. 4-9).

The no-GA controls (lanes 1 and 2) had no GA-like activity. Positive extraction controls (lanes 3, 4, and 5) demonstrated that the majority of the GA in the samples was extracted and retained biological activity. Lanes 6, 7, and 8 were internal standards in which GA_3 was added directly to the assay solutions. By comparing lane 9 with lanes 10, 11, and 12, the conclusion could be drawn that extracts from slender half-seeds contained GA-like concentrations as low as, or lower than either

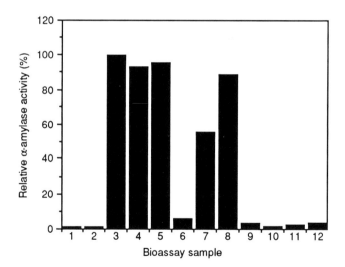

Figure 4-9. Bioassay of extractable GA-like substances in slender and normal aleurone layers of barley: α-amylase activity. Relative α-amylase activities secreted from Himalaya barley aleurone layers incubated in extracts of GA-like substances, or various control concentrations of GA_3 for 24 hours. Sample 1, no GA control; 2, no GA extraction control; 3,4,5, GA extraction controls, 10^{-6}, 10^{-7}, 10^{-8} M of GA_3, in the order given; 6,7,8, GA "standards," in the order 10^{-9}, 10^{-8}, 10^{-6} M of GA_3; 9,10,11,12, extracts from, in order, slender, whole Himalaya seeds, Himalaya half-seeds, and normal (*Hordeum vulgare* L. cv. Herta) half-seeds. (Lanahan and Ho, *Planta 175:* 107-114, 1988.)

extracts from whole Himalaya seeds, Himalaya half-seeds, or the siblings of the normal barley from which the slender mutant was obtained (Lanahan and Ho, 1988).

Thus, slender aleurone layers synthesize and secrete at least three hydrolytic enzymes without the requirement for GA treatment. Furthermore, the level of GA-like substances in slender half-seeds does not exceed that in normal half-seeds. This suggests that slender is a constitutive GA-response mutant, that is, those GA-induced activities that would normally be expressed only in response to GA, are expressed independently of GA levels (Chandler, 1988; Lanahan and Ho, 1988).

Because α-amylase, protease, and nuclease activities were affected similarly in slender aleurone layers, the mutation may disrupt an early regulatory event in the GA-controlled induction process. Constitutive hormone responses could also be explained by changes in the hormone receptor proteins. The observation that ABA retained its inhibitory effect on hydrolase secretion in slender indicated that ABA may act at a point after the action of the *sln-1* gene product. Alternatively, the mechanism of ABA suppression of hydrolase synthesis may be a distinct pathway of inhibition and not involve any component of the GA-induction pathway.

Case Study 4-3:   The Gibberellin-Insensitive Mutants.

Plant stature is a primary factor in survival because of advantages conferred in terms of preferential light interception, ease of pollination, and more effective seed dispersal. In economic terms, stature is a primary component of yield. Maximum production in crop plants frequently has a close relationship to stature, or to the success with which the plant breeder can manipulate it.

Control of plant height is achieved by varying the number and/or the extension rate of subapical internodes, and the process is amenable to intervention in either the division or elongation phases of the cell cycle. It should be appreciated that control varies from simple single-site events to complex allelic interactions at multiple loci and that a temporal control of gene expression (itself regulated by genetic factors) can be involved. In Case Study 4-2 examples were given of mutants in which this important aspect of the plant life cycle, growth in height, was affected by the relative absence of GAs in vegetative plant tissues. Another group of mutants that have modified growth responses to GAs are those that are relatively insensitive to this group of growth substances (Stoddart, 1987).

Designation as a GA-insensitive mutant has up to now rested on an altered response to applied GA. They are a large and heterogeneous group of mutants as they may influence any of the steps between reception of the GA signal and the manifestation of the GA response.

This group may include, therefore, mutants that are not directly involved with the GAs. For example, they include any mutant that makes the GAs nonlimiting for the response(s) being examined. At least potentially, these include factors such as insufficient substrates, limiting levels of other plant hormones and limited uptake or transport systems within the plant (Reid, 1986). In this Case Study is a description of the isolation, characterization, and use in physiological investigations of some GA-insensitive mutants of wheat as an example of this class of variants.

4-3.1.   Characteristics of the GA-insensitive Mutants of Wheat

The Norin-10 varieties of wheat have two genes for dwarfness, *Rht1* and *Rht2*, that result in statures around 70 percent of those typifying "tall" cultivars (Keyes et al., 1989). A third dwarfing factor, the "Tom Thumb" gene (*Rht3*) confers full dwarfness, with statures of 40 percent or less of those in tall controls. These plants are dark green and small leaved, but tillering patterns are not abnormal. All three genes exhibit dominance.

　　The Norin-10 genes resemble *Rht3* in their mode of action. Seedlings containing any of the three genes are insensitive to exogenous GA, showing none of the characteristic acceleration of growth rate exhibited by nondwarf cultivars after such treatment. The vegetative tissues of all dwarf lines reveal an interesting, and biologically significant, phenomenon. Seedlings of Norin-10 and related dwarf lines contain about 10 times as much extractable free GAs as do tall lines. Investigations of these lines demonstrated for the first time that no obligate link existed between extractable GA content and growth rate, and effectively terminated an era that sought to establish a causal relationship between rates of GA synthesis and growth. The *Rht* genes provide the plant physiologist with an important complementary tool to that furnished by the genetic blockages in the biosynthetic pathway and could be significant in establishing the mode of action of GAs in particular and endogenous growth regulatory processes in general (Stoddart, 1987).

4-3.2   The Use of GA-insensitive Mutants in Physiological Studies

4-3.2.1   *Response of some Dwarf and Tall Wheat Cultivars to Applied GA*

Dwarf wheat cultivars derived from Norin 10 have short stiff straw and utilize a smaller percentage of the total dry matter produced than the straw of taller forms. A number of dwarf cultivars of *Triticum aestivum* L.

were compared to several tall forms with respect to their response to applied GA and their endogenous GA content (Table 4-10; Radley, 1970).

The tall cultivars had a marked response to GA$_3$ in contrast to the dwarfs that showed only a slight response (Table 4-10). In addition to not responding to added GA these dwarfs were found to contain much higher GA-like activity than talls. These results suggested a block in the utilization of GA that also causes an accumulation of the hormone (Table 4-11). Thus, cultivars such as the Norin-10 type dwarf wheats could also be useful subjects for the study of GA stability (Radley, 1970).

4-3.2.2   Control of Endosperm Amylase Activity in Standard Height and Dwarf Wheats

Hydrolytic enzymes, such as α-amylase in cereal endosperm, are induced by endogenous GA, which occurs in the scutellum of the embryo. Many factors influence the level of amylase activity attained in response to GA and it is not known if plant height and GA response are controlled by pleiotropic genes or if the two characteristics are controlled by independent or linked genes. Six wheat genotypes were studied that allowed the problem of the possible connection between plant height and amylase activity to be addressed: Ramona 50 and Nainari 60 are of standard height; D6301 and Norin 10 are short statured; Oleson and D6899 are also short-statured but are significantly shorter than the previous two genotypes. Norin 10 has two genes for dwarfness (*Rht1, Rht2*), the same two as D6301. Oleson has the Norin genes plus one other of uncertain origin. D6899 has the Tom Thumb gene (*Rht3*) and neither of the Norin genes (Fick and Qualset, 1975).

The two standard-height genotypes, Ramona 50 and Nainari 60, had long coleoptiles and first leaves and high amylase activity when compared with the four short-statured genotypes (Table 4-12). D6301, Norin 10, Olesen, and D6899 had coleoptiles and first leaves reduced in length to about the same degree as their mature height in comparison

Table 4-10. Heights (cm) of wheat seedlings grown with or without 10 gml GA$_3$ for 14 days.

Cultivars	Seedling Heights	
	-GA	+GA
5 Talls	28.1	40.4
13 Dwarfs	20.1	21.8
Significant Difference	0.8	
(P=0.05)		

(Radley, *Planta 92:* 292-300, 1970.)

Table 4-11. GA-like activity in some dwarf and tall wheat cultivars.
[a] proximal half-grains of wheat + agar substrate after 3 days of germination,
[b] 14-day-old wheat seedlings, and
[c] extending stems and upper leaves of garden-grown plants [given as
pg GA$_3$ eq./grain for (a) or per g FW for (b) and (c)].

Cultivar	Grains[a]	Seedlings[b]	Upper[c] leaves	Stems[c]
Tall 1	0	360	438	662
Tall 2	8	306	564	276
Tall 3	14	250	270	662
Dwarf 1	52	3,590	2,094	6,776
Dwarf 2	55	3,360	1,352	--
Dwarf 3	30	3,855	632	5,140

(Radley, *Planta 92*: 292-300, 1970.)

with the standard-height varieties. Endosperm amylase activity was
lower for D6301, Norin 10, and Oleson than for Ramona 50 and Nainari
60, but the differences were not all significant. Most striking was the
approximately four fold lower amylase activity of D6899 when compared
with the short-statured and standard height genotypes. These results
showed clearly that, in this group of genotypes, there was no direct
relationship of plant height to endosperm amylase activity during seed
germination (Table 4-12). The increases in amylase activity due to GA
were about the same for the three short wheats that have dwarfing
genes from Norin 10, but D6899 showed a small and nonsignificant
response to GA. Thus, these short-statured wheats differed in their
sensitivity to GA during enzyme induction as the germination process
started, but were similar in terms of early seedling growth (Table 4-12).
    Because of the extremely low amount of amylase synthesized during
germination of D6899 (Table 4-12) and because a single major gene is
known to control the dwarfing effect, it was believed that genetic control
of the amount of amylase produced might also be simply inherited.
Therefore, amylase activity and seedling response to GA treatment
were studied in the F_3 generation of crosses between Ramona 50 and
D6899 using seeds from 21 random F_2 plants. Table 4-13 shows the
data for the parents and 21 random F_2 lines, grouped according to their
genotype as assumed from F_2 plant heights. Levels of amylase activity
for the seven F_3 lines, assumed to have the Ramona 50 genotype, and for
the four F_3 lines, assumed to have the D6899 genotype, were similar to
the parental values. Mean amylase activity of the 10 F_3 lines derived
from heterozygous F_2 plants was intermediate to the activities of the
high and low amylase lines. Similar patterns were noted for the three
groups of genotypes with respect to response to GA; that is, the tall
genotypes characterized by high amylase activity responded to GA,

Table 4-12. Mature plant height and comparisons of coleoptile length, first seedling leaf length, and α-amylase activity in endosperm of germinating seeds of six wheat varieties untreated and treated with GA.
[a,b] GA effects significantly different from the untreated controls at the 0.05 and 0.01 probability levels, respectively.

Variety	Plant height (cm)	Coleoptile length (mm) Untreated	GA effect	First seedling leaf length (mm) Untreated	GA effect	Amylase activity $(\Delta A \cdot T_v)/t \cdot v$ Untreated	GA effect
Ramona 50	108	83.6	15.9[b]	125.8	19.0[a]	214	56[b]
Nainari 60	107	67.0	14.5[b]	124.5	28.3[b]	224	67[b]
D6301	80	51.8	2.5	100.8	6.7	194	18
Norin 10	71	38.2	-1.1	78.5	-6.0	176	29[a]
Olesen	50	38.8	0.6	81.6	4.8	196	22[a]
D6899	49	38.0	0.7	72.0	-4.5	48	7
Standard error	1	1.7		5.0		12	

(Fick and Qualset, *Proc. Natl. Acad. Sci. USA.* 72: 892-895, 1975.)

whereas the short, low amylase genotypes did not (Table 4-13; Fick and Qualset, 1975).

Although the populations were small, these results (Table 4-13) showed close associations of GA response, amylase activity, seedling growth, and plant height in this cross and supported the single gene hypothesis. Previous results (Table 4-12), however, showed quite clearly that not all wheats have the same physiological bases for dwarfism. Nonetheless, the limited GA response shown by D6899, predominantly controlled by a single gene, offers a genetic approach to study biosynthetic pathways and gene action of growth processes in higher plants. This genotype should be useful for further investigations involving the induction of hydrolytic enzymes in response to GA (discussed later).

4-3.2.3 Evidence for a Regulatory Step Common to Many Diverse Responses to GAs

Various responses of plant tissues to GA have been well documented. The most intensively studied responses are the induction of hydrolytic enzymes in the aleurone layers of cereal grains and the elongation of stem internodes, leaves, hypocotyls, and epicotyls. Despite the fact that much is known about these GA responses a basic problem remains unsolved. Is there any single regulatory step that is common to all the diverse physiological responses to GA? It is generally accepted that

Table 4-13. Amylase activity, coleoptile length, response in coleoptile length from treatment with gibberellic acid (GA), and plant height for parents and F_3 lines from the cross Ramona 50 x D6899 wheat varieties. [a] GA effect significantly different from the untreated populations at the 0.01 probability level.

Identity	F_2 genotype	No. of F_3 lines [a]	Amylase activity	Coleoptile length (mm) Un-treated	Coleoptile length (mm) GA effect	F_1 plant height (cm)
Ramona 50	$a_1 a_1$		275	96.1	8.1^a	102
F_2 tall	$a_1 a_1$	7	262	87.0	10.2^a	101
F_2 intermediate	$a_1 a_1$	10	165	67.7	5.3^a	53
F_2 short	$a_2 a_2$	4	54	48.0	1.3	32
D6899	$a_2 a_2$		40	41.6	1.2	30

(Fick and Qualset, *Proc. Natl. Acad. Sci. USA.* 72: 892-895, 1975.)

both hydrolytic enzyme induction and the maintenance of long-term elongation growth depend on protein and RNA synthesis. The complexities of proteins and RNA synthesized in plant tissues, however, do not permit an easy search for regulatory molecules that may be common to different GA-mediated physiological processes.

Another approach to this problem is the use of genetic mutants that do not respond to GA to determine whether there are genetic loci that are capable of regulating all the diverse GA effects. Dwarf varieties of cereal plants, which do not respond to GA, appear to be suitable for this approach. One of these, D6899, did not respond to GA in coleoptile and leaf elongation tests, nor in the production of α-amylase during germination (Fick and Qualset, 1975; Section 4-3.2.1). Genetic analysis indicated that D6899 had a single locus mutation in the *Rht3* gene (Fick and Qualset, 1975).

The lack of GA response in a tissue could be attributed to many causes, such as the presence of inhibitors and a general slowdown in cellular metabolism. To determine whether a specific regulatory step is affected by the mutation in the *Rht3* gene, the following questions were asked (Ho et al., 1981). First, does D6899 wheat have gross aberrations in basic cellular metabolism that cause its failure to respond to GA? Second, are the diverse GA effects in the aleurone layers of cereal grains retarded in D6899 wheat? Third, does D6899 wheat have a higher endogenous level of inhibitors (eg., ABA) than the wild type, which prevents most of the GA effect?

When compared with the standard height wheat variety Nainari 60, D6899 did not have a slowdown in cellular metabolism such as respiration rate, protein content, rate of protein synthesis, or uptake of amino acids. The content of ATP was even higher in D6899. Nainari 60 and D6899 had essentially the same uptake and metabolism of GAs and their levels of endogenous inhibitors did not differ drastically. Thus, it did not appear that cellular metabolism or the level of inhibitors could explain the behavior of D6899 (Ho et al., 1981).

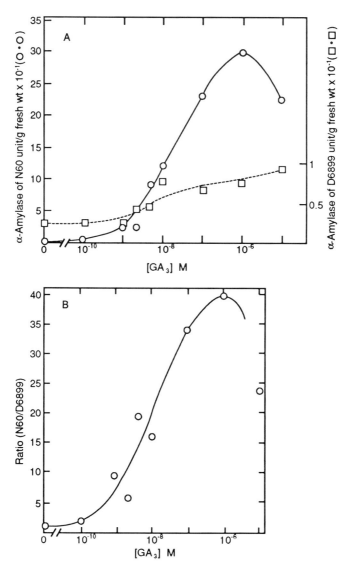

Figure 4-10. Dosage-response curve of GA_3-mediated α-amylase production in wheat aleurone layers. A Absolute values of α-amylase; Nainari 60 (○); D6899 (□); B Ratio of Nainari 60 to D6899. Note that the two curves in A are on different scales. (Ho et al., *Plant Physiol.* 67: 1026-1031, 1981; reproduced by permission of the ASPP.)

Table 4-14. Other GA_3 effects that are partially blocked in D6899 wheat aleurone layers.

Identity	Nainari 60			D6899		
	-GA	+GA	+GA/-GA	-GA	+GA	+GA/-GA
Protease	*unit / g fresh wt*					
Medium	1.7	6.3	3.71	1.9	1.9	1.00
Layer	2.3	4.3	1.87	2.5	2.3	0.92
Total	4.0	10.6	2.65	4.4	4.2	0.96
Phosphatase						
Medium	27.1	78.5	2.90	31.5	41.7	1.32
Layer	94.5	114.2	1.21	99.3	101.3	1.02
Total	121.6	192.7	1.58	130.8	143.0	1.09
Release of phosphate ion	*μmol Pi released / g fresh wt*					
	8.4	30	3.57	15	16	1.07
Release of reducing sugar	*mg glucose equivalent / g fresh wt*					
	0.35	2.54	7.26	0.26	0.28	1.04

(Ho et al., *Plant Physiol. 67:* 1026-1031, 1981; reproduced by permission of the ASPP.)

As shown in Fig 4-10, the amount of α-amylase in Nainari 60 increased to a maximum at 10^{-6} M GA_3, whereas in the aleurone layers of D6899 the response to GA_3 was much less and had a different pattern. Table 4-14 summarizes the other GA_3 effects that have been studied in these two types of wheat aleurone layers. GA_3 enhanced the production of protease in Nainari 60 but not in D6899. The release of protease, phosphate ion, and sucrose are likely to be membrane-related phenomena. GA_3 significantly affected all of these processes only in Nainari 60. The release of phosphatase has been shown to involve the transfer of preexisting enzyme molecules from cell wall to the incubation medium. Again, the GA_3-mediated process did not function effectively in D6899 (Table 4-14). In other studies, the effect of GA_3 on leaf elongation in the two genotypes was determined (Table 4-15). The leaves of D6899 were about half as long as those of Nainari 60 and elongated only slightly when GA_3 was applied.

A step common to GA stimulation of all these diverse affects (Fig. 4-10; Tables 4-14 and 4-15) must be responsible for the insensitivity to GA phenotype. The data suggest that a rate-limiting step, common to many of the diverse gibberellin responses, is partially blocked in D6899 wheat. Perhaps a defect in a GA-receptor would be such a common factor because the individual reactions themselves are so different from one another. It may be in the number or affinity of GA-recognition sites, but the definitive proof of this suggestion requires direct evidence on binding to GA receptors.

Table 4-15. Effect of GA$_3$ on leaf elongation growth in wheat. Plants were grown in vermiculite in half-strength Hoagland solution containing 1 mM CCC, a GA inhibitor. For GA treated plants 1 μM GA$_3$ was also included in the Hoagland solution. Plants were grown at 22°C and the length of the oldest leaf of each plant measured on the 10th day after germination. ^{a} Not determined.

	Length of Oldest Leaf Blade (Mean ± SE)		
	Nainari 60	Nainari 60 (seedlings from seeds whose endodosperm had been cut off)	D6899
		cm	
Experiment 1			
Control	11.4±0.6	ND^a	6.0±0.3
GA$_3$-treated	17.4±1.2	ND	7.4±0.2
Experiment 2			
Control	10.4±0.4	6.8±0.5	5.5±0.4
GA$_3$ -treated	19.7±0.7	12.0±0.8	6.0±0.6

(Ho et al., *Plant Physiol.* 67: 1026-1031, 1981; reproduced by permission of the ASPP.)

It is worth noting that the dwarfism in D6899 is caused by the reduced response to GA, whereas many other dwarfism cases in corn, rice, wheat, and pea, for example, are caused by lack of GA production. It is not understood why D6899 can survive without normal responses to GA. Probably there are reduced responses to GA in D6899 that still allow it to complete its life cycle and to yield viable progeny. Another possibility would be that responses to GA are not essential for the survival of wheat. The consequences of having reduced GA responses is merely a reduction in plant size, such as the length of leaves and the size of seeds (Ho et al., 1981).

4-3.2.4 The Effect of Low Temperature on GA$_3$ Sensitivity in Wheat

Low temperature has been shown to alter numerous physiological responses of plant tissues: the stratification of seeds, breaking of dormancy in buds of deciduous trees, and vernalization are all well documented cases. More specifically, there are a number of examples of interactions between GA responses and low temperatures in various species of grasses, including wheat. The results in Table 4-16 indicate one of these interactions. The low temperature induction of GA$_3$ sensitivity in aleurone tissue seems to be operative only in varieties having at least one of the three *Rht* alleles (Singh and Paleg, 1984). Varieties lacking *Rht* alleles showed no increase in α-amylase when subjected to low (5°C) temperature (Table 4-16).

Table 4-16. Effect of low temperature preincubation on the GA_3 sensitivity of wheat grain with various genetic backgrounds. Deembryonated seeds were preincubated for 20 hours at 5°C or 30°C and their GA_3 sensetivtiy was monitered by measuring the amount of a-amylase produced after 24-hour incubation at 30°C with 0.1 μg/ml GA_3. Increases in GA_3 sensitivity are expressed as percent increase in the amount of a-amylase produced as a result of 5°C preincubation over 30°C preincubation.

[a] 5°C preincubation treatment not significantly different from 30°C preincubation.

Variety		α-Amylase (SIC/Endosperm Preincubated at 30°C)	% Increase
Olympic	(No *Rht* allele)	26.4	nil[a]
Nainari 60	(No *Rht* allele)	27.5	nil[a]
Halbred	(No *Rht* allele)	24.6	nil[a]
Aroona	(*Rht1*)	14.8	88.36
Kite	(*Rht2*)	7.8	279.64
Tom Thumb	(*Rht3*)	7.3	132.34
Tardo	(*Rht3*)	3.5	555.62
Minister Dwarf	(*Rht3*)	5.2	207.22

(Singh and Paleg, *Plant Physiol.* 76: 139-142, 1984; reproduced by permission of the ASPP.)

The *Rht3*-containing GA_3-insensitive deembryonated wheat (*T. aestivum* L. var. Cappelle Desprez x Minister Dwarf) aleurone, that can be made responsive to GA_3 by low temperature (Table 4-16), can also be rendered GA_3-sensitive by preincubation with indole acetic acid (IAA). This response is concentration dependent, relatively sensitive, and similar in magnitude to that induced by low temperature. IAA has no effect on the wild type (Fig. 4-11; Singh and Paleg, 1986). The effects of low temperature preincubation and IAA pretreatment in enhancing the subsequent GA_3 response are not synergistic and both can produce similarly increased GA_3 responsiveness of the *Rht3* aleurone tissue when present alone or together during the preincubation period (Fig. 4-12).

The IAA-induced increase in subsequent GA_3 responsiveness was concentration dependent. IAA had no effect on the production of α-amylase in the absence of subsequent GA_3, and the amount of -amylase produced by IAA-pretreated, deembryonated *Rht3* seed was quite comparable to that produced by 5°C preincubated seed, which, in turn, was comparable to that produced by the tall selection (Fig. 4-12; Table 4-16; Singh and Paleg, 1986).

The suggestion that hormones can act in sequence in the control of growth and development is not new. The *Rht*-containing aleurone, however, was the first tissue that demonstrated an all-or-nothing

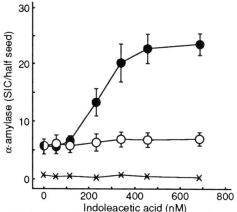

Figure 4-11. Effect of various concentrations of IAA on GA₃-induced α-amylase production by Rht3 (dwarf) de-embryonated seed. IAA was added during the 30°C 24h pre-incubation (●) or incubation (○) periods. Controls (x) to which IAA was added during pre-incubation had no GA₃ added subsequently. (Singh and Paleg, *Plant Physiol. 83:* 685-687, 1986; reproduced by permission of the ASPP.)

triggering response to two hormones that, acting clearly and completely in sequence, controlled the developmental response of the aleurone during the mobilization of endosperm reserves.

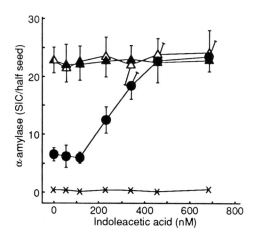

Figure 4-12. Comparison of effects of IAA and low temperature pre-incubation on GA₃-induced α-amylase production by *Rht3* deembryonated seed. IAA was added during the preincubation period to seed preincubated at 30°C (●), or during preincubation (▲) or incubation (△) periods to seed preincubated at 5°C. Controls (x) to which IAA was added during preincubation at 30°C, had no GA₃ added subsequently. (Singh and Paleg, *Plant Physiol. 83:* 685-687, 1986; reproduced by permission of the ASPP.)

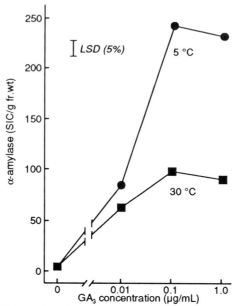

Figure 4-13. Effect of 20-hour preincubation at 5°C or 30°C on α-amylase production by isolated aleurone layers of *Rht3* dwarf wheat. (Singh and Paleg, *Aust. J. Plant Physiol. 12:* 269-275, 1985a; reproduced by permission of the CSIRO Editorial and Publishing Unit.)

This work with deembryonated seed makes it apparent that the embryo does not mediate the role of IAA in stimulating GA_3 sensitivity in the aleurone. In addition, if low temperature acts on the aleurone of dwarf (*Rht*) wheat by enabling the production of IAA, it is also clear that the embryo is not needed for these reactions, and that the aleurone itself (at least the dwarfed genotypes) has a greater range of responsive capacity than has hitherto been appreciated.

Thus, low temperature treatment has been shown to significantly increase the GA_3 sensitivity of wheat aleurone tissue containing the *Rht* gene. It has been suggested that this low temperature treatment, which cured or circumvented the genetic lesions manifested in the *Rht1*, *Rht2*, and *Rht3* genotypes, might involve an increase in hormone (GA_3) receptor sites (Singh and Paleg, 1984). Further examination was made of this hypothesis by studying the effects of low temperature treatment on the GA_3 sensitivity of wheat aleurone tissue from tall and dwarf lines isogenic with respect to all characteristics other than the *Rht3* gene (Singh and Paleg, 1985a).

As illustrated in Figure 4-13, a preincubation at 5°C, as compared with 30°C, of deembryonated seed of the *Rht* dwarf increased GA_3 sensitivity dramatically (Singh and Paleg, 1985a). One possible explanation for this increased GA sensitivity was that the low temperature preincubation shortened the lag time of α-amylase

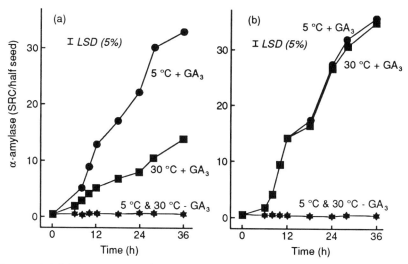

Figure 4-14. Time course of α-amylase production by deembryonated seed of (a) *Rht3*, dwarf and (b) *rht3*, tall wheat preincubated for 20 hours at 5°C or 30°C (GA$_3$ = 0.1 g/ml). (Singh and Paleg, *Aust. J. Plant Physiol. 12:* 269-275, 1985a; reproduced by permission of the CSIRO Editorial and Publishing Unit.)

production. Low temperature preincubation, however, had no effect on the lag time of α-amylase production by deembryonated seed of the dwarfs. Thus, the increased amount of α-amylase produced by the 5°C preincubated seed of the dwarf was due to a higher rate of α-amylase production and not to a shortening of the lag phase. Figure 4-14 further illustrates that low temperature preincubation raised the level of α-amylase production by the dwarf to the same level as that of the tall variety (Singh and Paleg, 1985a). Thus, low temperature can cure or reverse the genetic lesions manifest in the aleurone tissue containing the *Rht3* allele and the effect could be mediated by an increase in GA$_3$ receptor sites. These studies, however, are still a long way from demonstrating whether or not this is the case.

Another manifestation of low temperature treatment is the changes it brings about in the production of phospholipids in the aleurone tissue of dwarf wheat (Fig. 4-15; Singh and Paleg, 1985b). Deembryonated seed of the dwarf were subjected to a 12-hour imbibition at 30°C followed by preincubation at 5°C or 30°C for various lengths of time (0 to 24 hours). The aleurone tissue was separated and the phospholipids and their component fatty acid were analyzed. Imbibition of water for 12 hours at 30°C had no significant effect on the total phospholipids of the aleurone tissue (Fig. 4-15a). A further preincubation of the deembryonated seed of the dwarf for periods of 16 hours or longer at 5°C as compared with 30°C brought about significant increases in the level of the total phospholipids. In the case of the tall variety treated in the

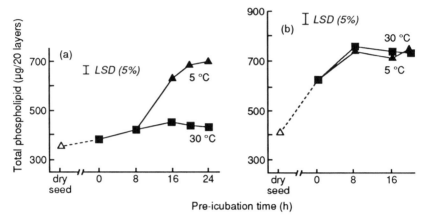

Figure 4-15. Effect of imbibition and preincubation temperature and time on the total phospholipid content of aleurone tissue of deembryonated wheat seed. Imbibition was carried out for 12 hours at 30°C. (a) *Rht3*, dwarf; (b) *rht3*, tall. (Singh and Paleg, *Aust. J. Plant Physiol.* 12: 277-289, 1985b; reproduced by permission of the CSIRO Editorial and Publishing Unit.)

same way as the dwarf, no difference could be detected in the total phospholipids (Fig. 4-15b).

Examination of individual phospholipids during the imbibition and preincubation periods also revealed that there was an almost complete lack of synthesis of the phospholipids during imbibition. A further preincubation at 5°C, however, caused a large stimulation in the synthesis of phosphatidylinositol (PI), phosphatidylcholine (PC), and phosphatidylethanolamine (PE) (Table 4-17; Singh and Paleg, 1985b). Preincubation at 30°C, for any of the periods examined, had no effect on the levels of any of the other constituent phospholipids of the aleurone tissue of the dwarf (Table 4-17).

The low temperature-induced increases in the levels of PI, PC, and PE were very highly correlated with the low temperature-induced increase in GA_3 sensitivity This was indicated by the highly significant values of r (the Pearson correlation coefficient) computed for the amount of α-amylase produced *after* 24 hours incubation with GA_3 versus the concentration of individual lipids in the tissue *before* to treatment with GA_3 (Table 4-18).

The simplest interpretation of the close relationships between phospholipid composition and hormonal sensitivity implicates and underlines the possibility that the GA_3 (hormone) receptor sites are membrane based. Two important differences, however, emerged from the comparison of the phospholipid behavior of the aleurone tissue of the tall variety with that of the dwarf. First, only the imbibition of water was required to initiate the synthesis of all the major phospholipids of the aleurone tissue in the tall variety (results not shown in this

Table 4-17. Phospholipid composition of aleurone tissue of *Rht3* (dwarf) wheat deembryonated seed preincubated at 5°C or 30°C for various lengths of time.
PI = phosphatidylinositol; PC = phosphatidylcholine; PE = phosphatidylethanolamine.
[a,b]Significantly different from the control at $P = 0.05$ and 0.01, respectively.

Experimental treatment	Phospholipid (µg lipid/20 layers)					
	PI	PC	PE	PI	PC	PE
	pre-incubated at 30°C			pre-incubated at 5°C		
Dry seed	26	197	69	26	197	69
12 h inhibition (control)	26	205	73	26	205	73
Preincubated for:						
8 h	34	223	72	32	233	70
16 h	40	260[a]	76	95[a]	357[b]	100[b]
20 h	32	257[a]	72	124[b]	380[b]	100[b]
24 h	35	249[a]	75	129[b]	382[b]	105[b]

(Singh and Paleg, *Aust. J. Plant Physiol. 12:* 277-289, 1985b; reproduced by permission of the CSIRO Editorial and Publishing Unit.)

excerpt). Thus, the *Rht* gene might be responsible for the observed aberrations in phospholipid metabolism in the dwarf. In view of the likely role of active phospholipid metabolism in the aleurone in relation

Table 4-18. Correlations between subsequent α-amylase production by aleurone tissue of *Rht3* and preceding individual phospholipids. The Pearson correlated coefficients were computed for the amount of α-amylase produced after 24 hours incubation with GA$_3$ (0.1 g/mL) versus the concentration of lipids in the tissue before treatment with GA$_3$. r, Pearson correlation coefficient: [a] correlation significant at $P = 0.01$; n.s. = correlation not significant; PG = phosphatidylglycerol; LPC = lysophosphatidylcholine; other abbreviations, see Table 4-17.

Correlation (r) for:						
PI	PC	PG		LPC		PE
0.977[a]	0.935[a]	0.521	n.s	0.062	n.s	0.846[a]

(Singh and Paleg, *Aust. J. Plant Physiol. 12:* 277-289, 1985b; reproduced by permission of the CSIRO Editorial and Publishing Unit.)

to GA$_3$ sensitivity, the present results make it probable that the *Rht* gene itself influences the GA$_3$ sensitivity of wheat aleurone through an involvement with phospholipid biosynthesis. Whether these observations can also account for the GA$_3$ insensitivity of vegetative tissue of *Rht*-containing wheat remains to be seen (Singh and Paleg, 1985b).

Although many of the studies illustrated here have so far proved to be inconclusive, it is also apparent that much can be learned about the physiology of plants through the use of such varieties as those that are GA insensitive. Enough has been learned through their use in a wide variety of investigations to be confident that further work with them as experimental material will provide significant advances in our understanding of hormone action.

Case Study 4-4:   The Occurrence of Mutants Involving Other Groups of PlantGrowth Substances

As seen in Case Studies 4-1 to 4-3, a wide range of mutants have been isolated with aberrations in the formation and role of plant growth substances. To these cases involving auxin and gibberellin mutants can be added Case Study 1-1 where the wilty (abscisic acid) mutants were described and discussed in regard to stomatal action and in Case Studies 6-1 and 6-2 which deal with mutations affecting gibberellins and abscisic acid.

Very few mutants have been isolated so far in higher plants that affect the other major groups of growth substances, the cytokinins and ethylene. In this short Case Study, some examples are given of recent investigations of mutants in which the metabolism of these substances appears to differ from wild-type.

4-4.1   Cytokinins

The "McIntosh" apple originated in Canada as a chance seedling discovered in 1811. Of particular interest to apple breeders was the discovery in 1965 of a nonbranching, highly spurred mutant now referred to as "McIntosh Wijcik." Unlike previous spur-type mutants of McIntosh, McIntosh Wijcik transmits its unique growth habit to a high proportion of its progeny by the compact (*Co*) gene (Looney et al., 1988).

The physiological basis of the spur-type growth habit in apple remains unknown. In McIntosh, the greater the degree of spurriness and upright growth habit, the greater the precocity of flowering and tendency toward biennial fruit bearing. Clearly, increased knowledge about the physiology of these growth-type mutants would be essential for them to be used most effectively in apple production.

Table 4-19. Total gibberellin and cytokinin concentrations (in nanograms of GA_3 or zeatin equivalent per gram dry weight of shoot tissue) in actively growing shoot tips of four strains of McIntosh apple.
[a] Mean separation within columns by Duncan's multiple range test, $P = 5$ percent.

McIntosh strain	Polar GAs (mainly GA_{19})	Cytokinins ($ng \cdot g^{-1}$ dry wt)
Summerland Red	24.1^{ba}	22.5^{a}
MacSpur	21.4^{b}	15.6^{a}
Morspur	23.1^{b}	14.4^{a}
Wijcik	6.8^{a}	50.4^{b}

(Looney et al., *J. Am. Hort. Sci. 113*: 395-398, 1988.)

Although shoot tips of Mcintosh Wijcik did not differ significantly from the other strains in total GA-like activity or concentration of less-polar GA-like substances, they displayed dramatically lower activity of GA_1 and GA_{19} (Table 4-19; Looney et al., 1988). Similarly, McIntosh Wijcik was the only strain to differ significantly in cytokinin content, displaying levels significantly higher than the other strains (Table 4-19).

The results of the measurements of endogenous cytokinin and GA levels suggested that both classes of growth regulators play a role in defining the spur-type growth habit in apple. It was suggested that the McIntosh Wijcik clone is both dwarf and spurry because it is relatively high in cytokinin and low in GA activity. It was predicted that those of its progeny with the *Co* gene also will display high cytokinin activity but their greater or lesser tendency to develop lateral branches will depend on their inherent growth potential or vigor. Vigor of growth is, in turn, likely to be controlled, at least in part, by the endogenous level of GA (Looney et al., 1988).

4-4.2   Ethylene

4-4.2.1   *Arabidopsis*

The gaseous compound ethylene is an endogenous regulator of growth and development in higher plants. Increases in the level of ethylene influence many developmental processes from seed germination and seedling growth to leaf abscission, organ senescence, and fruit ripening. A number of environmental stresses including oxygen deficiency, wounding, and pathogen invasion enhance ethylene synthesis. Stress-induced ethylene also elicits adaptive changes in plant development. Wounding and pathogen invasion may result in ethylene-mediated acceleration of senescence and abscission of infected organs and in the induction of specific defense proteins (Bleecker et al., 1988).

To gain further insight into the mechanism of ethylene action mutants with altered responses to ethylene in *Arabidopsis thaliana* were isolated. Ethylene-insensitive mutants were identified by taking advantage of the ethylene-mediated inhibition of hypocotyl elongation in dark-grown seedlings. Populations of M_2 mutagenized seed were placed in chambers through which ethylene (5 µl/L) in air was circulated. Seedlings that had grown more than 1 m after 4 days were selected as potential ethylene-insensitive mutants. A screen of 75,000 seedlings yielded three mutant lines that showed heritable insensitivity to ethylene (*etr*, a mutant allele of the *etr* gene; Bleecker et al., 1988). Dose-response curves for the effect of ethylene on hypocotyl elongation in the wild-type and the *etr* mutant line showed that although elongation of wild-type hypocotyls was inhibited by 70 percent with ethylene at 1 µl/L, hypocotyl elongation of the mutant line was unaffected by concentrations of up to 100 µl/L (Fig. 4-16).

The results summarized in Table 4-20 further indicated that a variety of ethylene responses, occurring in different tissues and at different stages in the life cycle of the wild type plant are affected by the *etr* mutation and, therefore, must share some common element in their signal transduction pathway. It was argued that the lesion produced by the *etr* mutation must occur early in the signal transduction chain because the biochemical bases for the various responses were presumably different. It was interesting that the apparent elimination of detectable ethylene responses did not drastically alter the growth and development of the plant, although the ethylene responses examined in these

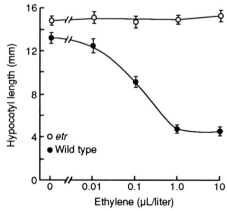

Figure 4-16. The effect of increasing ethylene concentrations on hypocotyl elongation of dark-grown wild-type (●) and *etr* (○) *Arabidopsis* seedlings. (Bleecker et al., *Science 241:* 1086-1089; copyright 1988 by the AAAS.)

Table 4-20. Responses to ethylene in wild-type and ethylene-insensitive (*etr*) *Arabidopsis.*

Parameter	Ethylene response (percent of air control)	
	Wild type	*etr* mutant
Organ elongation		
Root	17	111
Hypocotyl	36	104
Chlorophyll content		
Leaf	35	97
Stem	110	94
Peroxidase activity		
Leaf	303	102
Stem	421	92
Ethylene biosynthesis		
Leaf	15	96
Seed germination	664	100

(Bleecker et al., *Science 241:* 1086-1089; copyright 1988 by the AAAS.)

experiments represented many of the effects by which ethylene has been recognized as a regulator of plant growth and development.

Because all the ethylene responses that were examined were affected by the *etr* mutation, a single receptor for ethylene was hypothesized to be present in all tissues of *Arabidopsis*, and it was thought possible that the mutation directly affects this receptor. The results of binding studies using ^{14}C-labeled ethylene indicated that saturable binding of ethylene in *etr* mutants was one-fifth of that found in wild-type plants (Bleecker et al., 1988).

In a very recent study in which the response of *Arabidopsis* was used to identify ethylene mutants, variants with high and low levels of ethylene were identified as well as mutants that were insensitive to ethylene (Guzman and Ecker, 1990).

4-4.2.2   *Tomato*

In Case Study 4-1 the single gene mutant *dgt* of tomato was described. It was noted there that the phenotype of this mutant could be at least partially reverted to normal by exposure to low concentrations of ethylene, although it has been shown that the fundamental lesion of the

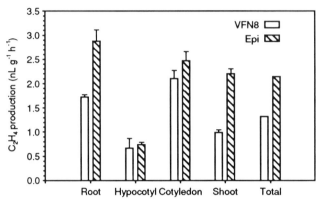

Figure 4-17. Ethylene production rates by VFN8 and *Epi* tomato root, hypocotyl, cotyledon, and shoot tissue 0.5 hours after excision. (Fujino et al., *Plant Physiol. 88:* 774-779, 1988; reproduced by permission of the ASPP.)

dgt mutant seems to be insensitivity to auxin. Recently, another single-gene tomato mutant derived from the same parent line (VFN8) as *dgt* was found that is characterized by a contrasting developmental pattern, that is, extreme leaf epinasty, thickened stems and petioles, an apparent reduction in anthocyanin production, a shortened and highly branched root system, and very erect growth. This *Epi* mutant has been hypothesized to be an ethylene overproducer or extremely sensitive to ethylene (Fujino et al., 1988). Indeed, ethylene production by shoot and root tissues of *Epi* were 122 percent and 67 percent higher, respectively, than the corresponding tissues of VFN8 (Fig. 4-17)

The higher concentration of internal ethylene in *Epi* suggested that there was either a difference in the concentration of the precursor of ethylene, 1-aminocyclopropane-1-carboxylic acid (ACC), or a difference in the ethylene-forming enzyme (EFE) activity between *Epi* and wild type. In fact, the ACC content of *Epi* was found to be more than six times that of VFN8, whereas the EFE activity was similar in both genotypes (Table 4-21). The reason for the elevated ethylene synthesis

Table 4-21. Internal ethylene concentration, ACC concentration and EFE activity of VFN8 and Epi genotypes of tomato.

Genotype	$[C_2H_4]$	[ACC]	EFE Activity
	$nL\ g^{-1}$	$nmol\ g^{-1}\ fr\ wt$	$C_2H_4\ nL\ g^{-1}\ h^{-1}$
VFN8	31± 5	0.48±0.2	76.0±10.8
Epi	162±44	3.2 ±0.7	77.9± 7.6

(Fujino et al., *Plant Physiol. 88:* 774-779, 1988; reproduced by permission of the ASPP.)

rates in *Epi* remains unknown. In the future, however, the mutant may provide a system for investigating the regulation of ethylene biosynthesis and the role of its target cell types in plant growth and development.

Case Study 4-5:   Gravitropic Mutants

In 1880, Darwin, assisted by his son Francis, authored a book entitled *The Power of Movement in Plants.* Darwin was led to conclude that the perception of gravity by roots was localized at the root tip, from which he hypothesized that some influence was transmitted to adjoining parts causing the root to bend. In spite of considerable investigative effort, there is still today no widely accepted explanation of the mechanisms regulating this root gravitropism. Investigators have, in more recent times, subdivided the gravitropic response into three components, or processes: perception of the stimulus, transduction of the message, and reorientation of the root.

Perception

During the perception phase, the root senses the direction of the gravitational force. Most workers are satisfied with the conclusion that this gravity-sensing mechanism for the root is located within the root cap. Efforts to determine how and where within the cap gravity is perceived have focused on special starch grain-containing cells termed *statocytes.* It is proposed that the direction of sedimentation of the starch grains and their asymmetrical distribution within the statocyte is under gravitational control. Repositioning the root in relation to gravity results in a movement of starch grains toward the lowermost, inner cell wall of individual statocytes (Fig. 4-18). It is suggested that the gravity-induced displacement of starch grains, now known to be enclosed in membrane-bound organelles called amyloplasts, is an

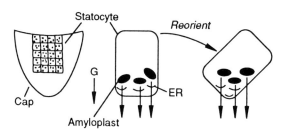

Figure 4-18. Diagram of statocytes within the root cap and the proposed distribution in relation to gravity (G) of amyloplasts and endoplasmic reticulum (ER) within the statocyte. (Feldman, *Physiol. Plant.* 65: 341-344, 1985.)

important component of the gravity-sensing mechanism (however, see later discussion in this Case Study) (Feldman, 1985).

Transduction

The second and intermediate phase of the gravitropic response involves communicating the direction of the gravitational force from the cap, where it is perceived, to the region of the root where curving will occur. Initial views of the nature of this signal were influenced by proponents of the Cholodny-Went hypothesis, which suggested that curvature resulted from the accumulation of one or more substances inhibitory to growth on the lower portion of the root. Two substances, IAA and ABA, are considered prime candidates for the root cap message. Data for the asymmetrical redistribution of IAA or ABA under gravistimulation, however, are equivocal. Others suggest that perhaps we need to abandon our notion that these substances "work" by directly bringing about an inhibition (or stimulation) of growth (Feldman, 1985).

Reorientation

Curvature, the final phase of the gravitropic response, results in a reorientation of the root in relation to the gravity vector. It is believed that a root curves downward because of difference in elongation rates between the upper and lower portions of the root, with the cells in the upper portion growing faster and resulting in a downward curvature of the root. But, there are at least six possible ways in which elongation could occur and lead to downward curving (Table 4-22).

It is most likely that curvature occurs because of a promotion of curvature on the uppermost side of a root but not enough roots have been examined to make this a certainty. From the few roots examined in detail, one would conclude that curving occurs because of a promotion of cell extension on the upper side, rather than an inhibition of growth on the lower surface of a gravireacting root. In Figure 4-19, an example

Table 4-22. Positive root gravitropism occurs when the initially uppermost side elongates more rapidly than the lowermost side. Six ways in which this might occurs. *Key:* (+) elongation stimulated, (-) elongation slowed, (0) no change in growth rate, (x) elongation arrested.

	1	2	3	4	5	6
Uppermost side	+	+	+	0	0	-
Lowermost side	0	-	x	-	x	x

(Feldman, *Physiol. Plant 65:* 341-344, 1985.)

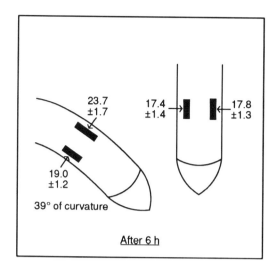

Figure 4-19. Lengths of cortical cells (in microns) in bending zones from roots positioned vertically or horizontally for 6 hours. (Feldman, *Physiol. Plant.* 65: 341-344, 1985.)

is given of the clear increase in length of cells on the uppermost side of a root stimulated gravitropically when compared with a control root growing vertically (Feldman, 1985).

4-5.1   The Isolation and Characterization of Gravitropic Mutants

The gravitropic bending of a plant organ is the culmination of a highly coordinated sequence of events, as described earlier. It is the relationship between these events that dictates whether an organ will be positively or negatively gravitropic or grow at an angle to the gravity vector. If the site of action of a mutation affects any step in the program, then the chances of the gravitropic response being impaired in some manner are high. Such a modification in the sequence could ultimately lead to a change in gravitropic orientation or sensitivity, or even the inability to respond to a gravitropic stimulus.

One of the earliest gravitropic mutants discovered was described in 1958 and termed *Pisum sativum ageotropum* (*agt*). More recently, the agravitropic nature of roots of the mutant has been confirmed in roots subjected to unilateral gravitropic stimulation (Fig. 4-20) and an anatomical study of the root carried out (Roberts, 1987).

A comparison of apparently agravitropic mutants with normal plants provides an opportunity to study the mechanism of gravity perception and the physiology of the graviresponse. In an attempt to

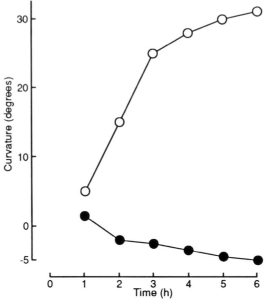

Figure 4-20. Gravitropic bending of pea roots from normal or *ageotropic* seedlings. (○) Normal; (●) *agt.* (Roberts, in *Developmental Mutants in Higher Plants,* H. Thomas, D. Grierson (eds.), Cambridge University Press, London, pp. 135-153, 1987; data redrawn from G.M. Olsen and T-H. Iverson, *Physiol. Plant 50:* 269-274, 1980.)

distinguish the various steps involved in gravitropic responses, a systematic and large-scale search for agravitropic mutants is underway in a species that is especially suitable for rapid isolation of mutants, *Arabidopsis thaliana* L.. The first group of *Arabidopsis* mutants was isolated on the basis of their resistance to 2,4-D and included mutations at two unlinked loci, *aux* and *Dwf.* Two alleles of the *aux* locus were distinguished; among these, *aux-1* gave agravitropic seedling roots with reduced sensitivity to exogenous auxins (Fig.4-21; Mirza et al., 1984; Roberts, 1987).

Wild-type roots of *Arabidopsis* responded to unilateral gravitropic stimulus by developing curvatures in the direction of the gravity vector (Table 4-23). The stimulation of wild-type roots at 135 degrees yielded significantly larger curvatures than at 90 degrees which, in turn, produced larger curvatures than at 45 degrees during a 22-hour period. The average rate of curvature of wild-type roots decreased with the time of stimulation when stimulated at 45, 90 and 135 degrees.

The development of wild-type hypocotyl curvatures was also greatest after stimulation at 135 degrees, with 45 degree stimulation producing the smallest response. Neither an asymmetry in the root cap nor a curvature was observed in *aux-1* roots when subjected to continuous gravitational stimuli. Unlike roots, *aux-1* hypocotyls responded to

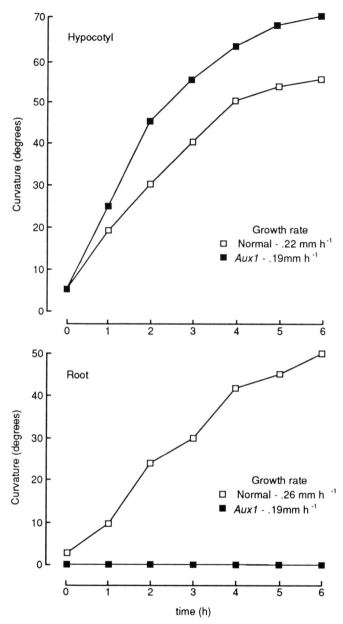

Figure 4-21. Gravitropic bending in normal or *aux-1 Arabidopsis thaliana* seedlings. (Roberts, in *Developmental Mutants in Higher Plants,* H. Thomas, D. Grierson (eds.), Cambridge University Press, London, pp. 135-153, 1987; redrawn from Mirza et al., 1984.)

gravity and developed greater curvatures after 22 hours than wild type. Curvature also developed more rapidly than in wild type (Table 4-23). The negative gravitropic curvature attained by *aux-1* hypocotyls showed

Table 4-23. Curvature rates(degrees/hour) of wild-type and *aux-1* roots (top) and hypocotyls (bottom) stimulated at 45, 90, and 135 degrees.

Geno- type	0-2 h			2-6 h			6-22 h		
	45°	90°	135°	45°	90°	135°	45°	90°	135°
WT	8.6	11.5	17.5	3.0	7.4	6.9	0.6	1.3	2.4
	±0.6	±1.0	±1.0	±0.5	±0.3	±0.4	±0.1	±0.1	±0.1
aux-1	0	-0.08	-0.08	-0.4	-0.08	-0.04	0.01	0.03	0.03
Geno- type	0-2 h			2-6 h			6-22 h		
	45°	90°	135°	45°	90°	135°	45°	90°	135°
WT	11.8	18.6	18.9	1.9	4.7	9.6	0.1	0.2	0.8
aux-1	14.3	22.7	23.7	3.0	7.4	15.2	0.1	0.7	1.3

(Mirza et al., *Physiol. Plant.* 60: 516-522, 1984.)

that the responses of roots and hypocotyls can be clearly separated by mutation and must be, to some extent, under different genetic control (Mirza et al., 1984).

4-5.2   Physiological Investigations Using Agravitropic Mutants.

4-5.2.1   The Role of Statocytes in Gravitropic Movement.

Several attempts have been made over the years to demonstrate that amyloplasts with starch grains act as statocytes and form an integral part of the graviperception system in roots of higher plants. Thus far, however, it has not been possible to elucidate the basic mechanism of the gravitropic responses, that is, the way in which the movements of statocytes initiate the reaction chain eventually leads to the curvature of the plant organ. To help understand this mechanism, physiological and ultrastructural comparisons of normal plants with plants (mutants) that apparently do not respond to gravity have been made (Olsen et al., 1984).

   In coleoptiles of the amylomaize corn mutant, the amyloplasts are much reduced in size in comparison with the wild-type corn, permitting a comparison of gravitropic responsiveness as related to lateral displacement of amyloplasts and lateral transport of auxin. The displacement of starch grains was followed after orientation of coleoptiles in the horizontal position. Differences between wild type and amylomaize were seen in the rate of displacement as well as in the final asymmetry achieved. Whereas the wild type achieved a maximum of 40 to 60

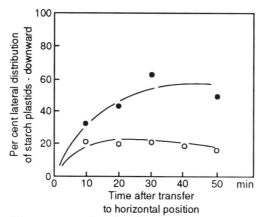

Figure 4-22. Time curves for the displacement of starch plastids after placement in the horizontal position. Plastids were scored in the upper and lower half of cells in several fields 5 to 7 mm below the tip. (●) Wild-type; (○) amylomaize. (Hertel et al., *Planta* 88: 204-214, 1969.)

percent displacement, the amylomaize did not exceed 20 percent (Fig. 4-22; Hertel et al., 1969).

The striking difference in starch grain properties in the two types of corn presented an opportunity to test the starch-statolith hypothesis. The effect of gravity on lateral movement of auxin was measured using cylinders of coleoptiles, comparing the lateral movement in three different orientations of the coleoptiles of the two corn strains (Fig. 4-23). In wild-type coleoptiles in the vertical position, 27 percent of the total radioactivity in the section moved to the receptor side. A comparable

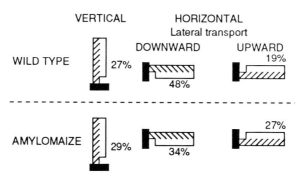

Figure 4-23. Lateral auxin transport diagram showing three coleoptile orientations. The donor blocks (black) are in direct contact with half of the coleoptile cross section. After the experiment the donor half and the receptor half are separated by a cut, extracted and assayed for IAA-^{14}C. As a measure of lateral transport, the average cpm/100 mg in receptor half are expressed as percent of total uptake (= average cpm/100 mg in receptor half + donor half). (Hertel et al., *Planta* 88: 204-214, 1969.)

amount of lateral movement was found in vertical amylomaize coleoptiles (29 percent). When the coleoptiles were held in the horizontal position, 48 percent lateral movement occurred in wild type in the direction of gravity, and 34 percent lateral movement occurred in amylomaize in the same orientation. Thus, lateral auxin movement in wild type and amylomaize occurred in the ratio of about 1 to 1 in the vertical orientation, and 1.4 to 1 in the direction of gravity in the horizontal orientation (Table 4-23).

The difference in gravity sensitivity between wild type and amylomaize could further be documented measuring gravitropic curvature (Table 4-24). The gravity effect on lateral IAA transport was seen to be greater in wild type than in amylomaize, and within either type the effect was stronger in the tip region than in the zone below. The starch sedimentability in the same locations was again greater in the wild type and in the tip regions than in the zone below. The responses of starch grain movement, lateral auxin transport, and curvature response between wild type and amylomaize were roughly proportional in each instance. These findings are readily compatible with the concept of the starch grains acting as gravisensors (Hertel et al., 1969).

In other studies, seedlings of *Arabidopsis thaliana* L., wild type and its three mutants *aux-1, aux-2* and *Dwf* were used and the movement of amyloplasts and nuclei measured (Fig. 4-24; Olsen et al., 1984). The sedimentation velocity calculated from Figure 4-24 showed that the pattern of movement of amyloplasts in wild type and *aux-2* was almost identical at 6.0 and 6.6 m hour, respectively. In *aux-1*, on the other hand, the sedimentation was significantly (<1 percent level) slower within the inversion period (2.4 m/hour).

Table 4-24. Gravity effects on (a) starch-grain distribution, (b) lateral transport of IAA-^{3}H, and (c) gravitropic curvature (measured at 90 min).

Material	Zone (mm)	(a) Starch distribution index	(b) Lateral transport (%) $\dfrac{\text{cpm downward} - \text{cpm upward}}{\text{cpm downward} + \text{cpm upward}} \times 100$	(c) Degrees curvature
Wild type	2- 7	44	31.7±6.3	intact.: 27.7±1.6
	8-13	29	26.5±3.3	decap.:15.8 ±1.6
Amylomaize	1-7	32	16.3±5.0	intact.: 17.9±1.4
	8-13	10	2.0±3.6	decap.:7.5± 1.4

(Hertel et al., *Planta 88:* 204-214, 1969.)

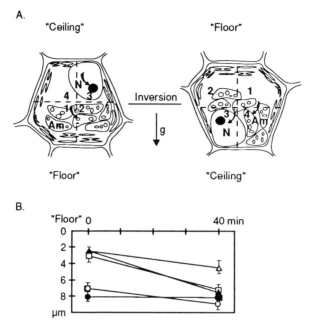

Figure 4-24. Movement of cell structures in statocytes of gravitropically normal and agravitropic *Arabidopsis* roots. A. Semischematic representation of a normal statocyte demonstrating the position of amyloplasts (Am) and nucleus (N) before and after 40 minutes of inversion. B. Movement of amyloplasts (□,△,▲) and nuclei (■,○,●) during 40-minute inversion of root tip. Wild type, ■ and □; aux-1,△ and ○; aux-2, ▲ and ●. (Olsen et al., *Physiol. Plant. 60:* 523-531, 1984.)

Thus, the most obvious difference between *Arabidopsis* wild type and *aux-1* was the retarded amyloplast movement in the latter. Whether these differences were of basic importance for the initial phase of the gravity perception in *Arabidopsis* it was not possible to say from these results. Probably one link in the whole gravitropic reaction chain was altered or missing and thereby caused agravitropic behavior, but it was not possible to say which link this may be, or where the link was to be found (Olsen et al., 1984).

The conclusion that it is the amyloplasts that are the site of graviperception is borne out further by yet another study:Etiolated hypocotyls from normal tomato plants showed a negative gravitropic response within 20 minutes of stimulation. In contrast, etiolated hypocotyls from the gravitropic mutant *Lazy-1*, did not reorient after gravistimulation (Fig. 4-25; Roberts, 1984). Anatomical studies revealed that etiolated hypocotyls from normal plants contained sedimenting amyloplasts located within endodermal cells. Such sedimenting amyloplasts were absent in *Lazy-1* tissue. It was hypothesized that the

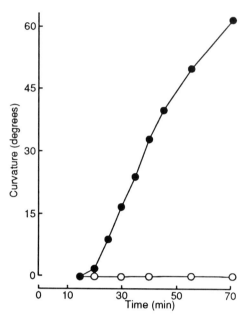

Figure 4-25. Curvature of hypocotyls from normal (●) and *Lazy-1* (○) tomato seedlings after a gravitropic stimulus. (Roberts, *Plant Cell Environ.* 7: 515-520, 1984; reproduced by permission of Blackwell Scientific Publications Ltd.)

hypocotyl of *Lazy-1* was agravitropic as it was unable to perceive a gravistimulus (Roberts, 1984).

Thus, several studies have shown that amyloplasts are essential for root graviresponsiveness. Cells of graviresponsive as well as nongraviresponsive roots of *Hordeum vulgare*, however, possess sedimented amyloplasts. Similar findings have been reported previously for graviresponsive and nongraviresponsive roots of other plants, indicating that the presence of sedimented amyloplasts does not ensure that a root is graviresponsive. Similarly, amyloplasts are not present in cells of some graviresponsive roots, indicating that amyloplasts need not even be present in cells for roots to be graviresponsive. Such observations, however, may reflect the fact that other mutations occur than those that affect amyloplasts directly and that have an impact on the gravitropic response (Moore, 1985, 1989). One example of such a mutant is described later.

4-5.2.2   Transduction of the Gravitropic Stimulus

The gravitropic bending response is thought to involve either an asymmetric distribution of a plant growth regulator or an asymmetric sensitivity to a plant growth regulator. In roots, the nature of this plant growth regulator is unclear. Most evidence favors a growth inhibitor. This could be IAA, ABA, or so far, an unidentified plant growth inhibitor.

In this present study, (Eason et al., 1987) in vitro culture methods were developed to allow the investigation and quantification of the graviresponse of both adventitious roots and isolated roots of the pea including the mutant *Pisum sativum ageotropum*. Mutant roots isolated in liquid culture showed the same growth potential as the roots of the normal pea. The application of basally applied ABA and IAA to isolated roots in agar culture inhibited root elongation but only in normal peas and could not restore a positive graviresponse in the mutants. Only asymmetrically applied auxin or replacement of the mutant root cap with one from a normal pea root was effective in restoring a positive graviresponse (Table 4-25).

The mutant roots grew preferentially gravinegatively, whereas the normal roots grew gravipositively. Decapitation of both genotypes induced a significant change in the growth of the root when compared with the intact roots, both genotypes growing almost horizontally. When root caps were removed and replaced the graviresponse was lessened but only significantly in the normal roots. Reciprocal transfers, however, had a profound effect. When mutant tips were placed on normal roots, the gravipositive response was considerably reduced, whereas, when normal tips were placed on mutant roots, the normal positive graviresponse was substantially restored (Table 4-25). Similarly,

Table 4-25. Effect of root cap transfers on graviresponse of pea root tips.

	Mean curvature degrees ± s.e.	Significance level
Wild type (JI 932)		
Intact	+58.5±3.5%	
Caps removed and replaced	+30.3±7.4	1%
Caps removed and replaced with JI 819 caps	+9.0±7.4	1%
Caps removed	+1.1±3.7	1%
'Ageotropum' (JI 819)		
Intact	-14.8±4.6	
Caps removed and replaced	-12.1±3.9	ns
Caps removed and replaced with JI 932 caps	+28.5±6.0	1%
Caps removed	-2.5±4.1	ns

(Eason et al., *J. Plant Physiol. 131:* 201-213, 1987; reproduced by permission of Gustav Fischer Verlag.)

when auxin at 1 μM was applied to mutant and wild-type roots it was effective at inducing curvature in both. No repair of the mutant by ABA, however, could be observed on either intact or decapitated roots (Eason et al., 1987).

These results seem to suggest that auxin, but not ABA, plays a role in the transmission of the signal resulting from the perception of gravity. Similarly, in the case of the *flacca* mutant of tomato, which is deficient in ABA, seedlings exhibit normal gravitropic behavior, results that run contrary to certain theories regarding the involvement of the growth inhibitor in the control of gravitropism (see Weyers, 1985).

4-5.2.3   *Root curvature*

Light and gravity both are important environmental factors in the growth of plant roots. Darwin noticed that in nature most of the seeds that germinate on the surface of the soil receive light and gravity acting perpendicular to the plane of root growth until roots penetrate the soil. How do roots react to these dual stimuli?

The agravitropic mutant *aux-1* of *Arabidopsis* provides an opportunity to study the responses of plant roots to light within the gravitational field of the earth. Advantage can be taken of the agravitropic nature of *aux-1* roots to study and distinguish light and gravity effects on the direction of horizontal root growth of *aux-1* and its isogenic wild-type (Mirza, 1987).

The mutant *aux-1* was induced and isolated in *Arabidopsis thaliana* L., ecotype Landsberg (*erecta* mutant), on the basis of its resistance to the herbicide 2,4-D. Roots of the mutant are agravitropic and negatively phototropic. When *aux-1* seedlings were grown on the surface of horizontal agar with illumination from above, the roots developed clockwise curvature as seen from above. In contrast, *aux-1* roots developed counterclockwise curvature (as seen from above) when illuminated from below (Table 4-26).

When both light and gravity vectors acted in the same direction, a synergistic effect (clockwise coiling) was produced on the horizontal curvature of wild-type roots, whereas an antagonistic effect was produced when light and gravity vectors acted in opposite directions (Table 4-26). In contrast, with horizontal *aux-1* roots, light and gravity, whether acting in the same or opposite directions, produced similar low levels of clockwise or counterclockwise curvatures, as would be expected by the light alone.

These results showed that, in addition to the recognized negative phototropic and positive gravitropic responses of the root, there were also horizontal clockwise curvatures perpendicular to the direction of light and gravity. Light and gravity interacted to determine the direction and magnitude of the horizontal curvature of roots. In nature,

Table 4-26. Mean curvature of horizontal wild-type (WT) and *aux-1* roots of *Arabidopsis* with illumination from above or below. Clockwise curvatures are assigned as (+) and counterclockwise curvatures as (-).

Illumination of *aux-1* from Above/Below	Mean Curvatures as Seen from Above
	degrees ± SEM
Above	(+) 104.05±5.12
Below	(-) 59.32±6.09

Genotype	Mean Curvatures as Seen from Above	
	Illumination from above	Illumination from below
	degrees ± SEM	
WT	(+) 507.31±35.96 (27)	(+) 181.78±15.43(28)
aux-1	(+) 51.25± 9.36 (31)	(-) 42.79± 4.59 (30)

(Mirza, *Plant Physiol. 83:* 118-120, 1987; reproduced by permission of the ASPP.)

both light and gravity act synergistically on germintating seeds as long as the developing roots are on the soil surface and would, therefore, be expected to induce clockwise coiling (Mirza, 1987).

The investigations discussed earlier suggest that *Arabidopsis* can be a source of useful mutations affecting a significant physiological phenomenon, gravitropism. This is borne out by the fact that the list of *Arabidopsis* mutants with defects in gravitropic behavior continues to grow (Bell and Maher, 1990; Wilson et al., 1990). Additional recent work indicates that phototropism can also be approached genetically through the use of this species. Through the use of mutants with aberrations in the phototropic response it may not only be possible to investigate the light-related phenomenon itself but also the relationship between phototropism and gravitropism (Khurana et al., 1989; Khurana and Poff, 1989).

Case Study 4-6:   Phytochrome Mutants

Plants use specific regulatory photoreceptors to perceive light signals and appropriately adjust their developmental pattern. A number of plant regulatory photoreceptors have been identified, among them blue light receptors and phytochrome. Not much is yet known about either the blue light receptors or the molecular events of blue light-induced responses. The phytochrome system is better understood (Vanderhoef and Kosuge, 1984).

Of its two forms, phytochrome(P_r) absorbs maximally at 660 nm and phytochrome(P_{fr}) at 730 nm. These two forms are reversibly interconvertible by light (Salisbury and Ross, 1985). The absorption spectra of phytochrome as found in etiolated seedlings (P_r) and after saturating irradiation with red light (P_{fr}) are shown in (Fig 4-26).

Because of overlapping absorption of P_r and P_{fr} below 720 nm, it is impossible to photochemically convert all of P_r to P_{fr}. The relative quantities of the two species obtained after saturating red light actually represents the absorbance of a mixture of 15 percent P_r and 85 percent P_{fr}. In addition to the phototransformations there is a dark reversion of P_{fr} to P_r.

The phytochrome protein is synthetized in growing seedlings de novo as P_r. There is essentially nothing known of the biosynthetic origin of the linear tetrapyrrole chromophore. Although there is a turnover of P_r in dark grown seedlings relatively large amounts of the protein accumulate in meristematic tissues, as much as 0.5 percent of the extractable protein in etiolated oat and rye seedlings. Exposure of dark grown seedlings to light (conversion of P_r to P_{fr}) initiates a destruction process that is apparently a proteolytic phenomenon. This destruction appears to be an important part of the mechanism by which the level of the active form of phytochrome is controlled in the cell. Under continuous light the level of phytochrome is only a fraction of that in etiolated seedlings.

The photoconversion of P_r to P_{fr} in the cell induces a large number of diverse morphogenic responses. Reconversion to P_r cancels these

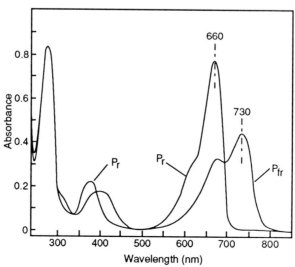

Figure 4-26. The action spectra of the P_r and P_{fr} forms of phytochrome.

inductions. Thus, phytochrome is considered to function as a reversible biological switch. The molecular mechanism by which P_{fr} induces the observed developmental changes, however, remains unclear (Smith, 1983).

The transduction chain from photoreception to the ultimate observed physiological response involves many steps, most of which are not light regulated. Until recently, photomorphogenesis was studied mainly by applying light treatments and observing their ultimate effects. With pigments, such as phytochrome, where photoreversible inductive effects require short periods of irradiation, this provided an elegant experimental approach. Complications arise, however, because of the coaction of several photoreceptive systems regulating the same processes or because of multiple effects induced by one photoreceptor. For the best characterized photoreceptor, phytochrome, the presence of two phytochrome pools (labile P [*l*P] and stable P [*s*P]) in higher plants provides an additional complication (Nagatani et al., 1989). The availability of genotypes (often as induced mutants) in which certain parts of the morphogenic pathway are eliminated, therefore, provides a useful additional tool for the study of photomorphogenesis (Koornneef and Kendrick, 1986).

4-6.1   Isolation and Characterization of Phytochrome Mutants

The isolation of mutants that exhibit specific phenotypic aberrations offers a potentially powerful means of elucidating the steps of sensory transduction chains. Mutations used to decipher these transduction sequences can be divided into two major categories, receptor and response mutations. A receptor mutation affects the abundance or functional integrity of the stimulus receptor and is, thus, the most useful class of mutation to examine the primary action mechanism. For receptor mutations, all phenotypic characters that are regulated by this receptor would be expected to be abnormal, whereas a response mutant would be expected to exhibit a more limited spectrum of aberrations. Koornneef and co-workers have used this rationale to isolate potential phytochrome receptor mutants in tomato (Parks et al., 1987).

In tomato, a number of mutants with an elongated hypocotyl have been described. With the exception of *procera (pro)*, they all have yellow-green leaves. Because the latter characteristic was the most conspicuous, these mutants were named *aurea (au)* and *yellow-green (yg)* (Koornneef et al., 1985). Tomato (*L. esculentum* Mill. cv. Moneymaker) mutants at the *au*, *yg-2* and *yg-6* loci phenotypically resemble each other in having reduced seed germination, reduced chlorophyll and anthocyanin contents, and an elongated hypocotyl in white light. Where such a complex phenotype results from single gene mutations, the genes involved more probably govern a common inducer of all these pathways. All properties peculiar to the mutants studied

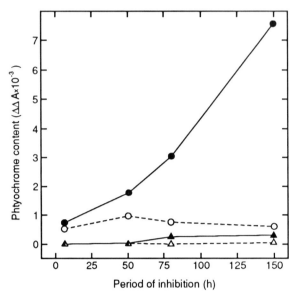

Figure 4-27.   Total phytochrome (●,▲) and far-red absorbing form of phytochrome (○,△) during imbibition of seeds and in young seedlings of wild-type (●,○) and seeds of the au^w mutant (▲,△) of tomato. (Koornneef et al., *J. Plant Physiol. 120:* 153-165, 1985; reproduced by permission of Gustav Fischer Verlag.)

have been shown either to be controlled by phytochrome or to be at least light regulated. This relationship, together with the observation that phytochrome levels in the mutants are greatly reduced (Fig. 4-27, for example), favors the hypothesis that the *au* and *yg* genes play a role in phytochrome synthesis or action (Koornneef et al., 1985). The type of spectroscopic measurements made in compiling data, such as those in Figure 4-27, however, could not preclude the possibility that the pigment merely possessed altered absorption characteristics in the mutant. The following measurements were, therefore, carried out to determine whether phytochrome was formed at all in the au^w/au^w (W616) mutant of tomato.

Four-day-old etiolated wild type tomato and mutant seedlings were used. The wild type displayed a normal P_r-P_{fr} spectrum with major peaks at 667 and 730 nm. The spectrum for the mutant, in contrast, revealed that these seedlings contained less than 5 percent (the detection limit) of the phytochrome level found for the wild type (Fig. 4-28; Parks et al., 1987).

A number of proteins from crude extracts of tomato were detected with heterologous polyclonal antibodies against phytochrome, in particular, a strong band with an apparent molecular mass of 116-kD (Fig. 4-29; Parks et al., 1987). This protein, although prominent in the wild-type tissue extract, was very faint in an extract of mutant tissue.

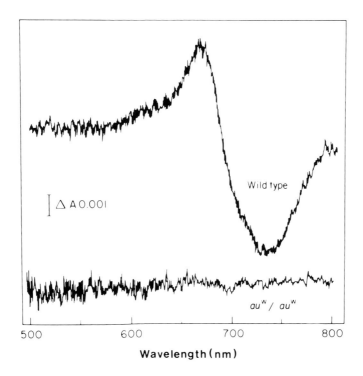

Figure 4-28. In vivo difference spectra of phytochrome from etiolated seedlings of mutant (*au^w^/au^w^*) and wild-type tomato. (Parks et al., *Plant Mol. Biol. 9*: 97-107, 1987; reprinted by permission of Kluwer Academic Publishers.)

Other protein bands, including the 116-kD band, were also recognized by one family of monoclonal antibodies directed against oat phytochrome (type 1B; Fig. 4-29, lane 10) but not by another type (type 1A; Fig. 4-29, lane 7), which served as a convenient nonimmune control. Additional diagnostic tests were performed to determine that, if any, of the immunodetectable bands represented tomato phytochrome. These confirmed that the 116-kD band was phytochrome (Parks et al., 1987).

There are three fundamental classes of aberrations that could explain why the *au* mutant is severely deficient in spectrally and immunochemically detectable phytochrome.

1. A mutation that results in a decreased production of active chromoprotein.

2. A mutation in the pigment which renders it unstable in vivo.

3. A mutation in the in vivo degradation pathway for phytochrome such that the molecule is now degraded rapidly in the P_r form without the normally necessary photoconversion to the P_{fr} conformation.

Genomic Southern blot analysis indicated that there were at least two and probably more phytochrome polypeptide structural genes in

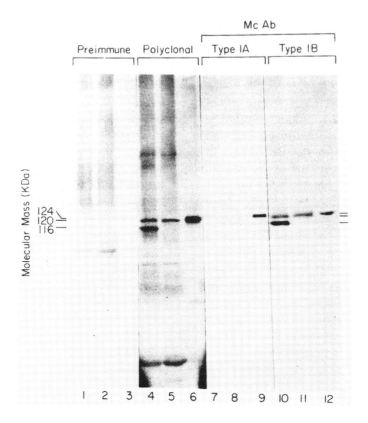

Figure 4-29. Immunochemical detection of phytochrome in crude extracts of etiolated tissue from wild-type and mutant seedlings of tomato on nitrocellulose. Lanes 1, 4, 7, and 10: cv. Moneymaker (parent wild type); lanes 2, 5, 8, and 11 - *aurea* locus mutant; lanes 3, 6, 9, and 12: 124-kD oat phytochrome (20 ng per lane). Three different primary antibodies were used, a rabbit polyclonal antiserum, and two murine monoclonal antibodies, Mc Ab Type 1A and 1B. (Parks et al., *Plant Mol. Biol. 9:* 97-107, 1987; reprinted by permission of Kluwer Academic Publishers.)

tomato.   RNA blot analysis showed that the *au* mutant contained normal levels of phytochrome mRNA and in vitro translation of *au* poly(A)+ RNA yielded a phytochrome apoprotein that was quantitatively and qualitatively indistinguishable on SDS gels from that synthesized from wild-type RNA.   These results indicated that the phytochrome deficiency in *au* was not the result of lack of expression of phytochrome genes but was more likely due to instability of the phytochrome polypeptide *in planta* (Sharrock et al., 1988).

Individual *au* mutant plants are capable of attaining maturity although they exhibit a partial etiolated appearance throughout their life cycle.   If phytochrome activity is necessary for plant survival, then possible explanations for the survival of mutant individuals might include:

1. That these mutants can survive with very low (<5 percent of weight) levels of etiolated-tissue type phytochrome,

2. That another type of phytochrome gene is active in mature tissue,

3. That a blue light photoreceptor can substitute for the lack of active phytochrome in these mutants.

In spite of the current uncertainties as to the precise identity of the lesions in question, the discovery of higher plant mutants, such as the *au* variants in tomato, that are severely deficient in phytochrome provides a valuable tool for investigating the mechanisms involved in light-mediated development at both the molecular and physiological levels (Parks et al., 1987).

4-6.2 Some Phytochrome Mutants and Their Physiological Characteristics

4-6.2.1 The Control of Hypocotyl Elongation in Arabidopsis

The inhibition by light of hypocotyl elongation in dark grown seedlings is the classic example of the high irradiance reaction (HIR). It is generally held that phytochrome is the pigment responsible for the HIR. An induced mutant of *Arabidopsis* with a hypocotyl (in white light) more than twice as long as that of the wild-type has been described, and given the gene symbol, *hy*.

Seed stocks used to generate the *hy* mutant were derived from the pure line "Landsberg-erecta" of *Arabidopsis thaliana* L. Heynh. For measurement of hypocotyl elongation, seeds were given a standard cold treatment followed by 15 minutes of red light (658 nm, 0.64 Wm⁻²) to induce germination. The seeds were kept in the dark for 24 hours and then placed in cabinets with monochromatic light. Hypocotyl lengths were measured on the seventh day after this treatment. Mutants were isolated after seeds were either treated with EMS or X-rays or fast neutrons. The M_2 seeds derived were scored for segregation of seedlings with elongated hypocotyls at day 8 after cold treatment. Phytochrome was determined spectrophotometrically in dark grown tissue of these mutants in which the inhibitory effect of far-red light was almost or completely absent. Forty-one independently induced mutants were found to be alleles at one of five loci, *hy-1* to *hy-5*. All mutants were monogenic recessives to varying degrees (Fig.4-30; Koornneef et al., 1980).

The quantity of spectrophotometrically detectable phytochrome varied greatly between various genotypes. The *hy-1* and *hy-2* mutants, for instance, had no pigment or only traces of it although concentrations below 5 to 10 percent would go undetected using the method of assay

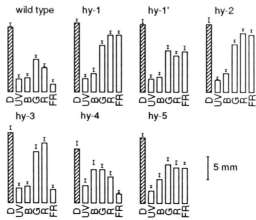

Figure 4-30. Hypocotyl lengths of wild-type *Arabidopsis* and a number of *hy* mutants in various spectral regimes. Shaded bars: dark controls. (Koornneef et al., *Z. Pflanzenphysiol. 100:* 147-160, 1980; reproduced by permission of Gustav Fischer Verlag.)

employed. Mutants at the other loci, *hy-3, hy-4*, and *hy-5*, showed about the same quantity as wild type (Table 4-27).

Therefore, it has been concluded that *hy-1* and *hy-2* genes regulate the synthesis of phytochrome, at least in the hypocotyl. Recent evidence suggests that these mutants do not represent mutations in a structural

Table 4-27. Phytochrome content of dark-grown, 1-week-old seedlings of a number of induced mutants of *Arabidopsis* and their length when grown under FR.

Locus	Phytochrome content xΔΔAx10³	Length in FR in mm	
		exp. on perlite	exp. on filter paper
+	-	2.8	1.8
hy-1	0.00	11.4	n.t.
hy-1	0.26	n.t.	8.8
hy-2	0.00	n.t.	12.6
hy-2	0.05	10.6	n.t.
hy-3	1.35	3.5	3.0
hy-3	1.21	2.7	n.t.
hy-4	0.97	3.1	1.9
hy-5	1.03	7.3	7.7

(Koornneef et al., *Z. Pflanzenphysiol. 100:* 147-160, 1980; reproduced by permission of Gustar Fischer Verlag.)

gene for phytochrome but may, rather, be dysfunctional in the biosynthesis and/or ligation of the chromophore to the protein moiety of the pigment (Parks et al., 1989). Because the activities of UV and B, in contrast to those of FR, are hardly diminished in the *hy-1* and *hy-2* mutants (Fig. 4-30, for example), it seems to follow that the short-wavelength part of the HIR in these mutants is mainly due to nonphytochrome pigments. In some of the other mutants, *hy-3* to *hy-5*, the amount of phytochrome is near normal (Table 4-27) and yet, especially in the case of *hy-5*, hypocotyl elongation is not very sensitive to far-red. Apparently, the reactivity to light in this genotype is blocked by some other factor than its capacity for phytochrome synthesis (Koornneef et al., 1980).

The general view that emerges is that the light-induced hypocotyl inhibition may depend on a complex system of photoreceptor pigments that are able to operate independently of one another. Plants of the genotypes described will develop into phenotypically almost, or completely, normal plants when cultivated under appropriate conditions in white light. Because it is likely that phytochrome has one or more functions during several stages of normal plant development, the mutants *hy-1* and *hy-2* probably should not be designated "phytochromeless."

4-6.2.2   *The Photoregulation of Gene Expression in Tomato*

Mutations at the *au* locus in tomato result in alteration of several phytochrome-controlled physiological properties such as seed germination, hypocotyl elongation, and chlorophyll and anthocyanin content. Here, the consequences of the *au* phytochrome deficiency for photoregulation of the chlorophyll a/b binding (*cab*) gene expression is given as an example (Sharrock et al., 1988).

Northern blots probed with a cDNA clone for corn *cab* showed a single band of hybridization at 1.2 kb in length (Fig. 4-31). The abundance of *cab* mRNA in wild-type tomato showed a very marked increase within 3 hours after a pulse of red light (Fig. 4-31c, lanes 1 and 2). A pulse of far-red light given immediately after the red pulse reversed this induction, resulting in a *cab* transcript level equivalent to that produced by irradiation with far-red light alone (Fig. 4-31c, lanes 3 and 4). The rapidity of this induction is a phytochrome-regulated response in tomato. In contrast to the wild-type tomato strain, the *au* mutant showed only a slight increase in *cab* mRNA abundance after a pulse of red light (Fig. 4-31c, lanes 5 and 6). Thus, the *au* mutant was almost completely lacking rapid phytochrome induction of *cab* mRNA, although the small increase that was observed was reversed by far-red light (Fig. 4-31c, lane 7).

The very low level of phytochrome-mediated induction of *cab* mRNA may at least in part explain the reduced chlorophyll content, reduced

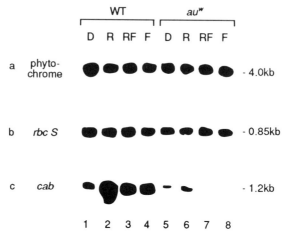

Figure 4-31. Northern blot analysis of poly(A)⁺RNA from wild-type (WT) tomato and the *au* mutant. RNA was isolated from etiolated (D), red light-irradiated (R), red then far-red irradiated (RF), and far-red light-irradiated (F) tissue of wild-type and *au*. Blots were probed for (a) phytochrome mRNA; (b) rbcS mRNA; (c) cab mRNA. (see text for discussion of lanes.) (Sharrock et al., *Mol. Gen. Genet. 213:* 9-14, 1988.)

thylakoid stacking, and yellow-green appearance of the *au* mutant (Sharrock et al., 1988), a conclusion confirmed by others, more recently (Ken-Dror and Horwitz, 1990).

4-6.2.3   Some Physiological Consequences of the Absence of Labile Phytochrome

Evidence has been gathered that there are two types of phytochrome, labile (*l*P) and stable (*s*P) (discussed earlier in this Case Study). It has been suggested, but not demonstrated unequivocally, that the small amount of phytochrome in the *au* mutant of tomato (<5 percent of that in the wild type) represents *s*P only (Adamse et al., 1988). Taking advantage of this hypothesis, the possible roles of the different types of phytochrome in elongation growth have been investigated recently.

On the one hand, when germinated in dark conditions and then transferred to continuous light of a broad spectrum, both the wild type and the *au* mutant showed similar hypocotyl growth in dark and far red. In red light and green light, in contrast to the wild type, no significant inhibition was observed for seedlings of the *au* mutant. In blue light and UV-A both the *au* mutant and the wild type were significantly inhibited compared to the dark controls, but a smaller response was exhibited in the case of the *au* mutant (Fig. 4-32; Adamse et al., 1988).

On the other hand, a pronounced end-of-day far-red response was observed in both wild type and the *au* mutant (Fig. 4-33). Elongation growth of many species (including tomato) increases as a result of a short irradiation with far-red at the end of the photoperiod in a white light/dark cycle. This effect (called the end-of-day response) is reversed by a subsequent white light pulse indicating that phytochrome functions as the photoreceptor (Adamse et al., 1988). In addition, when one cotyledon of deetiolated seedlings was covered with aluminum foil and the seedlings irradiated with white light from above both the wild type and the *au* mutant curved towards the uncovered cotyledon (37±5 and 46±7 degrees, respectively). This response of deetiolated seedlings due to red light absorption, presumably by phytochrome in the cotyledon, was exhibited by both the *au* mutant and wild type.

If the low level of phytochrome detected in light-grown plants of the *au* mutant is proposed to be *s*P, the difference between the responses of the *au* mutant and the wild type must, therefore, reflect the quantitative role of the phytochrome deficient in the *au* mutant. The deficiency of *l*P resulted in a loss of phytochrome control of hypocotyl elongation in etiolated seedlings of the *au* mutant (Fig. 4-32). In blue light and UV-A, however, both wild type and *au* mutant were inhibited, but the *au* mutant significantly less than the wild type. Apparently inhibition of

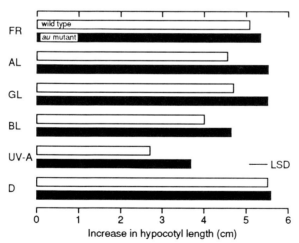

Figure 4-32. The increase in hypocotyl length of the *au* mutant of tomato and its isogenic wild type grown for 4 days in broad band light (2.1 mol/m²/ s ± 5 percent) at 25°C; far-red light (FR), red light (RL), green light (GL), blue light (BL), UV-A or darkness (D). Seedlings were grown for 3 days (wild type) or 4 days (*au* mutant) in D before transfer to the light. The bar indicates the least significant difference (LSD) at the 5 percent level between the dark control and the light treatments. (Adamse et al., *J. Plant Physiol. 133:* 436-440, 1988; reproduced by permission of Gustav Fischer Verlag.)

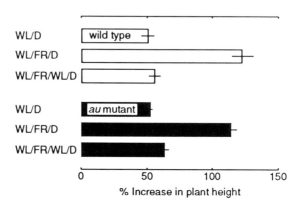

Figure 4-33. End-of-day far-red light (FR) response of stem elongation in *au* mutant of tomato and its isogenic wild type at 25°C. Plants grown for 28 days in daily white light (WL; 35 W/m² [PAR 160 mol/m²/s]; 14 hours)/dark (D; 10 hours) cycles were treated for 5 days with 20 minutes FR (20 mol/m²/s); 20 minutes FR immediately followed by 10 minutes WL; or D at the end of the daily light period. Results are presented as percentage increase in height of the plants during the end-of-day FR treatment. The actual increases in height were: wild type: D, 43 mm; FR, 76.3 mm; FR/WL, 45.2 mm; *au* mutant: D, 21.5 mm; FR, 46.5 mm; FR/WL, 22.7 mm. (Adamse et al., *J. Plant Physiol. 133*: 436-440, 1988; reproduced by permission of Gustav Fischer Verlag.)

hypocotyl elongation due to a BL/UV-A photoreceptor is retained in the *au* mutant, despite the lack of *l*P. This suggests either that inhibition of elongation growth by blue light does not require the presence of P_{fr} or it is extremely sensitive to the low level of P_{fr} present. The difference between the *au* mutant and its wild type in inhibition by blue light and UV-A represents a quantitative estimate of the role played by blue light and UV-A absorption by *l*P, deficient in the *au* mutant, in the photoinhibition of hypocotyl growth.

Both the wild type and *au* mutant retained pronounced end-of-day far-red responses (Fig. 4-33) and phototropic curvatures in white light . One hypothesis that could be derived from these observations is that at least these phytochrome-mediated responses are regulated by *s*P (Adamse et al., 1988).

Thus, by using variants such as the *au* mutant of tomato, it appears possible to dissect partially the multitude of physiological roles played by the different phytochrome types in the morphogenesis of higher plants. A similar approach taken recently with a new complementation group of *Arabidopsis thaliana* long hypocotyl mutant (*hy-6*) has led to the conclusion that phytochrome plays different roles in etiolated and green plants (Chory et al., 1989).

References

Adamse, P., Jaspers, P.A.P.M., Bakker, J.A., Wesselius, J.C., Heeringa, G.H., Kendrick, R.E., and Koornneef, M. 1988. Photophysiology of a tomato mutant deficient in labile phytochrome. *J. Plant Physiol. 133*: 436-440.

Behringer, F.J., Davies, P.J., and Reid, J.B. 1990a. Genetic analysis of the role of gibberellin in the red light inhibition of stem elongation in etiolated seedlings. *Plant Physiol. 94*: 432-439.

Behringer, F.J., Cosgrove, D.J., Reid, J.B., and Davies, P.J. 1990b. Physical basis for altered stem elongation rates in internode length mutants of *Pisum. Plant Physiol. 94*: 166-173.

Bell, C.J. and Maher, E.P. 1990. Mutants of *Arabidopsis thaliana* with abnormal gravitropic responses. *Mol. Gen. Genet. 220*: 289-293.

Bleecker, A.B., Estelle, M.A., Somerville, C.R., and Kende, H. 1988. Insensitivity to ethylene conferred by a dominant mutation in *Arabidopsis thaliana. Science 241*: 1086-1089.

Blonstein, A.D., Vahala, T., Koornneef, M., and King, P.J. 1988. Plants regenerated from auxin-auxotrophic variants are inviable. *Mol. Gen. Genet. 215*: 58-64.

Caboche, M., Muller, J-F., Chanut, F., Aranda, G., and Cirakoglu, S. 1987. Comparison of the growth promoting activities, and toxicities of various auxin analogs on cells derived from wild type, and a nonrooting mutant of tobacco. *Plant Physiol. 83*: 795-800.

Campell, B.R. and Bonner, B.A. 1986. Evidence for phytochrome regulation of gibberellin A$_{20}$ 3--hydroxylation in shoots of dwarf (*lele*) *Pisum sativum* L. *Plant Physiol. 82*: 909-915.

Chandler, P.M. 1988. Hormonal regulation of gene expression in the "slender" mutant of barley (*Hordeum vulgare* L.). *Planta 175*: 115-120.

Chory, J., Voytan, D.F., Olszewski, N.E., and Ausubel, F.M. 1987. Gibberellin-induced changes in the populations of translatable mRNAs, and accumulated polypeptides in dwarfs of maize, and pea. *Plant Physiol. 83*: 15-23.

Chory, J., Peto, C.A., Ashbaugh, M., Saganich, R., Pratt, L., and Ausubel, F. 1989. Different roles for phytochrome in etiolated, and green plants deduced from characterization of *Arabidopsis thaliana* mutants. *The Plant Cell 1*: 867-880.

Croker, S.J., Hedden, P., Lenton, J.R., and Stoddart, J.L. 1990. Comparison of gibberellins in normal, and slender barley seedlings. *Plant Physiol. 94*: 194-200.

Daniel, S.G., Rayle, D.L., and Cleland, R.E. 1989. Auxin physiology of the tomato mutant *diageotropica. Plant Physiol. 91*: 804-807.

Eason, W.R., Hall, G., and Wang, T.L. 1987. In vitro analysis of gravitropic mutants of pea. *J. Plant Physiol. 131*: 201-213.

Ephritikhine, G., Barbier-Brygoo, H., Muller, J-F., and Guern, J. 1987. Auxin effect on the transmembrane potential difference of wild-type, and mutant tobacco protoplasts exhibiting a differential sensitivity to auxin. *Plant Physiol. 83*: 801-804.

Estelle, M.A. and Somerville, C.R. 1987. Auxin-resistant mutants of *Arabidopsis thaliana* with an altered morphology. *Mol. Gen. Genet.* 206: 200-206.

Feldman, L.J. 1985. Root gravitropism. *Physiol. Plant.* 65: 341-344.

Fick, G.N. and Qualset, C.O. 1975. Genetic control of endosperm amylase activity, and gibberellic acid responses in standard-height, and short-statured wheats. *Proc. Natl. Acad. Sci. USA* 72: 892-895.

Fujino, D.W., Burger, D.W., Yang, S.F., and Bradford, K.J. 1988. Characterization of an ethylene overproducing mutant of tomato (*Lysopersicon esculentum* Mill. cultivar VFN8). *Plant Physiol.* 88: 774-779.

Fujioka, S., Yamane, H., Spray, C.R., Phinney, B.O., Gaskin, P., MacMillan, J., and Takahashi, N. 1990. Gibberellin A_3 is biosynthesized from Gibberellin A_{20} via gibberellin A_5 in shoots of *Zea mays* L. *Plant Physiol.* 94: 127-131.

Guern, J. 1987. Regulation from within: The hormone dilemma. *Ann. Bot.* 60: 75-102.

Guzman, P. and Ecker, J.R. 1990. Exploiting the triple response of *Arabidopsis* to identify ethylene-related mutants. *Plant Cell 2*: 513-523.

Hedden, P. and Phinney, B.O. 1979. Comparison of *ent*-kaurene, and *ent*-isokaurene synthesis in cell-free systems from etiolated shoots of normal and *dwarf-5* maize seedlings. *Phytochem.* 18: 1475-1479.

Hertel, R., De La Fuente, R.K., and Leopold, A.C. 1969. Geotropism, and the lateral transport of auxin in the corn mutant amylomaize. *Planta* 88: 204-214.

Hicks, G.R., Rayle, D.L., and Lomax, T.L. 1989. The *diageotropica* mutant of tomato lacks high specific activity auxin binding sites. *Science 245*: 52-54.

Ho, T-H. D., Nolan, R.C., and Shute, D.E. 1981. Characterization of a gibberellin-insensitive dwarf wheat, D6899. *Plant Physiol.* 67: 1026-1031.

Katsumi, M., Foard, D.E., and Phinney, B.O. 1983. Evidence for the translocation of gibberellin A_3, and gibberellin-like substances in grafts between normal, *dwarf-1*, and *dwarf-5* seedlings of *Zea mays* L. *Plant Cell Physiol.* 24: 379-388.

Ken-Dror, S. and Horwitz, B.A. 1990. Altered phytochrome regulation of greening in an *aurea* mutant of tomato. *Plant Physiol.* 92: 1004-1008.

Keyes, G.J., Paolillo, D.J., and Sorrells, M.E. 1989. The effects of dwarfing genes *Rht1*, and *Rht2* on cellular dimensions, and rate of leaf elongation in wheat. *Ann. Bot.* 64: 683-690.

Khurana, J.P. and Poff, K.L. 1989. Mutants of *Arabidopsis thaliana* with altered phototropism. *Planta 178*: 400-406.

Khurana, J.P., Ren, Z., Steinitz, B., Parks, B., Best, T.R., and Poff, K.L. 1989. Mutants of *Arabidopsis thaliana* with decreased amplitude in their phototropic response. *Plant Physiol.* 91: 685-689.

Koornneef, M., Rolff, E., and Spruit, C.J.P. 1980. Genetic control of light-inhibited hypocotyl elongation in *Arabidopsis thaliana* (L.) Heynh. *Z. Pflanzenphysiol.* 100: 147-160.

Koornneef, M., Cone, J.W., Dekens, R.G., O'Herne-Robers, E.G., Spruit, C.J.P., and Kendrick, R.E. 1985. Photomorphogenic responses of long hypocotyl mutants of tomato. *J. Plant Physiol. 120*: 153-165.

Koornneef, M. and Kendrick, R.E. 1986. A genetic approach to photomorphogenesis. In *Photomorphogenesis in Plants*, R.E. Kendrick, and G.H.M. Kronenberg (eds.), Martinus Nijhoff, Dortrecht, pp. 521-546.

Lanahan, M.B. and Ho, D.T-H. 1988. Slender barley: A constitutive gibberellin-response mutant. *Planta 175*: 107-114.

Looney, N.E., Taylor, J.S., and Pharis, R.P. 1988. Relationship of endogenous gibberellin, and cytokinin levels in shoot tips to apical form in four strains of 'McIntosh' apple. *J. Amer. Hort. Sci. 113*: 395-398.

Mirza, J.I. 1987. The effects of light, and gravity on the horizontal curvature of roots of gravitropic, and agravitropic *Arabidopsis thaliana* L. *Plant Physiol. 83*: 118-120.

Mirza, J.I., Olsen, G.M., Iversen, T-H., and Maher, E.P. 1984. The growth, and gravitropic responses of wild type, and auxin-resistant mutants of *Arabidopsis thaliana. Physiol. Plant. 60*: 516-522.

Moore, R. 1985. Morphometric analysis of the redistribution of organelles in columella cells in primary roots of normal seedlings, and agravitropic mutants of *Hordeum vulgare. J. Expt. Bot. 36:* 1275-1286.

Moore, R. 1989. Root graviresponsiveness, and cellular differentiation in wild type, and a starchless mutant of *Arabidopsis thaliana. Ann. Bot. 64*: 271-277.

Muller, J-F., Goujaud, J., and Caboche, M. 1985. Isolation in vitro of naphaleneacetic acid-tolerant mutants of Nicotiana tabacum, which are impaired in root morphogenesis. *Mol. Gen. Genet. 199*: 194-200.

Nagatani, A., Kendrick, R.E., Koornneef, M., and Furuya, M. 1989. Partial characterization of phytochrome I, and II in etiolated, and de-etiolated tissues of a photomorphogenetic mutant (*lh*) of cucumber (*Cucumis sativus* L.), and its isogenic wild type. *Plant Cell Physiol. 30*: 685-690.

Olsen, G.M., Mirza, J.I., Maher, E.P., and Iversen, T-H. 1984. Ultrastructure, and movements of cell organelles in the root cap of agravitropic mutants, and normal seedlings of *Arabidopsis thaliana. Physiol. Plant. 60*: 523-531.

Parks, B.M., Jones, A.M., Adamse, P., Koornneef, M, Kendrick, R.E., and Quail, P.H. 1987. The *aurea* mutant of tomato is deficient in spectrophotometrically and immunochemically detectable phytochrome. *Plant Mol. Biol. 9*: 97-107.

Parks, B.M., Shanklin, J., Koornneef, M, Kendrick, R.E., and Quail, P.H. 1989. Immunochemically detectable phytochrome is present at normal levels but is photochemically nonfunctional in the *hy-1*, and *hy-2* long hypocotyl mutants of *Arabidopsis. Plant Mol. Biol. 12*: 425-437.

Phinney, B.O. 1984. Gibberellin A_1, dwarfism, and the control of shoot elongation in higher plants. In *The Biosynthesis, and Metabolism of Plant Hormones*, A. Crozier, and J.R. Hillman (eds.), Society for Experimental Biology, Seminar Series 23, Cambridge University Press, London, pp. 17-41.

Pinthus, M.J., Gale, M.D., Appleford, N.E.J., and Lenton, J.R. 1989. Effect of temperature on gibberellin (GA) responsiveness, and on endogenous GA$_1$ content of tall, and dwarf wheat genotypes. *Plant Physiol. 90*: 854-859.

Radley, M. 1970. Comparison of endogenous gibberellins, and response to applied gibberellin of some dwarf, and tall wheat cultivars. *Planta 92*: 292-300.

Reid, J.B. 1986. Gibberellin mutants. In *A Genetic Approach to Plant Biochemistry*, A.D. Blonstein, and P.J. King (eds.), Springer-Verlag, New York, pp. 1-34.

Reid, J.B. 1990. Phytohormone mutants in plant research. *J. Plant Growth Regul. 9*: 97-111.

Reid, J.B. and Ross, J.J. 1988. Internode length in *Pisum*. A new gene, *lv*, conferring an enhanced response to gibberellin A$_1$. *Physiol. Plant. 72*: 595-604.

Roberts, J.A. 1984. Tropic responses of hypocotyls from normal tomato plants, and the gravitropic mutant *Lazy-1*. *Plant, Cell Environ. 7*: 515-520.

Roberts, J.A. 1987. Mutants and gravitropism. In *Developmental Mutants in Higher Plants*, H. Thomas, and D. Grierson (eds.), Cambridge University Press, London, pp. 135-153.

Salisbury, F.B. and Ross, C.W. 1985. *Plant Physiology*, Wadsworth, Belmont, CA.

Scott, I.M. 1990. Plant hormone response mutants. *Physiol. Plant. 78*: 147-152.

Sharrock, R.A., Parks, B.M., Koornneef, M., and Quail, P.H. 1988. Molecular analysis of the phytochrome deficiency in an *aurea* mutant of tomato. *Mol. Gen. Genet. 213*: 9-14.

Singh, S.P. and Paleg, L.G. 1984. Low temperature-induced GA$_3$ sensitivity of wheat. 1. Characterization of the low temperature effect on isolated aleurone of kite. *Plant Physiol. 76*: 139-142.

Singh, S.P. and Paleg, L.G. 1985a. Low-temperature-induced GA$_3$ sensitivity of wheat. III. Comparison of low temperature effects on alpha-amylase induction by aleurone tissue of dwarf, and tall wheat. *Aust. J. Plant Physiol. 12*: 269-275.

Singh, S.P. and Paleg, L.G. 1985b. Low-temperature-induced GA$_3$ sensitivity of wheat. IV. Comparison of low temperature effects on the phospholipids of aleurone tissue of dwarf, and tall wheat. *Aust. J. Plant Physiol. 12*: 277-289.

Singh, S.P. and Paleg. L.G. 1986. Comparison of IAA-induced, and low temperature-induced GA$_3$ responsiveness, and alpha-amylase production by GA$_3$ insensitive dwarf wheat aleurone. *Plant Physiol. 82*: 685-687.

Smith, W.O. 1983. Phytochrome as a molecule. In *Photomorphogenesis, Encyclopedia of Plant Physiology*, New Series, vol. 16A, W. Shropshire Jr., and H. Mohr (eds.), Springer-Verlag, New York, pp. 96-118.

Spray, C., Phinney, B.O., Gaskin, P., Gilmour, S.J., and MacMillan, J. 1984. Internode length in *Zea mays* L. The *dwarf-1* mutation controls the 3--hydroxylation of gibberellin A$_{20}$ to gibberellin A$_1$. *Planta 160*: 464-468.

Stoddart, J.L. 1987. Genetic, and hormonal regulation of stature. In *Developmental Mutants in Higher Plants*, H. Thomas, and D. Grierson (eds.), Cambridge University Press, Cambridge, pp. 155-180.

Thomas, H. and Grierson, D. 1987. *Developmental Mutants in Higher Plants*. Cambridge University Press, Cambridge.

Ursin, V.M. and Bradford, K.J. 1989. Auxin, and ethylene regulation of petiole epinasty in two developmental mutants of tomato, *diageotropica*, and *epinastic*. *Plant Physiol. 90*: 1341-1346.

Vanderhoef, L.N. and Kosuge, T. 1984. *The Molecular Biology of Plant Hormone Action: Research Directions for the Future.* Workshop Summaries II, American Society of Plant Physiologists, Rockville, MD., 39 pp.

Wareing, P.F. and Phillips, I.D.J. 1981. *Growth and Differentiation in Plants*. Pergamon Press, New York.

Weyers, J.D.B. 1985. Germination, and root gravitropism of *flacca*, the tomato mutant deficient in abscisic acid. *J. Plant Physiol. 121*: 475-480.

Wilson, A.K., Pickett, F.B., Turner, J.C., and Estelle, M. 1990. A dominant mutation in *Arabidopsis* confers resistance to auxin, ethylene and abscisic acid. *Mol. Gen. Genet. 222*: 377-383.

Zobel, R.W. 1973. Some physiological characteristics of the ethylene requiring tomato mutant *diageotropica*. *Plant Physiol. 52*: 385-389.

5 Mutants Affecting Flower and Fruit Development

In the "typical" life history of an herbaceous flowering plant there is usually an initial phase of vegetative growth, which sooner or later is followed by the reproductive phase, that is, flowering and seed production. In Chapter 5 there are four case studies dealing with the first of these processes, flowering. Chapter 6 is devoted to seed physiology.

The study of the induction and regulation of flowering is one of the great themes of plant developmental biology. One particular motif in this research area has been the study of the control of sex in the flower. This is a complex subject in higher plants because of the many reproductive strategies that exist: complete hermaphroditism with both sexes in one flower; sexes on separate plants; mixed male and female flowers on the same plant; and virtually all possible combinations of these. A variety of genetic and hormonal mechanisms have been discovered that govern these patterns of floral differentiation (Malmberg et al., 1985).

Male sterility is a deviant floral condition where no viable pollen is produced and the sterility is expressed in a variety of ways. In some cases, the male floral organs are completely absent, whereas in others, these reproductive structures are formed but they are either aborted or produce infertile pollen grains. The importance of male sterility in the production of hybrid seed is widely known. In tomato, several male sterile mutants exist that exhibit various degrees of abnormality in stamen development. Of interest is the single gene recessive *stamenless-2* (*sl-2/sl-2*) mutants, as stamen development in this mutant is hormone and temperature sensitive. The mutant produces twisted stamens that are laterally free, possess naked external ovules on the adaxial surface and contain shrunken and nonviable pollen. If mutant plants are treated with gibberellic acid (GA_3) or are grown under relatively low temperatures, they produce phenocopies of the normal and the stamens contain viable pollen. Alternatively, indoleacetic acid (IAA) treatment or high temperatures induce the formation of carpelloid stamens in which microsporogenesis does not occur. Through the use of a genetic system it should be possible to investigate the role of nutritional, hormonal, and environmental factors in a defined component of flower development (Rastogi and Sawhney, 1988a). Case Study 5-1 deals with some of the uses of this mutant in physiological investigations.

In many species, representing very diverse angiosperm families, self-fertilization is blocked by a mechanism termed self incompatibility. After self-pollination, interaction between the stigma or style and the pollen elicits a defined morphological response preventing normal pollen tube growth; fertilization is prevented because these pollen tubes are unable to penetrate the full length of the stigma and the style. The specificity of the incompatible interaction is determined by one or more genetic loci and requires that the alleles carried by the male and female parents are identical. In the crucifer *Brassica*, the self-incompatible reaction is localized at the stigma surface, and occurs within minutes after the initial contact between the pollen and the papillar cells on the outer surface of the stigma (Nasrallah et al., 1985b). In fact, in many plant species the control of self-incompatibility is attributed to a single locus known as the S locus. Most species have multiple alleles at this locus, designated as $S_1, S_2, \ldots, S_n$. The presence of identical alleles in pollen and pistil results in incompatibility. Thus, S_1 pollen fails on S_1 plants but presumably functions normally on $S_2 \ldots S_n$. In the absence of dominance, usually absent in species with gametophytic control, an S_1S_2 plant cannot be fertilized by either S_1 or S_2 pollen. Self-incompatibility is, in essence, a mating control device that physiologically restricts inbreeding in both higher plants and fungi. The effectiveness of self incompatibility systems in maintaining outbreeding is considered to be a major factor in the evolutionary success of flowering plants. The phenomenon has been investigated through the use of serological and recombinant DNA techniques on the S alleles and their hybrids (Nasrallah and Wallace, 1967), some of which are described in Case Study 5-2.

Polyamines have been implicated in several fundamental processes in plants, including embryogenesis, reproductive development, and virus multiplication (Galston and Kaur-Sawhney, 1988; Galston and Sawhney, 1990). Precise knowledge of their activities in vivo, however, is lacking (Balint et al., 1987). Polyamines arise from three amino acids; the polybasic acids arginine and ornithine furnish the main part of the carbon skeleton of the polyamines, and methionine contributes propylamino groups of the simple diamine, putrescine, to form, successively, the triamine, spermidine, and the tetramine, spermine (Fig. 5-1). Each of the amines formed can be catabolized by a diamine or polyamine oxidase. Putrescine is oxidized by diamine oxidase to form a monoaldehyde that then cyclizes spontaneously to form pyrroline (Galston, 1983). An analogous reaction catalyzed by one polyamine oxidase converts spermidine and spermine to 1,3-diaminopropane plus a pyrroline. Another polyamine oxidase produces putrescine from spermidine. Both the hydrogen peroxide and the aldehydes formed

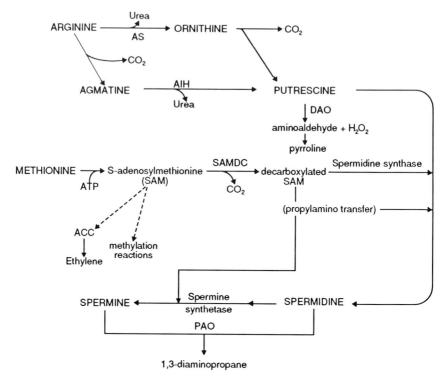

Figure 5-1. The biosynthesis and catabolism of some plant polyamines. AS = arginase; OrnDC = ornithine decarboxylase; ArgDC = arginine decarboxylase; SamDC = S-adenosylmethionine decarboxylase; AIH = agmatine iminohydrolase; DAO = diamine oxidase; PAO = polyamine oxidase.

from the oxidation of polyamines are cytotoxic. Methionine may contribute to the biosynthesis of polyamines through S-adenosyl-methionine (SAM). SAM is also the precursor of the plant hormone ethylene. Because ethylene tends to be a senescence inducer and polyamines, senescence inhibitors, the fate of SAM would seem to be crucial. Feedback controls exist that cause polyamines to inhibit ethylene formation and ethylene to inhibit polyamine formation (Galston, 1983). Thus, each pathway, once initiated tends to shut off the other (Fig. 5-1).

A probable regulatory role for polyamines in plants is indicated by their ubiquity, their abundance in meristematic and growing tissue, the regulation of their synthesis by physical and chemical factors that affect growth and development, and their dramatic effects on growth, senescence, and membrane stabilization when applied exogenously to selected plant systems. Different polyamines are related to different processes. Thus, stress of various kinds induces accumulation of putrescine. Elongation growth is also correlated with high putrescine, and cells in division are rich in spermidine and spermine. Although evidence is strong for modulation of various plant processes by polyamines, there is no good evidence for their translocatability, and, hence, they cannot be classed as hormones. They may possibly function as second messengers (Galston, 1983).

In animal cells, polyamine biosynthesis is controlled primarily by two enzymes. S-adenosylmethionine decarboxylase is necessary for the production of spermidine and spermine; ornithine decarboxylase is responsible for the production of putrescine. In higher plants there is an additional biosynthetic enzyme, arginine decarboxylase, that in parallel with ornithine decarboxylase, results in the production of putrescine (Fig. 5-1). A variety of physiological stimuli and stress reactions affect the activity of the decarboxylases in higher plants. Little is known, however, about the regulation of biosynthesis of these enzymes in plants, or how polyamine biosynthesis influences the development of the whole plant (Galston, 1983).

Unlike cultured animal cells, many plant cell suspensions are capable of regeneration into whole plants if they are exposed to the appropriate hormonal and nutritional conditions. For these reasons, plants offer an opportunity to study the effects on development of alterations, including genetic, in the polyamine pathway (Hiatt et al., 1986). Case Study 5-3 deals with a series of tobacco mutants altered in the development of flowers and selected for lesions in the polyamine biosynthetic pathway.

The main body of a fleshy fruit can be derived from a number of different parts of a flower. At the beginning of fruit development specific groups of cells derived from the mesocarp, pericarp, or receptacle undergo cell division and expansion. During the growth phase, when the fruit protects or supports the developing seeds, the cells are generally hard and unpalatable. At maturity, the fruit cells undergo a series of changes in ultrastructure, chemical composition, and physiological activity that makes them attractive to animals and birds and they are eaten. The seeds survive the digestive process and are dispersed some distance from the parent plant. The changes in color, texture, flavor, and aroma that serve to attract potential consumers constitute ripening as, for example, in the case of tomato (Table 5-1; Grierson et al., 1987).

Ripening is not a uniform process. Different fruits are derived from different cell types. There is a range of patterns of cell division and expansion, the cells accumulate different types of storage materials, and the biochemical conversions that occur during ripening also vary. In fruits, like tomatoes, dramatic changes in color are due to the production of β-carotene and lycopene in plastids. The variation in ripening patterns means that it is impossible to give a definition of the process in terms of biochemical pathways. The most appropriate definition is to consider ripening as a series of processes rendering the cells attractive to eat. In any specific fruit the changes may be defined precisely (e,g., Table 5-1). They invariably involve all the compartments of a cell and are highly coordinated. Although it is quite clear that the

Table 5-1. Ripening changes in tomato.

Metabolism	50-100 fold increase in synthesis and evolution of ethylene Increase in respiration Solubilization of starch Changes in metabolism of organic acids
Color	Loss of chlorophyll, dismantling of the photosynthetic apparatus and accumulation of carotenoids such as β-carotene and lycopene
Texture	Partial solubilizaton and modification of pectins and cellulose in cell walls
Flavour and aroma	Accumulation of sugars and changes in organic acid metabolism Destruction of alkaloids such as a-tomatine Production of a complex mixture of volatiles
Gene expression	Dissappearance of some mRNAs and proteins synthesized before ripening, such as those for chloroplast photosynthetic enzymes Appearance of new ripening-specific mRNAs directing the synthesis of enzymes catalysing specific ripening changes

(Grierson et al., *in Developmental Mutants in Higher Plants*, H. Thomas D. Grierson (eds.), Cambridge University Press, London, pp. 73-94, 1987.)

cells are alive and metabolically active when they are consumed, ripening may be considered as a specialized case of senescence. The developmental program that leads to ripening is only just beginning to be understood at the molecular level. Throughout the ripening process protein synthesis and turnover continue.   Many proteins disappear and new, but unidentified, ones accumulate. Thus, the process of fruit ripening also lends itself to genetic analysis through the generation of mutants with aberrations in this program of controlled senescence (Grierson et al., 1987). Among these are the *ripening inhibitor (rin)* mutants of tomato, the subject of Case Study 5-4.

Case Study 5-1:   Male Sterility in a Tomato Mutant

One promising tool in documenting the relationship between gene and form is the use of single-gene mutants that affect the development of form or a step in the development of form. One such area concerns the development and differentiation of floral organs.

Tomato (*Lycopersicon esculentum* Mill.) provides a large number of single-gene mutants affecting one, or more than one, organ in the flower. In addition, it can be cultured easily and, if necessary, also can be propagated vegetatively through cuttings. The mutant *stamenless-2* (sl_2/sl_2) in tomato has the unique feature of bearing naked external ovules (EO) on the stamens. Inheritance studies have shown that the sl_2 allele behaves as a single, recessive gene and the morphological abnormality is expressed only in the homozygous (sl_2/sl_2) condition. The phenotype of the heterozygote ($+/sl_2$) is identical to the homozygous dominant ($+/+$) form.

The *sl2* mutant differs phenotypically from the normal only in stamen structure. Stamens of sl_2/sl_2 flowers are shorter, possess fewer sporangia per anther, produce shriveled, nonviable pollen, are paler in color, and above all, bear naked ovules on the adaxial surface. Because the morphological characters produced by the sl_2 mutant have the unique feature of influencing both male and female characters on the same organ, the mutant represents an excellent subject for experimentation into the relationship between gene and character in the flower (Sawhney and Greyson, 1973).

5-1.1   The Hormonal and Temperature Control of Male Sterility in the *stamenless-2* Tomato Mutant

Male sterility in plants is extremely valuable for the production of hybrid seeds and for investigating the gene controlled mechanisms in stamen development. Generally, cytoplasmic male sterility (CMS) is used in producing hybrids, as CMS can be readily maintained in the F_1 generation. Alternatively, nuclear or genic male sterile lines can be useful for hybrid production, but maintenance of such lines is tedious. If, however, male-sterile mutant plants can be reverted by chemical or environmental treatments to produce normal stamens, such pollen can be used to pollinate mutant flowers and the seeds thus produced will generate all mutant plants. This approach can, therefore, be useful in maintaining pure male-sterile lines (Fig. 5-2; Sawhney, 1984).

Gibberellic acid at a concentration of $10^{-3}\mu M$ induced the formation of normal stamens in flowers that were initiated after the treatment.

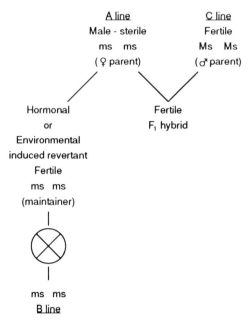

Figure 5-2. A proposal for the maintenance of a genic male-sterile line by the experimental induction of revertants and the production of mutant and hybrid lines. (Sawhney, in *Proceedings 8th International Symposium on Sexual Reproduction in Seed Plants, Ferns and Mosses,* Pudoc Publications, Wageningen, pp. 36-38, 1984. Adapted from R. Frankel and E. Galun, *in Pollination Mechanisms, Reproduction and Plant Breeding,* Springer-Verlag, New York, 1977.)

The stamens were yellow and pubescent, produced normal pollen, formed a staminal cone, and were as long as the normal stamens (Table 5-2). The pollen produced were viable and when used to pollinate the sl_2/sl_2 flowers formed fruits with normal seeds, which upon germination produced all mutant plants. IAA induced the production of carpellike stamens (CS) that showed no signs of microsporogenesis and possessed a basal ovary-like region that contained ovules (Table 5-2). The stamens were smaller than GA_3-treated sl_2/sl_2 and +/+ flowers, and the stamens did not fuse to form a staminal cone. Such stamens were functionally male sterile (Sawhney, 1984).

These observations (Table 5-2) showed that the male fertility of the sl_2/sl_2 mutant could be either restored by GA_3 treatment or the stamens could be induced to form a completely sterile structure by IAA. The timing of hormonal treatment was, however, a critical factor. The ability of sl_2/sl_2 flowers to produce normal characteristics when treated with GA_3 was restricted to the early stages of floral development (Sawhney and Greyson, 1979). Because normal pollen and viable seeds were produced after GA_3 treatment, it was clear that this treatment could be used to maintain the male-sterile line (the B-line in Fig. 5-2).

Table 5-2. Effects of GA_3 and IAA on the development of stamens of the *stamenless-2* mutant of tomato.
Concentrations: GA_3, 2 x 10 µl of 10^{-3}µM; IAA, 2 x 25µl of 10^{-4}µM. Y.P. = Yellow and pubescent (normal); c.s. = carpellike stamens. Means followed by the same letter (a, b, or c) in a vertical column not different at the 5 percent level of confidence.

Treatment	No. of Y.P.stamens/ flower	No. of c.s stamens/ flower	Stamen length (mm)	Staminal cone (I flowers)
GA_3	6.60^a	0.0^a	9.30^a	100.0
IAA	0.25^b	4.25^b	6.50^b	0.0
sl_2/sl_2 (control)	2.25^c	0.5^a	7.80^c	0.0
+/+(control)	6.36^a	0.0^a	9.80^a	100.0

(Sawhney, in *Proceedings 8th International Symposium on Sexual Reproduction in Seed Plants, Ferns and Mosses,* Pudoc Publications, Wageningen, pp. 36-38, 1984; data from V.K. Sawhney and R.I. Greyson, *Can. J. Bot. 51*: 2473-2479, 1973.)

Mutant plants grown under different temperature regimes also showed somewhat similar results to those in Table 5-2. Mutant plants grown under low temperatures produced normal stamens that were yellow and pubescent, were as long as +/+ stamens, and were fused to

Table 5-3. Temperature effects on the stamen development of the *stamenless-2* mutant of tomato.
Temperature regimes: LTR = low; ITR = intermediate; HTR = high. Y.P. = yellow and pubescent; c.s. = carpellike stamens. Means followed by the same letter (*a, b,* or *c*) in a column not different at the 5 percent level of confidence.

Genotype Temperature regime	No. of Y.P.stamens/ flower	No. of c.s.stamens/ flower	Stamen length (mm)	Staminal cone (% flowers)
sl_2/sl_2 (LTR)	6.16^a	0.13^a	9.92^a	66.6
" (ITR)	0.50^b	1.30^b	7.61^b	0.0
" (HTR)	0.00^b	5.46^c	5.72^c	0.0
+/+ (ITR)	6.70^a	0.00^a	9.41^a	100.0

(Sawhney, in *Proceedings 8th International Symposium on Sexual Reproduction in Seed Plants, Ferns and Mosses,* Pudoc Publications, Wageningen, pp. 36-38, 1984; data from Sawhney, *J. Hered. 74: 51-54, 1983.)*

form a staminal cone (Table 5-3). The pollen produced in these stamens was normal and viable and, if used to pollinate the mutant flowers, resulted in fruits with seeds that were all sl_2/sl_2. The high temperature conditions, on the other hand, caused the production of carpellike stamens in which microsporogenesis did not take place and instead ovules were produced at the basal end (Sawhney, 1984).

Thus, the male sterility of sl_2/sl_2 mutants could also be controlled by growing mutant plants under different temperature regimes (see also Sawhney, 1983). This, therefore, can be an alternative approach to the manipulation of hormones for the maintenance of pure sterile lines.

5-1.2. Some Physiological Characteristics of the *stamenless-2* Mutant

5-1.2.1 Soluble Protein Profiles in the Mutant and the Normal Plant

These observations showed that sl_2 is a "hormone- and temperature-sensitive" mutant and could serve as a useful system to investigate biochemical changes in the expression of male sterility. One approach to defining biochemically the effect of the mutation was to characterize the soluble proteins in the sl_2/sl_2 mutant and normal (+/+) stamens and other floral organs and to determine whether the temperature regulation of male sterility in this mutant was correlated with any specific proteins (Sawhney and Bhadula, 1987).

A protein analysis showed that the normal stamens contained nearly twice as much protein as those of the mutant at intermediate temperatures (12.26 for sl_2/sl_2 and 24.34 for +/+, g/mg fresh weight of protein). The sl_2/sl_2 stamens reverted to normal by low temperatures (15.86 g/mg) or GA_3 (18.44 g/mg) contained more protein than the untreated sl_2/sl_2 stamens, but less than those of the +/+ stamens. The carpellike stamens produced under high temperatures contained lower levels of protein (10.1 g/mg) than in any of the other conditions. The lower level of protein in the mutant versus normal stamens may have been primarily related to the proteins associated with pollen development; the reversion to normal stamens and pollen by GA_3 and low temperatures restored some of these proteins (Sawhney, 1984).

An SDS-polyacrylamide gel analysis showed that mutant stamens contained three bands of approximately 33, 28, and 23-kD that were either faint or absent in the +/+ stamens. The low temperature-reverted mutant stamens did not have these bands, but they were present in the high temperature stamens. In the GA_3-treated stamens,

Table 5-4. A developmental program of the normal (+/+) and *stamenless-2* (*sl₂/sl₂*) mutant stamens of tomato (*Lycopersicon esculentu*m).

Developmental stage	Normal	sl_2/sl_2 mutant
i	Microspore mother cells	Microspore mother cells
ii	Tetrads	Tetrads, abnormal tapetum
iii	Microspores	Microspores; some abnormal
iv	Vacuolate microspores exine deposition. Start of tapetum degeneration,	Some normal microspores; others devoid of a wall. Vacuolate tapetum.
v	Binucleate pollen, well developed exine.	Many degenerating microspores; some normal. Start of tapetum degeneration.
vi	Pollen with small vacuoles. Tapetum degeneration complete.	Mostly abnormal microspores Tapetum degenerating.
vii	Pollen same as in stage vi	Many microspores empty or with large vacuoles.
viii	Mature pollen, dehiscent anther	Mostly degenerated microspores, some normal pollen. Non-dehiscent anther.

(Bhadula and Sawhney, *J. Exp. Bot. 40:* 789-794, 1989; reproduced by permission of Oxford University Press.)

also, protein bands found in the mutant stamens were much fainter (Sawhney, 1984).

It appeared, therefore, that the differences in the protein levels or protein bands in the mutant and normal stamens were partly removed by the restoration of normality in the mutant stamens either by low temperature or by GA_3 application (see Sawhney and Bhadula, 1987). The possible roles of these proteins, however, has not been determined.

5-1.2.2   *Esterase and Amylolytic Activities*

Because esterases are implicated in pollen development and amylases, through the breakdown of starch, are believed to provide free sugars for developing pollen, a study on the activity and isozymes of these enzymes in the normal and sl_2/sl_2 mutant was conducted at eight stages of development (as defined in Table 5-4 taken from Bhadula and Sawhney, 1989).

Esterase Activity. Esterase activity was significantly higher in the normal than in mutant stamens at all stages of development (Fig. 5-3). In the normal, at stages i to iii, the esterase activity was high. It declined at stages iv and v and was further reduced at stages vi and vii. In the mutant, there was a gradual increase in esterase activity up to stage iv, but then it declined until maturity. In low temperature (LTR) reverted mutant stamens (MS-REV), the esterase activity was higher than those grown in intermediate temperatures (ITR), or in the greenhouse. The esterase activity of LTR-reverted mutant stamens was less than that of normal stamens at stages i to iii and at anthesis but not different at stages iv to vii (Fig. 5-3; Sawhney and Bhadula, 1988a).

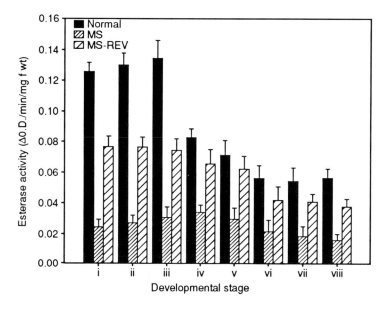

Figure 5-3. Esterase activity (OD/minute/mg fresh weight) normal, male sterile *stamenless-2* mutant (MS) and low temperature-reverted mutant (MS-REV) stamens at eight (i-viii) developmental stages. Vertical bars represent standard error. (Sawhney and Bhadula, in *Proceedings 10th International Symposium Sexual Reproduction,* M. Cresti, P. Gori,E. Pacini (eds), Springer-Verlag, Berlin, pp. 187-192, 1988.)

The overall activity of esterase was lower in the mutant than the normal throughout stamen development (Fig. 5-3). Esterases have been localized in the tapetum at early stages, and later on the microspores during exine deposition. Although esterases are likely involved in a variety of functions, it is possible that some forms of it have a role in the polymerization of sporopollenin in the tapetum and its deposition in the exine. Thus, the lower activity and low intensity of major isozymes of esterases in the sl_2/sl_2 mutant could well be related to the lack of exine deposition and eventual degeneration of many microspores (Bhadula and Sawhney, 1987; Sawhney and Bhadula, 1988b).

Amylolytic Activity. The total amylolytic activity of normal and mutant stamens was not different at stages i and ii (Table 5-4), but thereafter, the activity in normal stamens gradually increased through to stage vii, whereas in the mutant it did not change throughout development, and was significantly less than the normal. The mutant stamens also possessed fewer α- and β-amylase isozymes than the normal. The amylolytic activity and isozymes of amylases of LTR-reverted mutant stamens followed the same pattern as the normal (Fig. 5-4; Sawhney and Bhadula, 1988a).

The activity of amylases in the mutant was not different from the

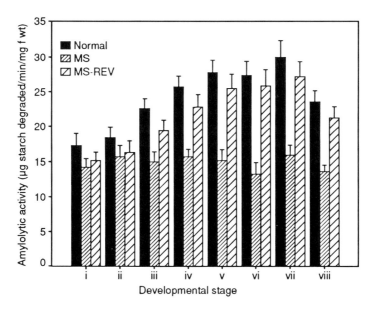

Figure 5-4. Amylolytic activity (g starch degraded/minute/mg fresh weight) of the normal, male sterile *stamenless-2* mutant (MS) and low temperature-reverted mutant (MS-REV) stamens of tomato at eight (i-viii, see Table 5-4 for explanation) developmental stages. Vertical bars represent standard error. (Sawhney and Bhadula, in *Proceedings 10th International Symposium Sexual Reproduction,* M. Cresti, P. Gori, E. Pacini (eds), Springer-Verlag, Berlin, pp. 187-192, 1988.)

normal at stages i and ii, but from stage iii onward, it was significantly greater in the normal stamens. It is well established that amylases break down starch and provide free sugars for the developing organs. The low activity and fewer isozymes of amylases in the mutant would, therefore, cause depletion of sugars, which are essential for pollen development. The consequent low level of free sugars could lead to pollen starvation, thus leading to their abortion.

Thus, through the use of the sl_2 mutant of tomato the roles of key enzymes in the normal development and performance of the stamen could be investigated. It may be that other biochemical events in stamen development could also be investigated using this or other mutants that are known to affect flower development in tomato and other higher plant species (Haughn and Somerville, 1988; Nester and Zeevaart, 1988; Martinez-Zapater and Somerville, 1990).

5-1.2.3   Flower Culture

Flower culture in vitro is a useful technique to eliminate the influence of vegetative parts, and is valuable in investigating the role of nutritional, hormonal, and environmental factors in flower development (Rastogi and Sawhney, 1989). Flower culture of normal and male sterile mutants should facilitate a comparative analysis of the nutritional and hormonal requirements of the two lines and also provide insights into the physiological control of normal stamen development.

In the sl_2/sl_2 mutant, it is of interest to recall that the exogenous application of GA_3 to developing floral buds in vivo results in the formation of normal stamens that contain viable pollen (previously discussed). In contrast, isolated buds grown in vitro did not produce normal stamens with viable pollen even at the highest GA_3 concentration (Rastogi and Sawhney, 1988a). These observations suggested that (1) gibberellins, although an important component, are not the only factor required for stamen development in the sl_2/sl_2 mutant and, (2) they presumably interact with other factors supplied by vegetative parts of the plant. The nature of these factors is not known. In addition, when normal and sl_2 stamen primordia were cultured in vitro in the presence of benzylaminopurine and GA_3 the mutant organs became more like the normal. In the case of both genotypes, however, microsporogenesis was lacking, results that show that whereas the expression of some of the mutant characteristics is independent of other floral parts, the differentiation of microsporogenous tissue in both genotypes is dependent on other floral organs. The results also suggest that although gibberellins are essential for stamen development, some other factors, yet unknown, are also required for stamen maturation (Rastogi and Sawhney, 1988b).

These few examples illustrate that in vitro floral culture can be an important tool in the investigation of flower development and that the technique can be made more effective through the addition of mutant material specifically altered in some aspect of flower morphogenesis.

Case Study 5-2:   Self-Incompatibility

All eukaryotic organisms have evolved barriers to intergeneric and interspecific crosses that act to maintain the identity and stability of the species. Mechanisms for promoting outcrossing are also important in maintaining genetic variability within a particular species. Animals fulfill this requirement by moving about to choose a mating partner. Flowering plants are not able to move but they have the same requirement to promote outcrossing. They have the additional problem that they are, in many cases, hermaphroditic; that is, the male and female sexual organs are carried on the same plant, often in close physical proximity that could facilitate self-mating and inbreeding. To prevent this self-mating, flowering plants have evolved another mechanism to promote outcrossing, which depends on a gene known as the *S* gene, or *self-incompatibility* gene. This gene acts to prevent fertilization involving gametes from the same plant and between certain individual plants of the same species. The effectiveness of self-incompatibility systems in maintaining outbreeding is considered to be a major factor in the evolutionary success of flowering plants (Bernatzky et al., 1987).

   Self-incompatibility systems are usually classified as heteromorphic or homomorphic. Heteromorphic species produce morphologically distinct types of flowers, whereas homomorphic plants produce morphologically identical ones. The distinctive features of the flowers of the heteromorphic, self-incompatible species are their relative style length and anther level, and these features present mechanical barriers to illegitimate matings (Bernatzky et al., 1987).

   There are two types of homomorphic self-incompatibilities: gametophytic and sporophytic. In gametophytic self-incompatibility the male gametophyte, the pollen, expresses its own *S*-genotype. In contrast, in sporophytic systems, pollen behavior is determined by the genotype of the pollen-producing plant, the sporophyte (Ebert et al., 1989). An example of the latter, in the Brassicaceae, is used to illustrate the progress that has been made through analysis of alleles of the *S*-gene and understand the homomorphic system of self-incompatibility.

Sporophytic Self-incompatibility in the Brassicaceae

Sporophytic self-incompatibility is apparently not as widely distributed as the gametophytic variety. It has been described for only two families, the Cruciferae (Brassicaceae) and the Compositae (Asteraceae). The genetic control is through a single S locus with multiple alleles. The incompatibility response is determined by the alleles carried by the pollen-producing plant (Fig. 5-5). In these systems, the arrest of growth of self-pollen tubes occurs at a very early stage, therefore, if the pollen does germinate, the tube growth is arrested on or just beneath the stigma surface.

Until recently, very little was known of the biochemistry by which female organs recognize specific pollen genotypes. A priori, the extensive variability at the S locus should also be manifest as biochemical variability. Several methods to detect such variability were attempted. Until about 1985, immunological techniques seemed the most promising, as illustrated in the following example (Nasrallah and Wallace, 1967).

Antisera were produced in rabbits against *Brassica oleracea* var. *capitata* stigmatic homogenates of S_1S_1, S_2S_2, and S_3S_3, and designated as AHS_1, AHS_2, and AHS_3, respectively. Test antigens were referred to as HS_1, HS_2, and HS_3, respectively. Homogenates of stigmas obtained from the three possible hybrids, S_2S_2, S_1S_1, and S_2S_3 were allowed to react with AHS_1, AHS_2, and AHS_3. In correlation with the presence of stigmatic S antigens, the F_1 genotypes rejected pollen from their respective parents while accepting pollen from the nonparental inbred (Table 5-5; Nasrallah and Wallace, 1967).

The question arose as to whether these antigens were present in other plant parts. Test antigens were prepared from styles, ovaries, anthers including pollen, embryos, and leaves, and were allowed to react against the sera. No S antigens were detected in any of these tissues. The results indicated that only stigmas contained detectable S antigens.

The detection of antigenic differences led to investigation of the inheritance of these differences and attempts were made to correlate the presence or absence of a given S antigen in F_2 plants with incompatibility behavior. The data of Table 5-6 indicate such correlation.

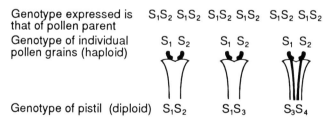

Figure 5-5. Behavior of pollen in the sporophytic incompatibility system. The pollen parent genotype is S_1S_2. When an allele in the pollen parent is matched with that of the pistil (eg., S_1S_2 or S_1S_3) pollen germination is arrested at the stigma surface. Where there is no match (S_3S_4), the pollen may germinate and grow through the style to the embryo sac. (Bernatzky et al., *Dev. Genet. 9:* 1-12; copyright © 1987 Wiley-Liss; reprinted by permission of Wiley-Liss, a Division of John Wiley and Sons, Inc.)

Table 5-5. Reaction of hybrid stigmas with antisera prepared against parental genotypes of *Brassica oleracea* L.

Antiserum	Absorbing antigen	Genotype of hybrid Stigma		
		$HS_1 S_2$	$HS_1 S_3$	$HS_2 S_3$
AHS_1	HS_2 or HS_3	+	+	-
AHS_2	HS_1 or HS_2	+	-	+
AHS_3	HS_1 or HS_2	-	+	+

(Nasrallah and Wallace, *Heredity 22:* 519-527, 1967.)

The individual F_2 plants were assayed for S_1 and S_2 antigens by testing stigmatic homogenates with absorbed AHS_1 and AHS_2. Cross-compatibilities with the parents were then determined with reciprocal crosses to both parents.

In the three S_1 x S_2 progenies, for example (Table 5-6), F_2 plants having S_2 antigen but no S_1 were found to be reciprocally cross-incompatible with the S_2S_2 parent and reciprocally compatible with the S_1S_1 parent. Plants having the S_1 antigen but no S_2 were reciprocally compatible with S_2S_2 and reciprocally incompatible with S_1S_1. Plants with both S_1 and S_2 stigmatic antigens exhibited cross-compatibilities with the parents that were identical to those of heterozygotes derived directly from hybridization. In such plants, S_2 was dominant over S_1 in the pollen. The pollen phenotypes of S_1S_2 was, therefore, S_2 irrespective of the pollen genotype. The data in Table 5-6 are in complete agreement with this interpretation.

Questions remained following this analysis: Are the S antigens polypeptides and does each S allele produce a unique antigenic specificity by changing the amino acid composition of this polypeptide? Are these antigens enzymes, inhibitors, or substances that control gene activity? Do the S antigens diffuse into pollen grains and, if so, where are they localized? What is the nature of the pollen specificity?

Table 5-6. Correlation of incompatibility genotype of F_2 plants of *Brassica oleracea* as determined serologically and by cross-pollination.

Cross (F_1)	Reaction with		Genotype determined by serology	Genotype determined by pollination
	AHS_1	AHS_2		
S_1xS_2	+	-	S_1S_1	S_1S_1
	+	+	S_1S_2	S_1S_2
	-	+	S_2S_2	S_2S_2

(Nasrallah and Wallace, *Heredity 22:* 519-527, 1967.)

In an attempt to answer at least some of these questions acrylamide gel electrophoresis and immunodiffusion were used to study the buffer soluble basic proteins of stigmatic homogenates of several S allele genotypes of cabbage. A number of useful observations were made: each S allele genotype had a different electrophoretic protein band pattern; each of the genotypes S_1S_1, S_2S_2, and S_3S_3 had a unique electrophoretically separable protein band; the electrophoretically separated S allele-specific proteins were shown to be heritable by the presence in heterozygous genotypes of the unique proteins of each of the two homozygous parents. From such observations it was hypothesized that S allele-specific proteins differ by amino acid substitutions and it was postulated that the basic S proteins act as regulators of enzymes needed for pollen germination and tube growth (Nasrallah et al., 1970).

In Brassicaceae, the pollen-stigma interaction of self-incompatibility is characterized by the arrest of pollen tube development at the stigma surface. The reaction occurs within minutes after contact between the male gametophytes and the tightly packed, elongated, unbranched surface cells of the stigma. In *Brassica oleracea*, these cells form a layer that consists of approximately 3,000 papillae that can accommodate a similar number of pollen grains at their surface. During flower development, the stigmas of *Brassica* are initially self-compatible and become self-incompatible only about one to two days before anthesis. The immature, self-compatible stigmas are fully receptive to pollen, thus permitting the production and maintenance of S homozygotes. The shift from self-compatibility to self-incompatibility has been correlated with the accumulation of S allele-specific antigens and, more recently, of S allele-specific glycoproteins.

Studies of the synthesis of S specific molecules in developing stigmas are necessary for defining the molecular steps that lead to the onset of the self-incompatibility response. To this end, the variation in the relative synthesis of S specific molecules was followed in extracts of stigmas harvested along one inflorescence. The inflorescence was divided into five developmental zones designated minus 3, minus 2, minus 1, 0 and +2, each consisting of a sequence of three flowers or flower buds. Stigmas from the minus 3, minus 2 and minus 1 zones were immature stigmas derived from flower buds that would normally reach anthesis in 3, 2, or 1 days, respectively; stigmas from the 0 and +2 zones were mature stigmas from flowers at anthesis, and 2 days after. The ^{14}C incorporation into S molecules was tissue specific in as much as it was detected in stigmas of all stages but not in the stylar controls. The relative incorporation into S molecules was found to increase with the onset of the self-incompatibility (SI) response and to reach a maximum in SI buds just before anthesis (Fig. 5-6; Nasrallah et al., 1985a). The S-specific molecules comprised 6 and 10 percent of total protein in stigmas from self-incompatible buds and from open flowers, respectively. The papillar cells would seem to be the site of this protein synthesis.

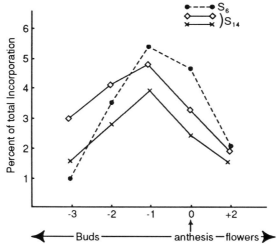

Figure 5-6. Relative incorporation of ^{14}C-amino acids into S specific molecules of stigmas of *Brassica oleracea* harvested along one inflorescence. (Nasrallah et al., *Planta 165:* 100-107, 1985a.)

The implication of these results is that the switch from self-compatibility (SC) to SI occurs when a certain threshold level of the S-specific molecules is attained. As the S-specific glycoproteins accumulate in the developing stigma, invasion of the papillar cells by pollen tubes is prevented, and later, with further accumulation, pollen germination may be inhibited as well.

Several lines of evidence suggest that these glycoproteins have an important role in incompatibility: The mobilities of these compounds vary in stigma extracts derived from *Brassica* strains with different *S*-locus alleles; these molecules are found in the stigma but not in the style or seedling tissue; the increased rate of synthesis of these *S*-locus-specific glycoproteins (SLSGs) in the developing stigma correspond with the onset of the incompatibility reaction in the stigma; mutations in genes unlinked to the *S* locus, which result in self-compatibility, are also associated with reduced levels of these molecules; the inheritance of the various forms of SLSG correlates perfectly with the segregation of *S* alleles in genetic crosses, indicating that the gene responsible for this polymorphism must be genetically located at or close to the *S* locus; when a complementary cDNA clone containing sequences encoding an SLSG from *Brassica oleracea* was isolated it was shown that the spatial and temporal distribution of the mRNA homologous to these sequences mirrors the appearance of the SLSG; and polymorphisms in homologous restriction fragments of *Brassica* genomic DNA segregate precisely with alleles at the *S* locus (Nasrallah et al., 1985b).

The sequences of the S glycoproteins from three homozygotes of *Brassica campestris* have been compared and show extensive homology. These glycoproteins have cysteine-rich clusters and six to seven

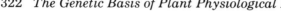
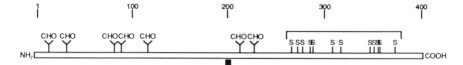

Figure 5-7. Schematic drawing of the S_8-glycoprotein of *B. campestris.* CHO, carbohydrate chain; S, cysteinyl residue. Heavy bar, zone variable among *Brassica* species. (Takayama et al., reprinted by permission from *Nature* vol. *326* pp. 102-105. Copyright © 1987 Macmillan MagazinesLtd.)

oligosaccharide chains (Fig. 5-7; Takayama et al., 1987). Most recently, this work has led to the isolation of the self-incompatibility genes themselves from papillar cells of *Brassica oleracea* and to the confirmation of the expression of the gene products in the cells that constitute the barrier to self-pollination (Nasrallah et al., 1988). To make matters more complicated, recent evidence suggests that more than one gene may be expressed in any one genotype leading to the production of more than one glycoprotein (Trick and Flavell, 1989).

In summary, the cruciferous genus *Brassica* has a self-incompatibility system that is under the sporophytic control of a single locus, *S*, and more than 50 *S* alleles have been reported. The existence of *S*-locus-specific proteins, termed S glycoproteins, has been confirmed. S glycoproteins appear only in the stigma of mature flowers and not in other parts of the plant. These glycoproteins, extracted from stigmas with different *S* alleles, are clearly distinguishable from each other by isoelectric focusing analysis, and their inheritance correlates perfectly with the segregation of *S* alleles in genetic crosses. Furthermore, S glycoproteins first appear in the stigma a few days before flowering, synchronously with the onset of the incompatibility reaction in the stigma. Thus, these *S*-allele-specific glycoproteins apparently play, directly or indirectly, an important role in the self-incompatibility system, although their structures and functions are still not completely known (Isogai et al., 1987). Through the use of the *Brassica* system of *S* alleles, much has already been learned about this particular type of control of a key developmental event.

Case Study 5-3:   Polyamines and Flowering in Tobacco

Tobacco has perfect, or complete, flowers with both male and female differentiation normally occurring in the same flower. In a vegetatively growing tobacco plant, the terminal apical meristem gives rise to the body of the plant; periodically, lateral initials are generated that differentiate into leaves. After the signal to flower has been received, the lateral initials develop into the parts of the flower - sepals, petals, stamens, and carpels - instead of leaves. This sequence is both temporal and spatial, with the sepals developing first and occupying the outermost position in the completed flower and the male stamen and the female carpel occupying the innermost and last developing positions. The tobacco flower has a fivefold symmetry; there are normally five of each flower part, except for the carpel. The sepals and petals each develop as fusions of their lateral initials, whereas the carpel develops as a composite fusion of the central meristem and the last developing initials.

Mutants have been isolated that alter the development of the tobacco flower; many of them seem to alter the expression of the initials that generate the flower parts, therefore, a variety of developmental switches are seen. They were isolated using the powerful somatic cell culture system of tobacco followed by plant regeneration. The mutants were selected for lesions in the polyamine biosynthetic pathway (Malmberg et al., 1985).

In the last 10 years, interest has developed in the study of the polyamine pathway (Fig. 5-1) in higher plants. Polyamines may be important growth regulatory substances in higher plants (Galston, 1983; Galston and Kaur-Sawhney, 1988). Additional stimulation has been given the discovery of several inhibitors of the pathway of polyamine synthesis (Fig. 5-8). Difluoromethylornithine (DFMO) and difluoromethylarginine (DFMA) block ornithine decarboxylase (OrnDC) and arginine decarboxylase (ArgDC), respectively; methylglyoxal-bis(guanylhydrazone) (MGBG) and dicyclohexyl-ammonium sulfate (DCHA) inhibit S-adenosyl-methionine decarboxylase (SamDC) and spermidine synthase (SpSyn), respectively. These and other inhibitors offer a potent tool to probe the pathway (Malmberg et al., 1985).

As Table 5-7 shows, there is considerable developmental regulation of the three enzyme activities, ArgDC, OrnDC, and SamDC, with respect to reproductive versus nonreproductive organs. Cell cultures, although theoretically undifferentiated, resemble flowers more than other tissues. Through the use of molecular probes it is suggested that the developmental regulation of OrnDC and SamDC in tobacco occurs at the pretranslational stage (Malmberg et al., 1985).

5-3.1 The Isolation and Characterization of Mutants Affecting Polyamines and Flowering

Mutants have been isolated from tobacco cell lines that are temperature sensitive or that are resistant to either MGBG, DFMO, or DCHA. In the case of MGBG, for instance, when plants were regenerated, every one that flowered had abnormal flowers. Some showed simple male and female sterilities without gross morphological changes, and others showed one or another developmental switch.

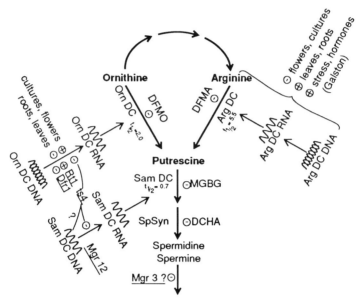

Figure 5-8. Polyamine pathway in tobacco showing the sites of action of some inhibitors of the pathway. OrnDC = ornithine decarboxylase; ArgDC = arginine decarboxylase; SamDC = S-adenosylmethionine decarboxylase; SpSyn = spermidine or spermine synthase; DFMO = difluoromethylornithine; DFMA = difluoromethylarginine; MGBG = methylglyoxal-bis(guanylhydrozone; DCHA = dicyclohexylammonium sulfate. Mutants (*Mgr12, ts4*, etc.) are discussed in the text. (Malmberg et al., *Cold Spring Harbor Symp. Quant. Biol. L:* 475-482, 1985.)

Wild-type cells were isolated from an anther-derived plant of *N. tabacum* cv. Wisconsin 38. After treatment with bromodeoxyuridine (BuDR) and light, several colonies were found that grew at the permissive

Table 5-7. Developmental regulation of enzyme activities. See Figure 5-8 for explanation of abbreviations. The data shown in the first three columns are the specific activities of the enzymes relative to the tissue in which the specific activity is the maximum. The fourth column (O/A ratio) is the ratio of the specific activities of OrnDC to ArgDC.

	OrnDC	ArgDC	SamDC	O/A ratio
Flower	79	20	100	2.24
Leaf	8	59	22	0.08
Stem	9	79	21	0.07
Root	20	100	24	0.11
Culture	100	31	31	1.83

(Malmberg et al., *Cold Springer Harbor Symp. Quant. Biol. L:* 475-482, 1985.)

temperature of 26°C but not at 33°C, notably the mutant *ts4* (Figs. 5-8 and 5-9; Malmberg, 1979, 1980). The mutant *ts4* regenerated into light green plantlets with short internodes that could be maintained only in shoot cultures. Another cell line, *Rt1*, a revertant of *ts4*, also regenerated to give plantlets that had short internodes but were dark green and produced flowers with a second row of petals in place of anthers. *Rt1* grew nearly as well as the wild type at 26°C and 33°C (Fig. 5-9; Malmberg, 1979).

In a search for the gene products that might be the source of temperature sensitivity, a striking effect was seen in the levels of the amino acid ornithine. In extracts from wild-type cells at both 26°C and 33°C and also in extracts for *ts4* at 26°C, the levels were virtually identical, but in *ts4* at 33°C there was a threefold elevation of ornithine level. Because polyamines are products of one pathway that starts with ornithine, the levels of polyamines were also examined in a similar experiment. Putrescine levels in *ts4* remained the same as in the wild type but spermine and spermidine were reduced approximately 50 percent in the mutant at low temperature and dropped further at the higher temperature. These results strongly implicated the enzymes OrnDC and SamDC as possible sites of the *ts4* lesion (Malmberg, 1980).

Results of enzyme assays showed that the *ts4* mutant had low levels of OrnDC and SamDC activity, whereas *Rt1* had restored levels of only OrnDC. This is the only enzyme activity in which *Rt1* resembles wild type more than it resembles *ts4*, suggesting that it is the site of metabolic alteration of the mutation (Table 5-8). It is difficult, however, to assign cause and effect in these mutants and revertants.

SDS-PAGE of total protein extracts from *wt*, *ts4*, and *Rt1* cells revealed a prominent band of approximately 30-kD in the *ts4* but not in

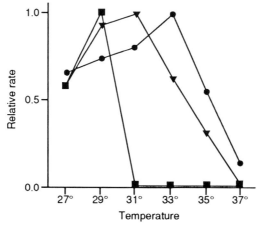

Figure 5-9. Growth of wt, and *ts4* and *Rt1* mutants of tobacco as a function of temperature. wt (●); *ts4* (■); *Rt1* (▼) (a revertant of *ts4*). (Malmberg, *Cell* 22: 603-609; copyright 1980 Cell Press.)

Table 5-8. Enzymes with altered specific activities in crude extracts from wt, *ts4*, and *Rt1* cells of tobacco and assayed either at 26°C or 33°C. The data are expressed as nm/hour/mg protein; in parentheses are the values relative to wild type at 26°C. PAL, phenylalanine ammonia lyase; NR, nitrate reductase; OrnDC, SamDC, see legend to Figure 5-8.

Enzyme	PAL		NR		SamDC	OrnDC
WT-26°C	30.2	(100)	333	(100)	.648 (100)	.604 (100)
WT-33°C	32.1	(106)	347	(104)	1.00 (155)	1.42 (235)
ts4-26°C	84.4	(179)	56.7	(17)	.026 (4)	.024 (4)
ts4-33°C	86.8	(287)	50.0	(15)	.078 (12)	.103 (17)
Rt1-26°C	56.2	(186)	63.3	(19)	.123 (19)	.411 (68)
Rt1-33°C	65.9	(218)	73.3	(22)	.220 (34)	1.12 (185)

(Malmberg, *Cell 22:* 603-609; copyright 1980 Cell Press.)

the *wt* or *Rt1* cells. When cells are grown in the presence of sublethal doses of MGBG, a 30-kD polypeptide is induced. Peptide mapping confirmed that this is the same polypeptide produced at high levels in the *ts4* cells (Malmberg and McIndoo, 1983). Further efforts were, therefore, made to isolate cell lines resistant to MGBG. One of these, *Mgr3*, regenerated quite rapidly into a whole plant. The plant was dark green and had very short internodes. Flowers initially appeared normal but then the stigma grew past the corolla and the stigma turned black. The anthers contained no viable pollen and died at the time the stigma turned black. Hundreds of anthers instead of ovules were found embedded in the placenta of the ovary. These and other abnormal structures could not have resulted from fertilization, as *Mgr3* was male sterile.

The levels of polyamines in *Mgr3* cultures and leaves were determined and compared with wild-type cultures and leaves (Table 5-9). In *Mgr3* cultures, the levels of putrescine were relatively low; whereas in leaves from regenerated mutant plants the levels of spermidine and spermine were quite high. The ratio of putrescine to

Table 5-9. Polyamine concentrations of *Mgr3* and wild type cells of tobacco

Cell Type	Concentration (nmol g^{-1})		
	Putrescine	Spermidine	Spermine
Wild-type culture	480(38)	430(53)	49(3)
Wild-type leaf	410(68)	430(24)	23(5)
Mgr3 culture	180(27)	520(27)	50(5)
Mgr3 leaf	570(36)	2,030(89)	160(7)

(Malmberg and McIndoo, reprinted by permission from *Nature* vol. *305* pp. 623-625. Copyright © 1983 Macmillan Magazines Ltd.)

spermidine in wild-type cultures and leaves was about 1:1, whereas in *Mgr3* cultures and leaves it was 1:3 or 1:4. In *ts4* and *Rt1* cells the ratio of putrescine to spermidine was about 2:1 (Table 5-9; Malmberg and McIndoo, 1983).

Several independent mutants have been isolated that are affected in polyamine synthesis and also have altered expression of floral meristems. This suggests that the two are related in some way. It may not be that the relationship is a direct and specific one, however, it may simply be that flower development is very sensitive to metabolic perturbations. Alternatively, it is possible that polyamines have a specific role in development: perhaps gradients of polyamines are involved in determining the developmental fate of floral meristems (Malmberg and McIndoo, 1983).

Following the results obtained with *Mgr3*, a number of cell lines resistant to MGBG were isolated. The partial female fertility of one line, *Mgr12*, permitted a meiotic genetic analysis of both the resistance and the abnormal floral development. In *Mgr12*, repeated attempts at fertilization with wild-type pollen eventually yielded eight F_1 seeds. These were germinated and grown into plants. A portion of a leaf from each of the eight plants was placed back into culture and tested for MGBG resistance. Seven of the eight plants were resistant, and one was sensitive to MGBG. When the plants flowered, it was observed that the seven resistant plants were male sterile. The single sensitive F_1 plant had normal flowers and was male fertile. The seven resistant plants were back crossed with wild-type pollen yielding a near normal number of seeds. Twenty-five of the back cross progeny have been tested for MGBG resistance, with a ratio of 12 resistant to 13 sensitive (Malmberg and McIndoo, 1984).

This meiotic genetic analysis of *Mgr12* strengthened the hypothesis of a physiological linkage between polyamine synthesis and the altered expression of floral meristems seen in this variant collection. The segregation data in the F_1 and backcross generations were consistent with *Mgr12* being a nuclear, dominant mutation.

5-3.2 Other Studies of Links Between Polyamine Synthesis and Floral Development

The observations based on mutational analysis hinted that polyamines may be correlated with whatever determines the developmental fate of individual meristems in a flower. There was no evidence; however, that the correlation was specific; it could have been that floral development was just highly sensitive to metabolic perturbations. Therefore, mutations in some other pathways than polyamine synthesis might also have led to alterations in the expression of floral development. Thus, other more recent studies, some of which are discussed later, have been directed to understanding the precise links between polyamine synthesis and floral development.

5-3.2.1   MGBG Treatment of Tobacco Suspension Cultures

As was learned earlier in this Case Study (Malmberg, 1980), *ts4* contained low levels of spermidine and spermine, as well as low OrnDC and SamDC activities. When the SDS-PAGE profile of *ts4* was compared to wild type, *ts4* was seen to overproduce a 30-kD peptide not detectable in wild type. A phenocopy of the overproduction of this polypeptide in *ts4* was observed in wild-type cultures exposed to sublethal levels of MGBG. A revertant of *ts4* was capable of further regeneration with the surprising result that upon flowering, the anthers were converted into a second row of petals. This led to the speculation that the defect in polyamine metabolism was in some manner related to the developmental switch. Subsequent regeneration of tobacco cell lines resistant to MGBG has supported this speculation; all MGBG-resistant plants are male sterile with a high frequency of abnormal floral development. The male sterility and abnormal floral development co-segregated with MGBG resistance suggesting that the two phenotypes are coincident (Hiatt et al., 1986).

Treatment of tobacco liquid suspension cultures with MGBG, an inhibitor of SamDC, resulted in a dramatic overproduction of a 30-kD peptide. MGBG treatment also resulted in a 20-fold increase in the activity of SamDC. Purification of SamDC from MGBG-treated cultures revealed that the overproduced 30-kD peptide and SamDC were identical. Other experiments revealed that the MGBG stabilized the enzyme to proteolytic degradation and did not increase its synthesis (Table 5-10). In addition to stabilizing SamDC, MGBG also resulted in the rapid and specific loss of ArgDC activity and the blockage of the increase in OrnDC activity after subculturing (Table 5-10).

Exogenously added polyamines had little effect on OrnDC, whereas SamDC and ArgDC activities rapidly diminished with added spermidine or spermine. Finally, inhibition of OrnDC was lethal to the cultures, whereas inhibition of ArgDC was only lethal during initiation of growth in suspension culture (Hiatt et al., 1986).

The simplest interpretation of these data is that polyamines are affecting the synthesis of various enzymes in the pathways leading to their synthesis. To understand the roles of the polyamines in development, it will be essential to understand more fully the nature of these controls. Thus, the study of these tobacco mutants has great value at the molecular, as well as at the morphological, level.

5-3.2.2   Polyamine Derivatives in Cultured Tobacco Cells Selected for Growth on Putrescine as Sole Nitrogen Source and in Tobacco Plants

Plants contain a number of metabolites that are made from polyamines, and it is unclear to what extent these may be responsible for observed

Table 5-10. Effect of MGBG on several decarboxylases in cultured tobacco cells.

	Specific activity, days after addition					
	Day 1	Day 2	Day 3	Day 4	Day 5	Day 6
	pmol CO_2 /mg protein /h					
A. Addition of fresh medium						
Arginine decarboxylase	4.3	18.8	16.3	4.2	3.2	1.1
Ornithine decarboxylase	34.1	15.2	7.8	4.6	1.1	1.2
S-Adenoxylmethionine decarboxylase	3.1	10.7	15.4	9.9	3.7	0.8
B. Addition of fresh medium containing 1 mM MGBG						
Arginine decarboxylase	0.09	0.07	0.11	0.05	0.10	0.03
Ornithine decarboxylase	1.1	1.2	0.9	0.85	0.76	0.79
S-Adenoxylmethionine decarboxylase	1.4	10.3	15.8	20.6	18.4	18.0

(Hiatt et al., *J. Biol. Chem. 261*: 1293-1298, 1986.)

polyamine-dependent functions. One prominent class of such metabolites includes polyamine amides of the hydroxycinnamic acids, in which one or both primary amino groups may be conjugated to *p*-coumaric, caffeic, ferulic, or sinapic acid. A feature of these compounds is that they accumulate rather specifically in the sex organs and associated tissues of virtually every plant species so far examined, where they are prominent nitrogenous and phenolic constituents. Furthermore, the assortment of hydroxycinnamic amides generally differs among species and between male and female organs in the same species. In maize, they are nearly absent from the anthers of *cms* (i.e., cytoplasmic male sterile) lines but are restored by the introduction of a nuclear gene that restores fertility to the pollen. In Araceae species, in which both fertile and sterile male and female flowers occur on the same plant, hydroxycinnamic amides are abundant in the fertile flowers of both sexes but are totally absent from the sterile flowers on the same plant. Thus, an important role in the reproductive development of higher plants seems likely for these polyamine metabolites (Balint et al., 1987).

One approach to polyamine function in plants is to explore their roles as intermediates in nitrogen metabolism of cultured cells. The XD cell line, which was derived from *N. tabacum* L. cv. Xanthi-nc, was found to be unable to grow when putrescine was supplied as the sole source of nitrogen. The cells remained alive, however, and developed a bluish cast due to the accumulation of hydroxycinnamoyl compounds. After repeated transfers to fresh putrescine-containing medium, the cells began multiplying at a high rate. The ability of these cells to grow on putrescine alone as a nitrogen source (called PUT cells) remained stable for several hundred generations and was not lost during prolonged growth on nitrate instead of putrescine (Balint et al., 1987).

When PUT cells were incubated with [^{14}C]putrescine, much of the radioactivity was found in [^{14}C]-4-amino-n-butyric acid (GABA), which appeared to be an obligatory intermediate in the assimilation of nitrogen from putrescine by PUT cells (Flores and Filner, 1985). Because N-caffeoyl-putrescine was the predominant hydroxycinnamoyl- putrescine in PUT cells and in tobacco flowers, a search was initiated for the stable metabolite of N-caffeoyl-GABA.

No N-caffeoyl-GABA was detected in extracts of XD cells that had been grown on GABA nitrogen (Table 5-11). These cells contained twice as much caffeoyl-putrescine as PUT cells. Thus, the ability to synthesize caffeoyl-GABA was a distinctive property of PUT cell that resulted from selection for the ability to grow on putrescine as sole nitrogen source.

The newly acquired enzymic activities implied by these findings (Table 5-11) were not likely to have arisen de novo but, rather, to have resulted from the activation of genes that are developmentally regulated in the intact plant. Thus, a localization of N-caffeoyl-GABA in tobacco plant tissues was attempted. The compound was found unequivocally in four tissues: flower buds, open flowers, leaves of the inflorescence, and green fruit (Table 5-11). Trace amounts were also possibly present in the apex and apical leaves of bolted plants and in the apex of vegetative plants. All lower leaves as well as apical leaves of vegetative plants were negative. This distribution closely paralleled that of caffeoyl-putrescine, the proposed precursor of N-caffeoyl-GABA (Balint et al., 1987). Both compounds reached their highest titers in flower buds but declined precipitously during subsequent development, implying that oxidation of hydroxycinnamoylputrescines may be quite active during the stage of germ cell formation.

Hydroxycinnamoyl-putrescines are major phenolic and nitrogenous components of the germ cells of higher plants and they are biochemical markers of fertility in maize and arums. Thus, the finding of a likely metabolite of caffeoyl-putrescine accumulating and disappearing in close parallel with it in reproductive tissues may provide a significant conduit for nitrogen and/or phenylpropanoid carbon for reproductive development in higher plants (Balint et al., 1987).

In any event, the polyamines and their derivatives are an intriguing group of secondary plant metabolites about whose roles in the plant much remains to be learned. The availability of mutants of the pathways leading to their formation provides an attractive and promising way to investigate their importance.

Table 5-11. Abundances of caffeoyl-GABA and caffeoyl-putrescine in cultured tobacco cells and tobacco plant tissues.
[a]"Lower" leaves were taken from the midheight of the plant in each case.

	Caffeoyl-GABA	Caffeoyl-putrescine
	nmol / g fresh wt	
PUT cells	2.70	890
XD cells on GABA	<0.01	1660
Flower buds	78.20	2430
Open flowers	14.00	1170
Green fruit	1.47	570
Floral leaves[a]	2.60	10
Lower leaves	(0.10)	<2
Bolted apex	(0.68)	49
Bolted apical leaves[a]	(0.43)	41
Bolted lower leaves	<0.03	<2
Vegetative apex	(0.37)	<2
Vegetative apical leaves	<0.03	<2
Vegetative lower leaves[a]	<0.03	<2

(Balint et al., *J. Biol. Chem. 262*: 11062-11031, 1987.)

Case Study 5-4:   The Ripening *(rin)* Mutant of Tomato

Tomatoes (*Lycopersicon esculentum* Mill.) are an obvious choice for biochemical and developmental studies on fruit ripening. After fertilization of the flower, fruit growth and maturation lasts for about 7 weeks. The ripening process itself extends over a period of 7 to 10 days and occurs in the absence of cell division or expansion (see Table 5-1). It is stimulated by ethylene, and synthesis of the gas by tomato fruit is one of the earliest signs of ripening. Using controlled environmental facilities, plants can be grown throughout the year, thus ensuring a continuous supply of material for experiments. Furthermore, the tomato has a good genetic map and a number of ripening mutants exist that are proving useful in dissecting the mechanism of ripening and its regulation.

Changes during tomato ripening begin with ethylene synthesis, which is followed by the appearance of polygalacturonase (PG) activity and the production of lycopene (Fig. 5-10; Grierson et al., 1987). The increase in ethylene synthesis is also accompanied by a rise in respiration, called the respiratory climacteric, that may actually be caused by

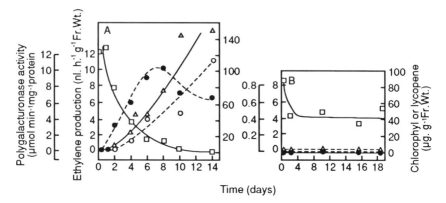

Figure 5-10. Ripening changes in normal and *rin* mutant tomatoes. Plants were grown in controlled environment chambers (16-hour light, 23°C; 8 hour dark, 15°C). Time 0 = 49 days after anthesis. Ethylene synthesis (●); polygalacturonase synthesis (○); chlorophyll (□); and lycopene (△). A. normal fruit; B. rin mutant. (Grierson et al., in *Developmental Mutants in Higher Plants,* H. Thomas, D. Grierson (eds.), Cambridge University Press, Cambridge, pp. 73-94, 1987.)

ethylene. The appearance of PG activity has been shown to be due to de novo enzyme synthesis. This enzyme accumulates in the cell walls where it plays a major part in solubilization of the pectin fraction leading to softening. Lycopene and β-carotene accumulate in the plastids as they are converted from chloroplasts to chromoplasts, a complex process. Initial changes in the chloroplasts, such as the dissolution of starch, may begin before ripening. This is followed by the loss of chlorophyll, which is associated with the disappearance of thylakoids. Throughout the ripening process synthesis and turnover continue. Many proteins disappear and new, but unidentified, ones accumulate.

Many mutations have been noted by plant breeders that affect ripening and their chromosomal location and map position determined. Among these, the ripening inhibitor (*rin*) mutation has an extreme effect on ripening behavior (Fig. 5-10; Grierson et al., 1987).

5-4.1   The Isolation and Characterization of *rin* Mutants of Tomato

Fruits of the *rin* mutant remain green, whereas normal fruits ripen and turn red. The mutant eventually turns a lemon color with little or no lycopene development. Genetic analysis showed that *rin* was a monogenic, recessive characteristic with no linkage associations except large sepal size.

Ethylene production of normal fruits begins to increase soon after harvest, whereas *rin* fruits maintain a low and constant production rate during ripening. In one example (Table 5-12) no surge of ethylene was detected in *rin* fruits monitored for up to 120 days from harvest (Herner and Sink, 1973). Apparently, the *rin* mutant lacks the genetic capacity for autocatalytic production of ethylene in the fruit in relation to ripening. When fruits were cut in half and ethylene production monitored, both *rin* and normal fruit exhibited a capacity for wound-induced ethylene production. The biochemical reason behind this distinction is intriguing and also unknown (Herner and Sink, 1973).

A number of enzymes change in activity during the respiratory climacteric that is associated with ripening in tomatoes. Two of the most obvious changes that occur are carotenoid synthesis and fruit softening, which is due to the activity of pectolytic enzymes. One of these enzymes, PG, is absent in normal green fruit but increases from the point of color change (Fig. 5-11; Tucker et al., 1980). There is a sequential appearance of two isoenzymes, polygalacturonase 1 and 2, during ripening. PG activity appeared at about the time fruit started to show visible signs of ripening and extractable activity increased with the stage of ripeness. The differential effects of heat enabled an estimate to be made of the relative amounts of the two isozymes of PG. Heating an extract for 5 minutes at 65°C inactivated the PG2 but only 10 to 15 percent of PG1. Thus, PG1 appeared first but reached only a low, stable level. PG2 appeared slightly later but increased dramatically

Table 5-12. Internal ethylene concentrations of normal and *rin* tomato fruits at designated stages of color development.

Tomato Fruit	Time from Harvest	Internal C_2H_4
	days	$\mu l/L$
Normal		
Green	4	0.18±0.03
Breaker	6	2.15±0.76
Orange	8	11.80±2.71
Red-orange	10	24.70±0.81
Red	18	17.92±2.81
rin		
Green	4	0.29±0.01
Green-yellow	30	0.15±0.04
Yellow-green	63	0.16±0.06
Yellow	102	0.22±0.05

(Herner and Sink, *Plant Physiol. 52:* 38-42, 1973; reproduced by permission of the ASPP.)

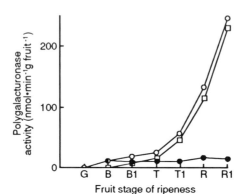

Figure 5-11. Total polygalacturonase (PG) activity and isoenzyme levels extracted from individual Ailsa Craig tomato fruit at various stages of ripeness. Isoenzyme activities were determined by heat inactivation. After heating for 5 minutes at 65°C the remaining activity could be attributed to 90 percent of the original PG1.   G = green; B = breaker (appearance of coloration); T = turning (appearance of orange coloration); T1 = late turning (totally orange fruit); R = ripe (red but firm); R1 = late ripe (red but slightly soft).   Total activity (○); PG1 (●); PG2 (□). (Tucker et al., *Eur. J. Biochem.* *112:* 119-124, 1980.)

as the fruit ripened. The *rin* mutant contained very little detectable PG activity at any stage of development (Tucker et al., 1980).

Several lines of evidence suggest that the *rin* mutation affects some part of the regulatory mechanism governing expression of ripening genes. Thus, although *rin/rin* fruit do not synthesize ethylene during ripening, they do produce the gas in response to wounding. Furthermore, although the fruits do not synthesize PG or its mRNA, an

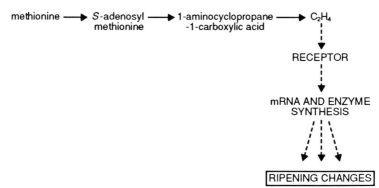

Figure 5-12. Ethylene and ripening in tomato. Ethylene is synthesized from *S*-adenosyl methionine by ACC synthase and the ethylene-forming enzyme (see Case Study 4-4). The gas stimulates its own synthesis and also promotes ripening. (Grierson et al., *Phil. Trans. R. Soc. Lond. B. 314:* 399-410, 1986.)

Table 5-13. Expression of some ripening-related mRNAs from tomato.

clone (pTOM)	protein M_r	mRNA expressed in			
		unripe fruit	ripe fruit	leaves	roots
5	48,000	-	+	-	-
6	55,000	-	+	-	-
13	35,000	-	+	-	-
36	52,000	-	+	-	-
75	28,000	+	+	+	+
99	44,000	-	+	-	-
137	57,000	+	+	+	+

(Grierson et al., *Phil. Trans. R. Soc. Lond. B. 314: 399-410, 1986.*)

immunologically similar enzyme is produced in the flower abscission zones. It seems probable that the mutation affects part of the response mechanism that switches on the "ripening genes" (Fig. 5-12; Grierson et al., 1986).

The changes in mRNA content that occur during tomato ripening were studied by translation in vitro and cDNA cloning. A number of ripening-related cDNA clones have been characterized. Five of the seven ripening-related clones identified were homologous to mRNAs, which were expressed only in ripe tomatoes, whereas two were also expressed in unripe fruit, leaves, and roots (Table 5-13; Grierson et al., 1986). To investigate the timing of these mRNAs in relation to ripening their expression was followed in fruit allowed to ripen while attached to the plant. The five ripening-specific mRNAs increased in concentration

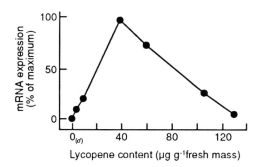

Figure 5-13. An example of the expression of mRNA homologues to a cDNA clone during ripening in tomato. The stage of ripening was determined by monitoring lycopene content of fruits. (Redrawn from Grierson et al., *Phil. Trans. R. Soc. Lond. B. 314:* 399-410, 1986.)

several hundredfold during ripening.   None of the mRNAs was particularly abundant, in each case representing less than 0.05 percent of the total mRNA.   All of the mRNAs appeared at an early stage of ripening.   Maximum expression was generally observed when tomatoes reached the fully orange stage, although the mRNAs persisted in soft red fruit (Fig. 5-13; Grierson et al., 1986).   Northern analysis of 6-week RNA samples, corresponding to the time of peak PG mRNA expression during normal ripening, showed a low level of hybridization only in the *rin* mutant (Fig. 5-14), thus confirming that one of the genes with much reduced expression in this mutant is that for the enzyme PG (Knapp et al., 1989).

5-4.2   The Use of the *rin* Mutant of Tomato in Physiological Studies

5-4.2.1   Polygalacturonase and Cellulase Enzymes and the Respiratory Climacteric

Fleshy fruits have been classified into general categories, climacteric and nonclimacteric, depending on changes in respiration during ripening

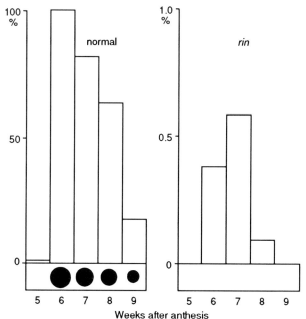

Figure 5-14. Levels of expression of PG mRNA in normal and *rin* fruit of tomato.   Note the hundredfold difference in scale for the *rin* histogram. (Knapp et al., *Plant Mol. Biol. 12:* 105-116, 1989; reprinted by permission of Kluwer Academic Publishers.)

and the response to exogenous ethylene. The sudden upsurge called the *climacteric rise* in respiration is often regarded as a turning point in the life of the fruit when development and maturation are complete and before senescence and deterioration have begun.

Because the *rin* mutant of tomato, a climacteric fruit, behaves like a nonclimacteric fruit, it is a system that can be used to investigate the physiological changes that occur before the respiratory climacteric. The tomato plants used were of the Rutgers and *rin* varieties. Changes in PG activity during fruit development in Rutgers and *rin* were compared (Fig. 5-15; Poovaiah and Nukaya, 1979). In Rutgers fruits, a detectable increase in PG activity was observed about 6 days before the respiratory climacteric. In *rin*, however, no change in PG activity was detected up to 100 days postanthesis. Cellulase activity started to increase in Rutgers fruits before the respiratory climacteric and continued to increase thereafter (Fig. 5-16). Similar changes in cellulase activity were also observed in the nonclimacteric *rin* fruits.

These results (Figs. 5-15 and 5-16) clearly illustrate the important point that PG plays a key role in fruit ripening, whereas cellulase is not directly involved in the process.

5-4.2.2 *Is Polygalacturonase Alone Responsible for Tomato Fruit Softening?*

The cell wall of tomato fruit is similar to other plant cell walls, and softening is thought to result from cell wall modifications that occur

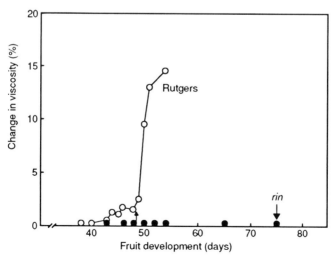

Figure 5-15. Changes in PG activity during fruit development in Rutgers and *rin* tomatoes. Arrow indicates onset of climacteric in Rutgers fruits. (Poovaiah and Nukaya, *Plant Physiol. 64*: 534-537, 1979; reproduced by permission of the ASPP.)

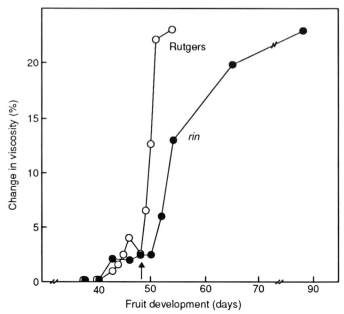

Figure 5-16.   Changes in cellulase activity during fruit development in Rutgers and *rin* tomatoes.   Arrow indicates onset of climacteric in Rutgers fruits. (Poovaiah and Nukaya, *Plant Physiol. 64:* 534-537, 1979; reproduced by permission of the ASPP.)

during ripening (discussed previously).   It has been proposed that polyuronide degradation is the primary determinant of tomato fruit softening and the agent of that degradation, the enzyme PG.   The evidence for this hypothesis is compelling.   First, coincident with the onset of fruit softening there is a dramatic increase in PG activity, mRNA concentration, and relative rate of gene transcription (see above). Second, there is a rough correlation between levels of PG activity and the rate of tomato fruit softening in different cultivars and in ripening-impaired mutants. Third, there is little correlation between the activity of other cell wall degrading enzymes (eg., cellulase discussed earlier) and the rate of tomato fruit softening (Giovannoni et al., 1989).

To clarify the role of PG in polyuronide degradation, tomato fruit softening, and other parameters of ripening, the strategy has been developed to modify PG gene expression in vivo and assess its physiological function using the pleiotropic tomato mutant *rin*. As we have seen, the fruits of *rin* do not soften, produce only basal levels of ethylene, do not accumulate carotenoid pigments, and have greatly reduced levels of PG, relative to wild-type. Thus, the *rin* mutation does not simply represent a lesion in the PG gene; the *rin* and PG loci are located on different chromosomes. The failure of *rin* fruit to activate PG gene expression results from a block at the level of gene transcription (Giovannoni et al., 1989).

The strategy in this study was to induce PG gene expression in *rin* fruit and to examine the effect on polyuronide degradation, fruit softening, and other parameters of ripening. For this purpose, a chimeric gene was constructed consisting of the PG structural gene fused to the regulatory sequence of another ripening-associated gene of unknown function, *E8*, but whose expression is tightly coordinated with that of the PG gene in wild type. Whereas in *rin* fruit PG gene transcription is severely inhibited, the relative rate of *E8* gene transcription is 60 percent of wild-type level. In addition, *E8* gene transcription, and not PG, can be activated by ethylene (propylene) in both unripe wild type and *rin* fruit. The strategy, then, was to introduce the chimeric gene into the *rin* mutant [*rin*(*E8*/PG)], induce its expression and follow the physiological consequences of its expression (Giovannoni et al., 1989) when compared to wild-type, the *rin* mutant itself, and the *rin* mutant transformed with plasmid not containing the chimeric (C) gene [*rin*(C)].

In the presence of propylene, used to activate the *E8* promoter, PG activity and polyuronide degradation could be caused to increase in the *rin*(*E8*/PG) transformant but not in *rin* or *rin*(C). Over the same time period and under the same treatment, however, only the wild-type fruits increased their compressibility (a measure of softening of the fruit) (Fig. 5-17; Giovannoni et al., 1989).

These results (Fig. 5-17) indicate that PG is the primary determinant of cell wall polyuronide degradation but that this degradation is not sufficient for the induction of softening (illustrated here), elevated rates of ethylene biosynthesis, or lycopene accumulation in *rin* fruits (not shown here). Additional events are required before the full range of ripening responses can be invoked in *rin* fruits.

5-4.2.3 *Studying Ethylene Action Using the* rin *Mutant of Tomato*

One hypothesis for the mechanism of ethylene action in the ripening process is that it controls the expression of specific genes. Exposure of unripe tomato fruit to exogenous ethylene has been shown to increase both the concentration of specific mRNAs and the transcription rate of specific genes that are expressed during ripening. As already discussed, a number of normal ripening processes (e.g., chlorophyll degradation, synthesis of carotenoid pigments, breakdown of cell wall components, climacteric respiration, increased ethylene biosynthesis) are inhibited by the *rin* mutation. All other aspects of *rin* plant growth and development, with the exception of calyx size, appear to be normal. Therefore, the *rin* mutant could be used to investigate the role of ethylene in ripening (Lincoln and Fischer, 1988).

The expression of four ethylene-inducible genes (*E4, E8, E17*, and *J49*) were compared in wild-type (Ailsa Craig) and an isogenic *rin* mutant during fruit development. *E4* and *E17* mRNA levels remained very low throughout *rin* fruit development (Fig. 5-18). The level of *E8* mRNA was significantly suppressed, and accumulated to only 30 percent of the level attained during wild-type fruit development. *J49* mRNA

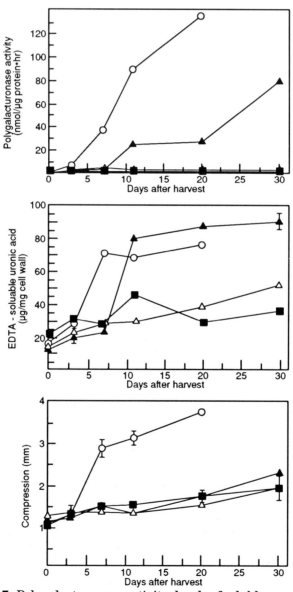

Figure 5-17. Polygalacturonase activity, levels of soluble uronides and the compressibility of wild type (WT), *rin*, and transgenic *rin* fruits of tomato treated with propylene.   WT (○); *rin* (■); *rin*(C) (△); *rin*(E8/PG) (▲). (Giovannoni et al., *Plant Cell 1:* 63-63: 1989; reproduced by permission of the ASPP.)

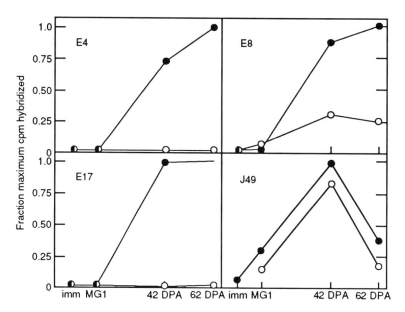

Figure 5-18. Accumulation of specific mRNAs during tomato fruit development. One microgram of mRNA from wild type fruit (●) or *rin* fruit (○) at the indicated stages was dotted onto nitrocellulose filters and hybridized with cloned ^{32}P-labeled DNA probes. After autoradiography, each dot was excised, and the extent of hybridization was determined by liquid scintillation spectrometry. (Lincoln and Fischer, *Plant Physiol. 88:* 370-374, 1988; reproduced by permission of the ASPP.)

accumulation during fruit development was only slightly depressed by the *rin* mutation. From these results it could be concluded that the *rin* mutation inhibited the accumulation of some but not all ethylene-inducible mRNAs during fruit development. Other experiments with isolated nuclei from normal and *rin* tomato plants indicated that the *rin* mutation inhibited the transcriptional activation of some, but not all, ethylene-inducible genes during fruit development (results not shown) (Lincoln and Fischer, 1988).

Thus, it would seem that the *rin* mutation caused ethylene to remain at basal levels throughout fruit development. These reduced levels of the hormone inhibited transcriptional activation of the *E4* gene and significantly reduced *E8* and *J49* gene transcription, although *J49* mRNA showed normal patterns of accumulation. In contrast, the *E17* gene exhibited relatively normal patterns of gene transcription but *E17* mRNA accumulation was inhibited (see Lincoln and Fischer, 1988, for details).

These results suggest that during fruit ripening, gene expression is regulated by both ethylene levels and sensitivity to ethylene, at both transcriptional and posttranscriptional levels (Lincoln and Fischer, 1988).

5-4.2.4   *Flower Abscission in Mutant Tomato Plants*

Abscission is believed to result from the coordinated degradation of the middle lamella and wall material of cells within the abscission zone. One enzyme believed to be involved in the degradation process is cellulase, which increases during abscission. Not only is the exact site of cellulase action in the abscission zone unknown but it also seems unlikely to be the only enzyme involved in the process. This conclusion has led to the investigation into possible roles for other enzymes, including the pectin-degrading enzymes such as PG (Tucker et al., 1984).

Flower explants from normal and *rin* plants were exposed to ethylene. In each case abscission was observed first 3 to 4 hours after exposure to ethylene. All the explants from normal and *rin* plants had abscinded within 6 hours. There was no detectable PG activity in normal or *rin* abscission zone tissues until 6 to 12 hours after exposure to ethylene. PG activity was found only in the abscission zone and not in the pedicel tissues proximal and distal to the zone (Fig. 5-19A). In contrast to PG, cellulase activity was found associated with proximal, distal, and abscission zone tissue but in no case until more than 12 hours after exposure of the tissues to ethylene (Fig. 5-19B). The appearance of the enzyme activity in *rin* tissues was delayed when compared with the normal (Fig. 5-19B).

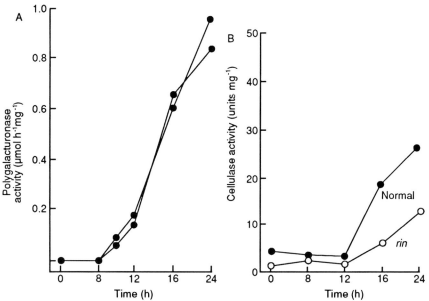

Figure 5-19. Polygalacturonase (A) and cellulase (B) activities associated with explants of abscission zone tissue of normal tomato (●), and the *rin* (▲) mutant at various times after exposure to ethylene. (Tucker et al., *Planta 160*: 164-167, 1984.)

From these results, it can be said that there are problems in correlating either PG or cellulase activity with abscission in tomato flower explants except that of the two, PG is the only one to be found exclusively in the abscission zone and not in adjacent tissues. Rises in the activities of both enzymes occur only after abscission is observed to be complete or well advanced and not at the time of initiation of abscission (Tucker et al., 1984). If the hypothesis is to be maintained, that one or more of these hydrolytic enzymes is involved in flower abscission in tomato, then an explanation has to be found for the apparent disparity between the timing of the abscission process and the appearance of the enzymes in the abscission zone. It may be a matter of the sensitivity of the assay for either one, or both, of the enzymes.

References

Balint, R., Cooper, G., Staebell, M., and Filner, P. 1987. *N*-caffeoyl-4-amino-*n*-butyric acid, a new flower-specific metabolite in cultured tobacco cells, and tobacco plants. *J. Biol. Chem. 262*: 11062-11031.

Bernatzky, R., Anderson, M.A., and Clarke, A.E. 1987. Molecular genetics of self-incompatibility in flowering plants. *Dev. Genet. 9*: 1-12.

Bhadula, S.K., and Sawhney, V.K. 1987. Esterase activity, and isozymes during the ontogeny of stamens of male fertile *Lycopersicon esculentum* Mill., a male sterile *stamenless-2* mutant, and the low temperature-reverted mutant. *Plant Science 52*: 187-194.

Bhadula, S.K., and Sawhney, V.K. 1989. Amylolytic activity and carbohydrate levels during the stamen ontogeny of a male fertile, and a 'gibberellin-sensitive' male sterile mutant of tomato (*Lycopersicon esculentum*). *J. Exp. Bot. 40*: 789-794.

Ebert, P.R., Anderson, M.A., Bernatzky, R., Altschuler, M., and Clarke, A.E. 1989. Genetic polymorphism of self-incompatibility in flowering plants. *Cell 56*: 255-262.

Flores, H.E., and Filner, P. 1985. Metabolic relationships of putrescine, GABA and alkaloids in cell and root cultures of Solanaceae. In *Primary and Secondary Metabolism of Plant Cell Cultures*, K-H. Neumann, W. Barz, and E. Reinhard (eds), Springer-Verlag, Heidelberg, pp. 174-185.

Galston, A.W. 1983. Polyamines as modulators of plant development. *BioScience 33*: 382-388.

Galston, A.W., and Kaur-Sawhney, R. 1988. Polyamines as endogenous growth regulators. In *Plant Hormones and Their Role in Plant Growth, and Development*, P.J. Davies (ed.), Kluwer Academic Publishers, Dortrecht, Netherlands, pp. 280-295.

Galston, A.W., and Sawhney, R.K. 1990. Polyamines in plant physiology. *Plant Physiol. 94*: 406-410.

Giovannoni, J.J., DellaPenna, D., Bennett, A.B., and Fischer, R.L. 1989. Expression of a chimeric polygalacturonase gene in transgenic *rin* (ripening inhibitor) tomato fruit results in polyuronide degradation but not fruit softening. *Plant Cell 1*: 53-63.

Grierson, D., Maunder, M.J., Slater, A., Ray, J., Bird, C.R., Schuch, W., Holdsworth, M.J., Tucker, G.A., and Knapp, J.E. 1986. Gene expression during tomato ripening. *Phil. Trans. R. Soc. Lond. B. 314*: 399-410.

Grierson, D., Purton, M.E., Knapp, J.E., and Bathgate, B. 1987. Tomato ripening mutants. In *Developmental Mutants in Higher Plants*, H. Thomas, and D. Grierson (eds.), Cambridge University Press, Cambridge, pp. 73-94.

Herner, R.C., and Sink, K.C. Jr. 1973. Ethylene production, and respiratory behavior of the *rin* tomato mutant. *Plant Physiol. 52*: 38-42.

Haughn, G.W., and Somerville, C.R. 1988. Genetic control of morphogenesis in *Arabidopsis. Dev. Genet. 9*: 73-89.

Hiatt, A.C., McIndoo, J., and Malmberg, R.L. 1986. Regulation of polyamine biosynthesis in tobacco. Effects of inhibitors, and exogenous polyamines on arginine decarboxylase, ornithine decarboxylase, and S-adenosylmethionine decarboxylase. *J. Biol. Chem. 261*: 1293-1298.

Isogai, A., Takayama, S., Tsukamoto, C., Ueda, Y., Shiozawa, H., Hinata, K., Okazaki, K., and Suzuki, A. 1987. S-locus-specific glycoproteins associated with self-incompatibility in *Brassica campestris. Plant Cell Physiol. 28*: 1279-1291.

Knapp, J., Moureau, P., Schuch, W., and Grierson, D. 1989. Organization and expression of polygalacturonase and other ripening related genes in Ailsa Craig "Niverripe", and "Ripening inhibitor" tomato mutants. *Plant Mol. Biol. 12*: 105-116.

Lincoln, J.E., and Fischer, R.L. 1988. Regulation of gene expression by ethylene in wild type and *rin* tomato (*Lysopersicon esculentum*) fruit. *Plant Physiol. 88*: 370-374.

Malmberg, R.L. 1979. Temperature-sensitive variants of *Nicotiana tabacum* isolated from somatic cell culture. *Genetics 92*: 215-221.

Malmberg, R.L. 1980. Biochemical, cellular, and developmental characterization of a temperature-sensitive mutant of *Nicotiana tabacum* and its second site revertant. *Cell 22*: 603-609.

Malmberg, R.L., and McIndoo, J. 1983. Abnormal floral development of a tobacco mutant with elevated polyamine levels. *Nature 305*: 623-625.

Malmberg, R.L., and McIndoo, J. 1984. Ultraviolet mutagenesis and genetic analysis of resistance to methylglyoxal-bis (guanylhydrazone) in tobacco. *Mol. Gen. Genet. 196*: 28-34.

Malmberg, R.L., McIndoo, J., Hiatt, A.C., and Lowe, B.A. 1985. Genetics of polyamine synthesis in tobacco: Developmental switches in the flower. *Cold Spring Harbor Symp. Quant. Biol., L*: 475-482.

Martinez-Zapater, J.M., and Somerville, C.R. 1990. Effect of light quality and vernalization on late-flowering mutants of *Arabidopsis thaliana. Plant Physiol. 92*: 770-776.

Nasrallah, J.B., Doney, R.C., and Nasrallah, M.E. 1985a. Biosynthesis of glycoproteins involved in the pollen-stigma interaction of incompatibility in developing flowers of *Brassica oleracea* L. *Planta 165*: 100-107.

Nasrallah, J.B., Kao, T-H., Goldberg, M.L., and Nasrallah, M.E. 1985b. A cDNA clone encoding an S-locus-specific glycoprotein from *Brassica oleracea. Nature 318*: 263-267.

Nasrallah, J.B., Yu, S-M., and Nasrallah, M.E. 1988. Self-incompatibility genes of *Brassica oleracea*: Expression, isolation, and structure. *Proc. Natl. Acad. Sci. USA 85*: 5551-5555.

Nasrallah, M.E., Barber, J.T., and Wallace, D.H. 1970. Self-incompatibility in plants: Detection, genetics, and possible mode of action. *Heredity 25*: 23-27.

Nasrallah, M.E., and Wallace, D.H. 1967. Immunogenetics of self-incompatibility in *Brassica oleracea* L. *Heredity 22*: 519-527.

Nester, J.E., and Zeevaart, J.A.D. 1988. Flower development in normal tomato, and a gibberellin-deficient (*ga-2*) mutant. *Am. J. Bot. 75*: 45-55.

Poovaiah, B.W., and Nukaya, A. 1979. Polygalacturonase and cellulase enzymes in the normal Rutgers and mutant *rin* tomato fruits and their relationship to the respiratory climacteric. *Plant Physiol. 64*: 534-537.

Rastogi, R., and Sawhney, V.K. 1988a. Flower culture of male sterile *stamenless-2* mutant of tomato (*Lycopersicon esculentum*). *Am. J. Bot. 75*: 513-518.

Rastogi, R., and Sawhney, V.K. 1988b. in vitro culture of stamen primordia of the normal and a male sterile *stamenless-2* mutant of tomato (*Lycopersicon esculentum* Mill.). *J. Plant Physiol. 133*: 349-352.

Rastogi, R., and Sawhney, V.K. 1989. in vitro development of angiosperm floral buds and organs. *Plant Cell, Tissue Organ Culture 16*: 145-174.

Sawhney, V.K. 1983. Temperature control of male sterility in a tomato mutant. *J. Hered. 74*: 51-54.

Sawhney, V.K. 1984. Hormonal and temperature control of male-sterility in a tomato mutant. In *Proceedings 8th International Symposium on Sexual Reproduction in Seed Plants, Ferns, and Mosses*. Purdoc Publ., Wageningen, Netherlands, pp. 36-38.

Sawhney, V.K., and Greyson, R.I. 1973. Morphogenesis of the *stamenless-2* mutant in tomato. 1. Comparative description of the flowers and ontogeny of stamens in the normal and mutant plants. *Am. J. Bot. 60*: 514-523.

Sawhney, V.K., and Greyson, R.I. 1979. Interpretations of determination and canalisation of stamen development in a tomato mutant. *Can. J. Bot. 57*: 2471-2477.

Sawhney, V.K., and Bhadula, S.K. 1987. Characterization and temperature regulation of soluble proteins of a male sterile tomato mutant. *Biochem. Genet. 25*: 717-728.

Sawhney, V.K., and Bhadula, S.K. 1988a. Histological and biochemical analyses of microsporogenesis in the normal and male sterile, *stamenless-2* mutant of tomato. In *Proceedings 10th International Symposium on Sexual Reproduction*, M. Cresti, P. Gori, and E. Pacini (eds.), Springer-Verlag, Berlin, pp. 187-192.

Sawhney, V.K., and Bhadula, S.K. 1988b. Microsporogenesis in the normal and male-sterile *stamenless-2* mutant of tomato (*Lycopersicon esculentum*). *Can. J. Bot. 66*: 2013-2021.

Takayama, S., Isogai, A., Tsukamoto, C., Ueda, Y., Hinata, K., Okazaki, K., and Suzuki, A. 1987. Sequences of *S*-glycoproteins, products of the *Brassica campestris* self-incompatibility locus. *Nature 326*: 102-105.

Trick, M., and Flavell, R.B. 1989. A homozygous *S* genotype of *Brassica oleracea* expresses two *S*-like genes. *Mol. Gen. Genet. 218*: 112-117.

Tucker, G.A., Robertson, N.G., and Grierson, D.   1980.   Changes in polygalacturonase isozymes during the 'ripening' of normal and mutant tomato fruit. *Eur. J. Biochem. 112*: 119-124.

Tucker, G.A., Schindler, C.B., and Roberts, J.A. 1984. Flower abscission in mutant tomato plants. *Planta 160*: 164-167.

6   Mutants Affecting Seed Dormancy and Development

The formation of a seed in the life cycle of higher plants is a unique adaptation. It incorporates embryo development with various physiological processes that are meant to ensure the survival of the plant in the next generation. These adaptations include the accumulation of nutritive reserves, an arrest of tissue growth and development, and finally, dessication. To survive long periods in this dry state, until environmental conditions are favorable to resume development into a seedling, numerous seeds have also acquired different mechanisms of dormancy.

From a more basic viewpoint, seed development is a convenient experimental system to study the underlying mechanisms of physiological and molecular regulation of cell and tissue development. The stages of seed development and germination involve both spatial and temporal regulation of cell and tissue growth and function. The sequence of events begins with rapid endosperm and embryo growth and differentiation after fertilization, followed by the transition from a state of high metabolic activity and growth to a quiescent state, and finally, a switch back to active growth of the embryo to form a seedling (Fig. 6-1; Quatrano, 1988).

During embryo maturation, there is a buildup of nutrient reserves, an arrest of tissue growth, a development of desiccation tolerance, and the acquisition of dormancy mechanisms. After desiccation, activation of the embryo results in initiation of a meristematic type growth pattern forming a seedling and of the utilization of nutrient reserves to support this development. All are associated with the "seed strategy," that is, mechanisms to ensure survival of the young seedling. Of course, there are exceptions (Quatrano, 1988). Because these processes are distinct and separable one can use them to study experimentally some stages in seed development. The objective in this chapter is to describe a few examples in which mutants have been used to investigate some of the prominent events in higher plant seed physiology.

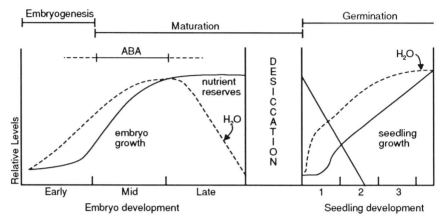

Figure 6-1. A generalized graph showing the relative levels of nutrient reserves, water content and growth during the embryogenic, maturation, and germination stages of embryo and seedling development. Times stated for seedling development are in days. Desiccation separates the end of maturation from the initiation of seedling growth, and ABA levels are temporally correlated with the onset of maturation and prevention of precocious germination during midembryo development. (Quatrano, in *Plant Hormones and Their Role in Plant Growth and Development*, P.J. Davies (ed.), pp. 494-514, 1988; reprinted by permission of Kluwer Academic Publishers).

Hormones are thought to play an important role in these processes because the amounts and activities of various hormones change dramatically during this developmental sequence. Cytokinins, auxins, gibberellins (GA), and abscisic acid (ABA) are found in relatively high concentrations in extracts from seeds at different developmental stages. One must critically ask, however, whether the changes in amounts or activities of hormones are correlated with changes in embryo growth, and do they play a causal role in embryo development? Based on correlative evidence only, a regulatory role for ABA has been suggested in the termination of seed growth, the start of rehydration, the prevention of precocious germination, the stimulation of food reserves, the induction of the developmental arrest, and the induction of primary dormancy (Koornneef, 1986).

The basic framework of the hormone hypothesis of seed dormancy originates from Luckwill (see Karssen and Lacka, 1986), who suggested that the dormancy of apple seeds depends on the interaction between naturally occurring growth-inhibiting and growth-promoting substances. In recent years, the interaction has often been described as a balance between simultaneously occurring promotive hormones, such as GAs and cytokinins, and the inhibitory hormone ABA. The hypothesis is based mainly on the effects of exogenous growth regulators on germination and dormancy, in which dormancy can be maintained, imposed, and released by these chemicals singly or in combination. Such effects are considered to reflect the action of naturally occurring hormones within seeds. Environmental factors, such as light and temperature conditions, are thought to operate on dormancy by causing changes in the balance between promotors and inhibitors (Karssen and Lacka, 1986).

Recent reviews, however, unanimously judge the present experimental evidence insufficient to justify the view that endogenous hormones control dormancy and germination, let alone that the control occurs in the form of a balance between promotors and inhibitors. There is a lack of correlation between hormone levels and dormancy. Improvement might be expected from a general adaptation of recent advances of hormone assays, in particular, the use of immunoassays that might enable studies of hormone compartmentation and redistribution within seed tissue. A lack of correlation between dormancy and hormone levels, however, might still exist because regulation is based on the sensitivity to hormones rather than on their levels. Such an hypothesis cannot be studied easily because endogenous hormone levels always interfere in the test for the sensitivity to applied growth regulators. Mutants that lack the capacity to synthesize a hormone or group of hormones are unique tools with which to circumvent that dilemma. Moreover, they provide the possibility of answering the question whether or not the presence of a hormone is required for a certain developmental process (Karssen and Lacka, 1986). In Case Study 6-1, discussion centers especially on a group of ABA-deficient mutants and the ways in which they have been used to investigate the role of hormones in seed dormancy.

Another way to investigate the causal relationships between ABA and developmental arrest is to make use of plants whose embryos are viviparous (i.e., are not arrested), but germinate directly upon completion of embryonic growth. Such systems could be used to investigate whether the failure of an embryo to undergo developmental arrest is due to the absence of ABA or an inability to respond to the hormone. Genetic vivipary has been reported, notably in *Avena nuda* L. and *Zea mays* L. The eight viviparous mutants of maize have great potential for studies of the events involved in the induction of developmental arrest because their embryogeny is normal until they germinate (Robichaud et al., 1980). These and other viviparous mutants are the subject of Case Study 6-2.

Grain filling is governed by the interdependent processes of photosynthesis, phloem loading, assimilate transport, phloem unloading, and utilization of sugars in starch synthesis. There exists a series of eight maternal-effect, spontaneous, shrunken endosperm mutants of barley (*Hordeum vulgare* L.) designated seg1- seg8. According to the inheritance pattern, the genes involved are nuclear, not cytoplasmic, but these recessive mutants are unlike most cereal endosperm mutants in that they do not express xenia (phenotype depends on genotype of the seed-bearing plant and is independent of pollen source). Therefore, it is unlikely that enzymes of the starch synthesis pathway are affected (Peterson et al., 1985). They can be categorized into two groups, the chalazal necrosis mutants (*seg1*, *seg3*, *seg6* and *seg7*) and the abnormal endosperm mutants (*seg2*, *seg4*, *seg5* and *seg8*). In the first group, the most likely reason for the maternal inheritance pattern is the death of the maternal chalaza and nucellar projection tissues that, by virtue of their position, must comprise the sugar transport pathway. Their abnormal development and early death prematurely cuts off the supply of sugars to the endosperm. The reason for maternal inheritance of the abnormal endosperm mutants is less clear as the maternal tissue appears to develop normally and characteristic irregularities occur in the endosperm tissue itself.

The mutations may become useful as probes to study endosperm development, sucrose translocation, and source-sink relationships (Peterson et al., 1985). Because of this possible future significance in physiological investigations of a most important aspect of seed development, a short consideration of these mutants is included here as Case Study 6-3.

Biochemical genetic analysis to elucidate gene-enzyme-phenotype relationship is a study of major significance. Such analyses have played a central role in understanding gene function and determination of rate-limiting steps in biosynthetic pathways. In maize, mutants with altered quantity or quality of starch in the endosperm tissue have been of particular importance in this regard. Because their phenotype readily identifies the area of metabolic lesion, gene-enzyme relationships have been intensively examined in these mutants. Four starch mutations are discussed in Case Study 6-4 as are their associated enzymatic lesions: the *waxy* locus with starch granule-bound glucosyl transferase; the *shrunken-2* and *brittle-2* mutations with ADP-glucose pyrophosphorylase; and the *shrunken-1* locus with sucrose synthetase (Chourey, 1981).

Case Study 6-1:   Mutants Affecting Seed Dormancy

Numerous reports have shown that exogenous growth regulators applied singly or in combination maintain, impose, or release dormancy in seeds. Convincing evidence that such effects reflect the action of naturally occurring hormones within the seeds is hard to produce. Attempts to correlate changes in dormancy with changes in endogenous hormone levels mostly fail to present reliable evidence for a causal relationship. Hormone mutants offer unique possibilities to study the role of hormones in seed dormancy without disturbance of the intact system (Karssen et al., 1987).

Thus far, abscisic acid-deficient mutants have been recognized because they showed excessive wilting and/or reduced seed dormancy.

These effects were shown to be reversed by exogenous ABA and correlated with greatly reduced endogenous ABA levels. All ABA-deficient mutants so far described have been monogenic recessives. For example, the tomato mutants with mutations at three different loci, *sitiens (sit)*, *flacca (flc)*, and *notabilis (not)*, are of this type (see Case Study 1-1).

In *Arabidopsis thaliana* and in tomato, GA-deficient mutants have more recently been found that do not germinate without exogenous GA. By selecting for germinating seeds among the progeny of mutagen-treated, nongerminating *(ga-1) Arabidopsis* seeds, mutants were isolated that were reverted at least for the nongermination trait. Genetic analysis showed this reversion to be due to a mutation in a suppressor gene segregating independently from the *ga-1* gene. Later, the suppressor gene mutation was shown to have all the characteristics of ABA deficiency (Koornneef, 1986). Such mutants are the main subject of this Case Study.

6-1.1 The Isolation and Characterization of ABA-deficient Mutants

In *Arabidopsis thaliana*, GA-deficient mutants were isolated by selecting for seeds that did not germinate without an application of GAs. The seedlings that developed from GA-treated seeds showed dwarf growth that could be reverted to normal growth by GA sprays. Deficiency for all GAs was demonstrated in the recessive mutants *ga-1, ga-2* and *ga-3* (Karssen et al., 1987; Koornneef and van der Veen, 1980). The *ga-1* mutants were then subjected to a mutagenic treatment and in their progeny seeds were selected that germinated independently of GA but still developed into dwarfs. Genetic analysis showed that this partial reversion to wild type was due to a mutation in a suppressor gene segregating independently from the *ga-1* gene. Subsequently, this suppressor gene was obtained as a homozygous recessive in a *Ga-1/ Ga-1* background. It was denoted *aba* as it showed all the characteristics of an ABA-deficient mutant (Karssen et al., 1987; Koornneef and van der Veen, 1980). Other mutants have also been isolated through their insensitivity to ABA. These are referred to as *abi* mutants.

Studies with *aba* and *abi* mutants clearly showed for the first time that ABA was absolutely required for the induction of seed dormancy in *Arabidopsis*. It is shown in Figure 6-2 that immature seeds of both an ABA-deficient *(aba-1)* mutant and an ABA-insensitive *(abi-3)* mutant developed full germination capacity, whereas wild-type seeds only showed some precocious germination around day 14. The wild-type and *abi-3* seeds contained high levels of ABA half way through seed development. The *aba* seeds contained practically no ABA during all stages of seed development (Tables 6-1 and 6-2). The germination of both *aba-1* and wild type could be completely inhibited by exogenous applied ABA (Fig. 6-3; Koornneef et al., 1982).

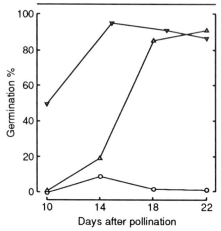

Figure 6-2. Precocious seed germination of different *Arabidopsis thaliana* genotypes in light at 24°C. Seeds were isolated from the siliques at the indicated times after pollination. Germination was recorded after 7 days. Wild-type (○); aba-1 mutant (△); abi-3 mutant (▽). (Karssen et al., in *Developmental Mutants in Higher Plants,* H. Thomas, D. Grierson (eds.), Cambridge University Press, Cambridge, pp. 119-133, 1987; data from Karssen et al., *Planta 157*: 158-165, 1983.)

The level of endogenous ABA was determined in dry ripe seeds and in siliques with seeds during development of several *aba* mutants of *Arabidopsis* (Table 6-2). The ABA in seeds, particularly during seed development, was found to be much reduced or below the level of detection. It appeared that ABA-types were deficient in ABA during various stages of their development (Koornneef et al., 1982). The rather unaffected sensitivity of the ABA-type to exogenously applied ABA (Fig. 6-3) seemed to exclude the possibility that the ABA-receptor sites were affected in the mutant. Very likely, the *aba* gene regulates the biosynthesis of endogenous ABA at all stages of the development of the *Arabidopsis* plant.

Table 6-1. Endogenous ABA levels in unripe siliques containing seeds of different genotypes of *A. thaliana.*

Genotype	ABA, ng $(g fr.wt.)^{-1}$	
	8-11 DAP	14-16 DAP
wild type	169	98
aba mutant	10	10
abi-3 mutant	279	144

(Karssen et al., in *Developmental Mutants in Higher Plants,* H. Thomas and D. Grierson (eds.), Cambridge University Press, Cambridge, pp. 119-133, 1987.)

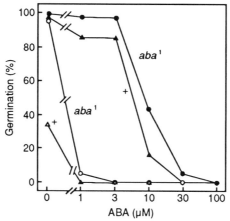

Figure 6-3. Germination in white light of wild type (+) and the *aba-1* mutant of *Arabidopsis thaliana* at different concentrations of ABA, scored 3 days (○,△) and 7 days (●,▲) after incubation. (Koornneef et al., *Theor. Appl. Genet. 61*: 385-393, 1982.)

The fact that the absence of seed dormancy of ABA-deficient mutants segregates as a single recessive gene in the progeny of heterozygous plants shows that dormancy in ripe seeds related to ABA metabolism is mainly determined by the genotype of the embryo and not by the genotype of the mother plant including the seed coat (Koornneef et al., 1982).

Table 6-2. Endogenous ABA content of ripe seeds and of developing siliques with seeds of *Arabidopsis thaliana*. The ripe seeds were extracted within 1 month of harvest and the developing seeds were extracted 10 days after pollination.
[a.] Seed weight 20 g/seed.
[b.] 93 percent or more of total ABA in siliques is present in the seeds.
nt= not tested.

	Ripe seeds harvested in:				Siliques with developing seeds harvested in:	
	April 1979		October 1979		January 1980	
	ng/g fresh weight	pg/ seed[a]	ng/g fresh weight	pg/ seed[a]	ng/g fresh weight	pg/ seed[b]
+/+ (wild-type)	71	1.42	27	0.54	117	10.5
aba^3/aba^3(G4)	8	0.16	7	0.14	12	1.0
aba^1/aba^1(A26)	nt	nt	<1	<0.02	nt	nt
aba^4/aba^4(A73)	nt	nt	<1	<0.02	nt	nt

(Koornneef et. al., *Theor. Appl. Genet. 61*: 385-393, 1982.)

6-1.2   Physiological Studies Using ABA-deficient Mutants

6-1.2.1   *The use of ABA mutants in seed physiology*

ABA-deficient lines of *Arabidopsis* are characterized by symptoms of withering, increased transpiration, and a lowered ABA content in ripe seeds and leaves. Therefore, the phenotype was called ABA-type and the recessive mutant allele, *aba*. Dormancy of the seeds of the ABA-type is strongly reduced but the sensitivity of the ABA-type seeds to applied ABA is only slightly less than that of the wild type (see earlier discussion for details). Reciprocal crosses between wild type and ABA-type have been used to study the origin of ABA in the seeds and the location of the dormancy mechanisms (Karssen et al., 1983).

In the F_1 seeds from the ABA-type x wild-type crosses, the genotypes of the maternal and embryonic tissues were different. If these crosses are made reciprocally, the development of seeds with the *Aba/aba* genotype *in embryo* can be studied in both a wild-type and an ABA-type mother plant. The development of such seeds were compared with F_1 seeds of *aba/aba* x *aba/aba* and *Aba/aba* x *aba/aba* crosses. In the latter case the embryos have genotypes *Aba/aba* and *aba/aba* of equal proportion. Precocious germination was tested during development of these different F_1 seeds (Table 6-3);

Dormancy was absent when both parents were ABA-type, but it developed fully in the heterozygous F_1 seeds irrespective of the genotype of the mother plant. The capacity for precocious germination in F_1 seeds of the *Aba/aba* x *aba/aba* crosses followed a course that happened to match perfectly the curve constructed from the addition of halved values of the curves for *Aba/Aba* x *aba/aba* and *aba/aba* x *aba/aba*

Table 6-3. ABA content and germination of immature (10 and 16 days) and mature (26 days) F_1 seeds of different crosses with *Arabidopsis* wild-type and *aba* mutant.

		Days after pollination				
		10		16		ripe
♀ x ♂	F_1	ABA ng/g	Germ%	ABA ng/g	Germ%	Germ%
Aba/Aba x *aba/aba*	*Aba/aba*	598	39	59	29	0
aba/aba x *Aba/Aba*	*Aba/aba*	35	44	36	49	0
aba/aba x *aba/aba*	*aba/aba*	8	64	1	99	95

(Koornneef, in *A Genetic Approach to Plant Biochemistry*, A.. D. Blonstein and P. J. King (eds.), Springer-Verlag, Wein, chapter 2, 1986.)

crosses (Fig. 6-4; Karssen et al., 1983). Therefore, it was concluded that the development of dormancy with regard to the *aba* gene was regulated by the genotype of the embryo and was not a maternal effect.

The genotype of the mother plant, however, was not completely without effect. Germination tests, for instance with mature seeds, showed that heterozygous F_1 seeds generated on an ABA-type plant required a shorter after-ripening period in dry storage to increase germination in the light than did seeds from a wild-type plant (Fig. 6-5; Karssen et al., 1983). Dormancy was not induced if the ABA level in the seeds was raised artificially. A spray with 100 µM of ABA solutions did not induce dormancy in ABA-type seeds even if the seeds were sprayed repeatedly at regular intervals. On the contrary, the seeds developed the capacity for precocious germination at an earlier time (Fig. 6-6).

The experiments clearly showed that endogenous ABA in developing fruits and seeds of *A. thaliana* has two origins. The first fraction is regulated by the genome of the mother plant (maternal ABA). It is responsible for the peak in ABA content half way in seed development

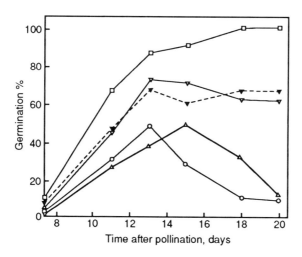

♀	♂		F_1
□	*aba / aba* x *aba / aba* – – – – –➤		*aba / aba*
▼	*Aba / aba* x *aba / aba* – – – – –➤		*Aba / aba* (½ + *aba / aba* (½)
○	*Aba / aba* x *aba / aba* – – – – –➤		*Aba / aba*
△	*aba / aba* x *Aba / Aba* – – – – –➤		*Aba / aba*

Figure 6-4. Germination of isolated developing F_1 seeds from different crosses of wild type and ABA-deficient mutants of *Arabidopsis*. The curve indicated with ▼ is the calculated average of curves □ and ○. Germination was counted after 14 days. (Karssen et al., *Planta 157*: 158-165, 1983.)

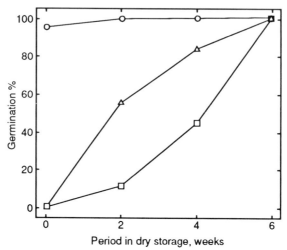

Figure 6-5. Change with time of dry storage in germination of mature F_1 seeds of *Arabidopsis* generated from the crosses *aba/aba* x *aba/aba* (□); *aba/aba* x *Aba/Aba* (△); and *Aba/Aba* x *aba/aba* (○). Germination was counted after 7 days. (Karssen et al., *Planta 157*: 158-165, 1983.)

(see also Fig. 3 in Koornneef, 1986). The second fraction is regulated by the genome of the embryo (embryonic ABA). This fraction is present during maturation, as well as at earlier stages of development (Karssen et al., 1983). The probability that ABA and dormancy induction are not

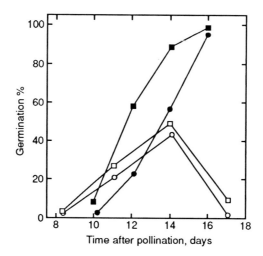

Figure 6-6. Changes with time after pollination in germination of isolated seeds from wild type (○,□) or ABA-type (●,■) *Arabidopsis*. During seed development plants were sprayed at intervals of 4 days with distilled water (○,●) or 100 μM ABA (□,■). Germination was counted after 14 days. (Karssen et al., *Planta 157*: 158-165, 1983.)

causally related is very low because these physiological genetic analyses have shown beyond doubt that a single gene is involved in the mutation. Very likely, the *aba* gene regulates the biosynthesis of endogenous ABA at all stages of development of the *Arabidopsis* plant. Dormancy was only induced in seeds of *A. thaliana* if the genome of the embryo contained the dominant *Aba* allele and, thus, the embryonic ABA fraction. Maternal ABA was not related to dormancy induction. Dormancy developed in spite of its absence (*aba/aba* x *Aba/Aba*). On the other hand, if maternal ABA was present the pattern of dormancy induction correlated still perfectly with the genotypes of embryo and endosperm. If it can be argued that ABA sprays resemble maternal ABA, then the results in Figure 6-6 reinforce these conclusions. It is assumed that their penetration into the embryo is prevented either by a permeation barrier or by an active metabolic breakdown. It seems that to induce dormancy ABA has to be synthesized close to its site of action in the embryo. The action mechanism of embryonic ABA is unknown (Karssen et al., 1983).

6-1.2.2 *The Role of Endogenous Gibberellins During Fruit and Seed Development in* Arabidopsis thaliana.

Fruit development has been related to seed development in several plant species. In many species the weight of the fruit tissues increases with an increasing number of well developed seeds and in some cases it has been possible to kill or remove the seeds and replace them with growth substances that restored normal fruit development. In addition, plant hormones, auxins and GAs in particular, are able to induce parthenocarpic fruits in several plant species. Together with the fact that seeds contain relatively high levels of hormones, and of GAs in particular, these data indicate that the hormones synthesized in the seeds are transported to, and affect the development of, the surrounding maternal fruit tissue. This does not exclude hormone synthesis by the maternal tissue itself (Barendse et al., 1986).

To distinguish between the role of the embryo and its surrounding maternal tissue a genetic approach can be followed when homozygous recessive mutants are available. Such a genetic approach has been applied to study the role of ABA in the induction of dormancy in developing seeds of *A. thaliana* as we have seen in the foregoing sections of this Case Study. In the same species mutants have been described with an absolute GA requirement for both seed germination and elongation growth. The mutants either have no GA synthesis or very little of it. Mutants at the *ga-1* locus were used to analyze the effects of the mutation, that is, GA biosynthesis on silique growth.

A time-course study on silique development after pollination was carried out after reciprocal crosses between wild type and the *ga-1* mutant (Fig. 6-7; Barendse et al., 1986). For all genotypes silique

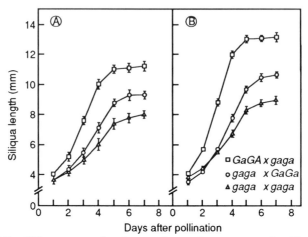

Figure 6-7. Silique growth after reciprocal crosses and self-pollinations with wild type (*GaGa*) and a *ga-1* mutant (*gaga*) of *Arabidopsis*. A. Average number of seeds per silique were, respectively: 41.7 ± 2.7 (□); 41.8 ± 1.9 (○); 42.1 ± 1.9 (△); and B. 60.5 ± 1.5 (□); 60.3 ± 0.9 (○); 59.1 ± 1.7 (△). The GA-sensitive mutant (*gaga*) was treated once with GA 4+7 1 week before pollination. (Barendse et al., *Physiol. Plant. 67*: 315-319, 1986.)

development was recorded for siliques containing a range of 30 to 50 seeds and 50 to 70 seeds. At both seed contents silique growth took place during the first 6 to 7 days after pollination in all genotypes. The differences in the ultimate silique length was the result of different growth rates of the different genotypes in this period. In addition, the growth rate of the silique was affected by the number of seeds per silique, that is, more seeds per silique enhances silique growth (Barendse et al., 1986). More recent evidence using the *ga-1* mutant of tomato indicates that a minimum number of seeds is required for successful fruit development and that this number is related to the presence or absence of the *Ga* allele (Groot et al., 1987).

The average number of ovules present in siliques was about the same in the mutant and the wild type. Seed set, however, was highest in the self-pollinated wild type and lowest in the mutant (Table 6-4;

Table 6-4. The effect of genotype on ovule number and seed set in wild type and a *ga-1* mutant of *Arabidopsis*.

Mother plant	Average number of ovules per silique	Percentage seed set after pollination with	
		Wild type	*ga-1* mutant
Wild type	50.7±1.1	72.9±1.2	36.4±1.2
ga-1 mutant	48.1±0.8	58.2±1.8	26.6±1.6

(Barendse et al., *Physiol. Plant. 67: 315-319, 1986.*)

Table 6-5. Ent-kaurene synthesizing capacity of cell-free preparations from 4 to 5-day-old siliques of wild type and a GA-sensitive mutant (*ga-1*) of *Arabidopsis*.

Genotype	Average fresh weight of siliques (mg)	[^{14}C]-*ent*-kaurene Bq (mg protein)$^{-1}$	
		exp. 1	exp. 2
Wild type	2.8	20.1	24.1
ga-1 mutant	2.5	0.7	0.1

(Barendse et al., *Physiol. Plant.* 67: 315-319, 1986.)

Barendse et al., 1986). Seed set in the crosses was intermediate, being higher in the crosses with the mutant as mother plant than in the reciprocal cross. Thus, both genotype of mother plant and of pollen affect seed set.

The *ent*-kaurene (GA) content of wild-type siliques during the rapid growth period was much higher in the wild type than in the mutant (Table 6-5). Thus, this gene appears to regulate an early step in the GA biosynthesis pathway (see Figure 4-5).

These results suggest a role for endogenous GAs in silique growth. GAs synthesized by the seed were responsible for growth of the surrounding maternal tissue. Silique length, however, was shown to be also affected by the genotype of the mother plant and is not solely determined by embryonic GAs. Thus, GA biosynthesis can originate both from maternal (fruit) tissue and from the developing embryo. Also, because in the GA-deficient genotypes a positive correlation was shown between seed number and silique length, there must be other factors produced by the developing seeds, apart from GAs, that are relevant to silique growth (Barendse et al., 1986).

6-1.2.3 *The Role of GA During Seed Development in Arabidopsis and the Hormone Balance Hypothesis.*

During 6 to 8 weeks of dry storage at room temperature or 4 to 8 days of cold stratification at 2°C wild-type seeds of *A. thaliana* developed the capacity to germinate in the light. Dark germination was not affected. Dormancy breaking treatment also did not promote the germination of *ga-1* seeds in water. When germination was tested in GA_{4+7} solutions it was seen that during cold stratification and dry storage the capacity to germinate in 10 µM GA_{4+7} rose in a similar way in seeds of wild type and *ga-1* mutant (Fig. 6-8). The test showed that cold stratification increased the sensitivity of *ga-1* seeds to GA_{4+7} about 100-fold. Thus, the relief of dormancy in *Arabidopsis* seeds involves a sensitization to GAs that occurs independently of GA biosynthesis (Karssen et al., 1987).

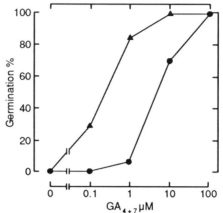

Figure 6-8. Germination of *ga-1* mutant seeds of *Arabidopsis* at 24°C in darkness in a range of GA_{4+7} concentrations with (▲) or without (●) a preceding 7-day incubation at 2°C. (Karssen et al., in *Developmental Mutants in Higher Plants,* H. Thomas, D. Grierson (eds.), Cambridge University Press, Cambridge, pp. 119-133, 1987).

Apart from cold stratification, light also enhances the GA sensitivity of *ga-1* seeds. In light, *ga-1* seeds require about 50 times less GA_{4+7} than in darkness (Fig. 6-9). In light, wild-type seeds are independent of exogenous GA. Such a change has never been observed in *ga-1* seeds. These results have been interpreted as an indication that light stimulates GA biosynthesis. It cannot be excluded, however, that also in wild-type seeds light only enhances GA sensitivity. Germination in water and

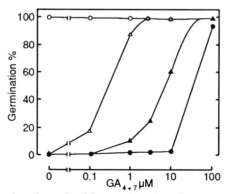

Figure 6-9. Germination of wild type (○,●) and *ga-1* mutant (△,▲) seeds of *Arabidopsis* at 24°C in a range of GA_{4+7} concentrations in constant white fluorescent light (○,△) or darkness (●,▲). (Karssen et al., in *Developmental Mutants in Higher Plants*, H. Thomas, D. Grierson (eds.), Cambridge University Press, Cambridge, pp. 119-133, 1987.)

light will then occur when the seeds have been made sensitive to the level of GA that is present in the seeds in darkness.

Revision of the Hormone Balance Hypothesis

These and other results using GA - and ABA-deficient mutants suggested that the hormone balance hypothesis may require revision. Thus, for example, freshly harvested mature *aba* seeds did not germinate in darkness. A few days of cold stratification sufficed to remove this feeble form of dormancy. In contrast, in seeds of the double mutant, *ga-1/ga-1, aba/aba*, the dark inhibition was more persistent, indicating a dependency on endogenous GA of ABA-deficient seeds. This was confirmed when the GA requirement of non-after-ripened seeds of the *aba, ga-1* and double mutants and wild type was compared in darkness. ABA-deficient seeds, indeed, required a little less GA_{4+7} than seeds of the double mutant, but both genotypes required about 100 times less GA_{4+7} than seeds of the genotypes with dominant *Aba* alleles. Thus, ABA action during seed development increased the GA requirement during germination (Fig. 6-10).

Therefore, the classic hypothesis of dormancy control that supposes a balance between simultaneously acting promotive and inhibitory hormones has to be substituted by a kind of "remote control" of the GA requirement during germination by the ABA of seed development through the intermediate dormant condition.

In conclusion, these studies with ABA-and GA-deficient mutants of *A. thaliana* leave no doubt that ABA and GA play an active role in the regulation of dormancy and germination, respectively. They also indicate that the environmental factors, light and low temperature, primarily act on the sensitivity of seeds to GA. It has also been shown that hormone mutants are very useful to determine when and where hormones are active.

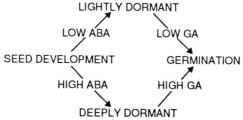

Figure 6-10. Schematic presentation of the control of seed dormancy and germination by ABA and GA in *Arabidopsis*. (Karssen and Lacka, in *Plant Growth Substances*, M. Bopp (ed.), Springer-Verlag, Heidelberg, pp. 315-323, 1986.)

Case Study 6-2:   The Viviparous Mutants

Vivipary, or premature germination, in maize (*Zea mays* L.) is expressed by the development of seedlings from immature embryos while still attached to the parent plant. The alteration in normal seed development can result from mutation at any one of several genes. Some viviparous mutants also affect carotenoid biosynthesis in the endosperm and developing seedling. Immature embryos from these mutants have reduced levels of ABA and have been postulated to represent blocks in ABA biosynthesis.

In another mutant case there is a unique pleiotropic effect in which carotenoid biosynthesis is not affected and the ABA content in developing embryos is normal; growth of embryos in culture in such cases is not inhibited by ABA. Instead, anthocyanin biosynthesis in the aleurone layer of the endosperm is blocked by the mutation. The biochemical basis for this viviparous phenomenon is not known (Wilson et al., 1973; Robertson, 1975; Dooner, 1985).

6-2.1   The characterization of viviparous mutants

The viviparous mutants of *Zea mays* are characterized by precocious germination 30 to 36 days after pollination, indicating a failure of the normal dormancy-inducing mechanism. In addition to premature germination, all viviparous mutants cause obvious changes in seed and/ or seedling phenotype. These pleiotropic effects allow the identification of homozygous viviparous seed on segregating ears to be made as early as 14 days postpollination. There are two major classes of these

Table 6-6. Growth of maize embryos cultured for 20 days with and without ABA in the medium.
[a.] *vp 1* growth was retarded but not inhibited at 3 mg percent ABA.
[b.] *vp 2* and *vp 9* seedlings are albino but grew at near normal rate.
- No visible growth.

Genotype	Growth	
	+ABA	-ABA
vp1 /vp1	+[a]	+
normal segregates	-	+
vp2 /vp2	-	+[b]
normal segregates	-	+
vp9/vp9	-	+[b]
normal segregates	-	+

(Smith et al., *Maize Genet. Coop. Newsletter 52*: 107-108, 1978.)

Table 6-7. ABA content of 20 day old maize embryos as determined by gas chromatography.

Genotype	ABA (mg)			% of normal
	Bound	Free	Total	
vp1/vp1	0.008	0.021	0.029	90.6
vp1/vp1	0.003	0.009	0.012	37.5
normal	0.006	0.026	0.032	

(Smith et al., Maize Genet. Coop. Newsletter 52: 107-108, 1978.)

mutants in maize, class I and II (Table 6-6; Robertson, 1975; Smith et al., 1978).

Class I mutants have normal carotenoid content in the endosperm and produce normal green seedlings and plants when homozygous. They fail to produce anthocyanin, however, in the aleurone in the presence of the genes required for aleurone color. The *vp1* mutant, which belongs to this class, produces normal amounts of ABA (Table 6-7), and growth of homozygous *vp1* embryos in culture is not inhibited by ABA at concentrations that prevent growth of normal embryos (Table 6-6). Such mutants may be defective for a specific ABA receptor (Smith et al., 1978).

Class II mutants fail to develop carotenoids in endosperm and leaf tissue, and the seedlings are albino when grown under normal illumination. Studies with *vp9* illustrated that these mutants were ABA deficient (Table 6-7). Embryo ABA content was 25 to 30 percent of that found in nonviviparous embryos of the same ears at various stages of development. In addition, the presence of ABA in the culture medium completely inhibited growth of *vp2* and *vp9* embryo explants (Table 6-6). Thus, the class II mutants probably lacked ABA (Smith et al., 1978).

6-2.2   The Use of Viviparous Mutants in Physiological Studies

6-2.2.1   *Possible Pathways of ABA Biosynthesis*

ABA is a sesquiterpenoid derived from mevalonic acid (MVA). The most likely pathway for ABA synthesis is that in which a C-40 precursor, such as violoxanthin, a xanthophyll that is derived from carotenes, is cleaved. Xanthoxin is probably the ABA precursor derived from violoxanthin (Fig. 6-11; Koornneef, 1986).

All the carotenoid-deficient mutants of maize have reduced but significant levels of ABA in embryo tissue, the values being variable and age dependent. Published data range from 7 to 70 percent of that of

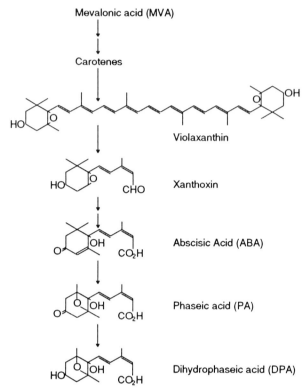

Mevalonic acid (MVA)

Carotenes

Violaxanthin

Xanthoxin

Abscisic Acid (ABA)

Phaseic acid (PA)

Dihydrophaseic acid (DPA)

Figure 6-11. The proposed "indirect" biosynthetic pathway for ABA and the major pathway for its catabolism. (Koornneef, in *A genetic Approach to Plant Biochemistry,* A.D. Blonstein, P.J. King (eds.), Springer-Verlag, Wien, pp. 35-54, 1986.)

the wild type.   However, ABA is not detectable in *w3, vp5* and *vp7* seedlings and roots (Fig. 6-12; Koornneef, 1986).

The blocks in carotenogenesis suffered by the mutants just discussed provide strong arguments for the C-40 pathway of ABA synthesis. Until very recently, however, no detailed biochemistry had been applied to these mutants (but see later in this Case Study). On the other hand, the green wilty mutants of tomato have been used to investigate the biochemistry of ABA synthesis. These mutants apparently represent, if the C-40 pathway operates, mutations in the last part of the ABA pathway (Koornneef, 1986).

Proponents of the indirect pathway have assumed that violaxanthin is the most likely C-40 precursor of ABA (Fig. 6-11 and 6-13).   Only three steps are required to convert the *cis*-xanthoxin to ABA.

1. Oxidation of the side-chain aldehyde to give the corresponding carboxylic acid.

2. Isomerization to form the γ-hydroxy, α,β unsaturated ring.

3. Oxidation of the 4' alcohol to a ketone.

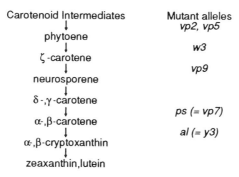

Figure 6-12. Carotenoid biosynthesis in maize and the metabolic blocks of different viviparous mutants. (Koornneef, in *A genetic Approach to Plant Biochemistry,* A.D. Blonstein, P.J. King (eds.), Springer-Verlag, Wien, pp. 35-54, 1986.)

Thus, there are only three enzyme-catalyzed steps that could be subject to genetic lesion (Fig. 6-13; Taylor, 1987).

If any ABA-deficient mutants can be shown to impair the conversion of labeled 2-*cis*-xanthoxin to ABA, then it should be easy to pinpoint the

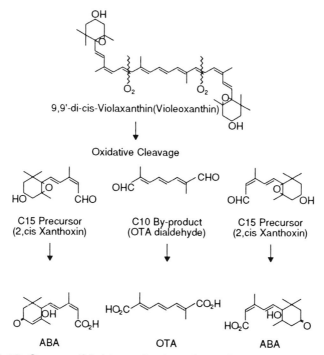

Figure 6-13. One possible biosynthetic pathway for ABA from a carotenoid precursor. OTA = 2,7-dimethyl-2,4,6-octatrienedioic acid. (Taylor, in *Developmental Mutants in Higher Plants,* H. Thomas, D. Grierson (eds.), Cambridge University Press, Cambridge, pp. 197-217, 1987.)

lesion. Elevated levels of C-10 byproducts would also be a clear indication that the mutants are acting "downstream" of the cleavage step. On theoretical grounds, the most likely initial byproduct of ABA biosynthesis would be the C-10 aldehyde, 2,7-dimethyl-2,4,6-octatriene dialdehyde, which might be oxidized in vivo to the corresponding dicarboxylic acid, 2,7-dimethyl-2,4,6-octatrienedioic acid (OTA). Assuming that any byproducts of ABA biosynthesis are not instantly metabolized, their concentration would be a reflection of the rate of cleavage of the C-40 precursor. Measuring byproduct levels could, therefore, provide important information on the rate of operation of this key step in the pathway (Fig. 6-13; Taylor, 1987). "Upstream" mutations affecting ABA biosynthesis at the C-40 level may be more difficult to define. However, any mutant having reduced levels of biosynthetic C-10 byproducts or C-15 intermediates would be a possible candidate for the upstream category.

Further confirmation of the relative sites of action of different mutants could be provided by an analysis of byproduct levels in double mutant combinations. Any upstream mutant should be dominant (epistatic) over the downstream mutants. By virtue of acting earlier in the pathway, the upstream mutants should ensure that the relatively low concentrations of the byproducts are found in double mutant homozygotes.

So far, OTA and OTA-dialdehyde have not been detected in any extracts of tomato tissue. Both ODA (2,7-dimethyl-2,4-octadienedioic acid) and OEA (2,7-dimethyl-4-octenedioic acid), however, have been detected in plant tissues. The reason for the occurrence of ODA and not OTA is not known (Table 6-8; Linforth et al., 1987; Taylor, 1987).

Table 6-8. Characteristic levels of ABA and the C-10 dicarboxylic acids (OTA, ODA and OEA, see the text) in ABA-deficient tomato genotypes grown in a controlled environment at 95 percent RH for 6 weeks. ND = Not detected.

Genotype	ABA $(ng.g^{-1}F.wt)$	OTA $(ng.g^{-1} F.wt)$	ODA $(ng.g^{-1} F.wt)$	OEA $(ng.g^{-1} F.wt)$
+	110	ND	74	231
not	52	ND	25	157
flc	23	ND	115	166
sit	9	ND	130	159
flc/sit	10	ND	135	177
flc/not	4	ND	30	128
sit/not	3	ND	47	175

(Taylor, in *Developmental Mutants in Higher Plants,* H. Thomas, D. Grierson (eds), Cambridge University Press, Cambridge, pp. 197-217, 1987.)

The *flc* and *sit* mutations have the same effect of elevating ODA while reducing ABA. Double mutants also elevate ODA. Thus, both are downstream mutants leading to an increased rate of cleavage of the C-40 precursor resulting in abnormally high levels of the C-10 byproduct. It can, therefore, be predicted that *flacca* and *sitiens* plants would be less efficient in converting radiolabeled xanthoxin to ABA than corresponding isogenic controls. In contrast, *not* is an upstream mutant because it gives rise to a reduction in ODA, a phenotype that is dominant over both *flc* and *sit* (Table 6-8).

6-2.2.2   *Xanthoxin as an Intermediate in ABA Biosynthesis*

Since its discovery as a photochemical degradation product of violaxanthin and the demonstration that it can be converted to ABA, xanthoxin has been implicated as an intermediate in ABA biosynthesis. Past studies, however, have done little to clarify the position of xanthoxin as an endogenous precursor of ABA. Attempts to measure xanthoxin in plants and correlate its levels with those of ABA have been fraught with technical difficulties.

Mutant plants exhibiting ABA deficiency appear to fall into two categories. Those in which ABA deficiency correlates with lack of carotenoids and those exemplified by the wilty mutants of tomato, potato, and *Arabidopsis*, possessing an apparently normal complement of carotenoids. If ABA is of apocarotenoid origin, then this latter category of mutant may exhibit metabolic blocks at one or more steps of the oxidative pathway from the carotenoids to ABA. Based on a working hypothesis that xanthoxin lies on this pathway, the metabolism of [^{13}C]xanthoxin has been investigated in leaves of wild-type and wilty mutant plants of tomato (Parry et al., 1988).

There are several routes whereby xanthoxin could be converted to ABA (Fig. 6-14). ABA-1',4-*trans*-diol (ABA-*t*-diol) is a possible intermediate and has been suggested as an ABA precursor. This, along with other possible intermediates, was investigated.

The results of feeding xanthoxin or *t*-xanthoxin to stressed leaves of wild-type and mutant leaves of tomato are given in Table 6-9. Wild-type and *not* leaves converted about 12 and 16 percent, respectively, of the applied xanthoxin to ABA, while *flc* and *sit* converted <1 percent to ABA (Table 6-9; Parry et al., 1988). The stimulation of ABA production, and the high ^{13}C content of ABA extracted from stressed *not* leaves fed with xanthoxin indicated that the production of xanthoxin, or a similar compound, was very likely limiting ABA biosynthesis. The *not* mutation, therefore, might have been affecting a step before xanthoxin or even the perception or transduction of the water stress signal. The results for *flc* and *sit* were more consistent with a block occurring after xanthoxin. In *flc*, incorporation of xanthoxin into ABA and *t*-xanthoxin into *t*-ABA were both affected, whereas in *sit* only incorporation into ABA seemed impaired.

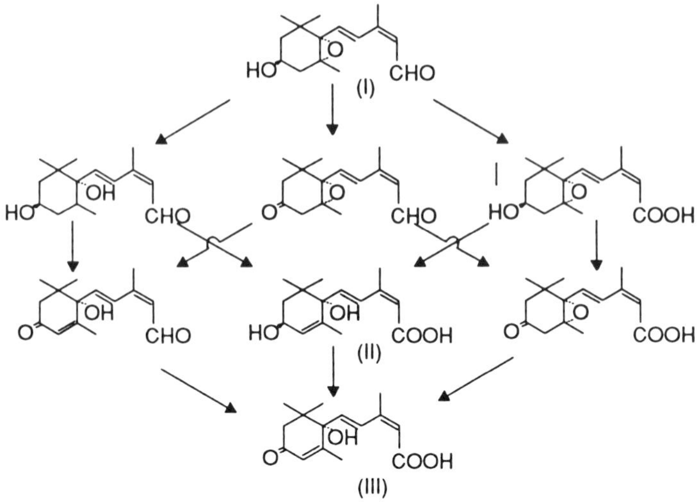

Figure 6-14. Possible routes for the conversion of xanthoxin to ABA. I = Xanthoxin; II = ABA-*t*-diol; III = ABA. (Parry et al., *Planta 173*: 397-404, 1988.)

These results (Table 6-9) clearly demonstrated that the *flc* and *sit* mutants were unable to convert xanthoxin into ABA and were consistent with xanthoxin being an endogenous ABA precursor in plants and with xanthoxin and ABA being derived by cleavage of a xanthophyll-like

Table 6-9. Conversion of [¹³C]xanthoxin and *t*-xanthoxin to ABA and *t*-ABA by stressed leaves of *L. esculentum* cv. Ailsa Craig wild type, *not*, *flc*, and *sit* mutants.
n.d. = not detected.

Tissue	¹³C isomer fed	Isotopic composition of extracted ABA (%¹³C)		% incorporation of applied Xanthoxin	
		ABA	t-ABA	ABA	t-ABA
wild type	Xan	29.9	53.5	12.0	0.4
	t-Xan	6.0	49.0	1.6	0.3
not	Xan	80.0	n.d.	16.5	-
	t-Xan	26.8	60.9	1.6	1.2
flc	Xan	23.6	n.d.	0.7	-
	t-Xan	4.7	3.9	0.1	0.1
sit	Xan	15.4	60.6	0.4	0.2
	t-Xan	n.d.	32.8	-	2.0

(Parry et al., *Planta 173*: 397-404, 1988.)

molecule (Parry et al., 1988). Other recent evidence has led to partial confirmation of the results of Parry et al. (1988), but also to important additional conclusions about the biosynthesis of ABA and the role of xanthoxin in the pathway.

When deuterium-labeled ABA-aldehyde was fed to various tomato genotypes, normal and *not* mutant plants incorporated substantial amounts of label into ABA (Taylor et al., 1988). This conclusion supported that drawn by Parry et al. (1988) that *not* is a mutation upstream in the pathway of ABA biosynthesis. In contrast, the *flc* and *sit* mutants reduced ABA-aldehyde to ABA-alcohol rather than oxidizing it to ABA, which suggests that these two lesions both act to block the last step of the ABA biosynthetic pathway (Fig. 6-15). It was suggested that the two mutant loci could be involved in coding for different subunits of the same dehydrogenase enzyme (Taylor et al., 1988). More recent evidence suggests that not all wilty mutants are blocked at this step in ABA biosynthesis (Duckham et al., 1989).

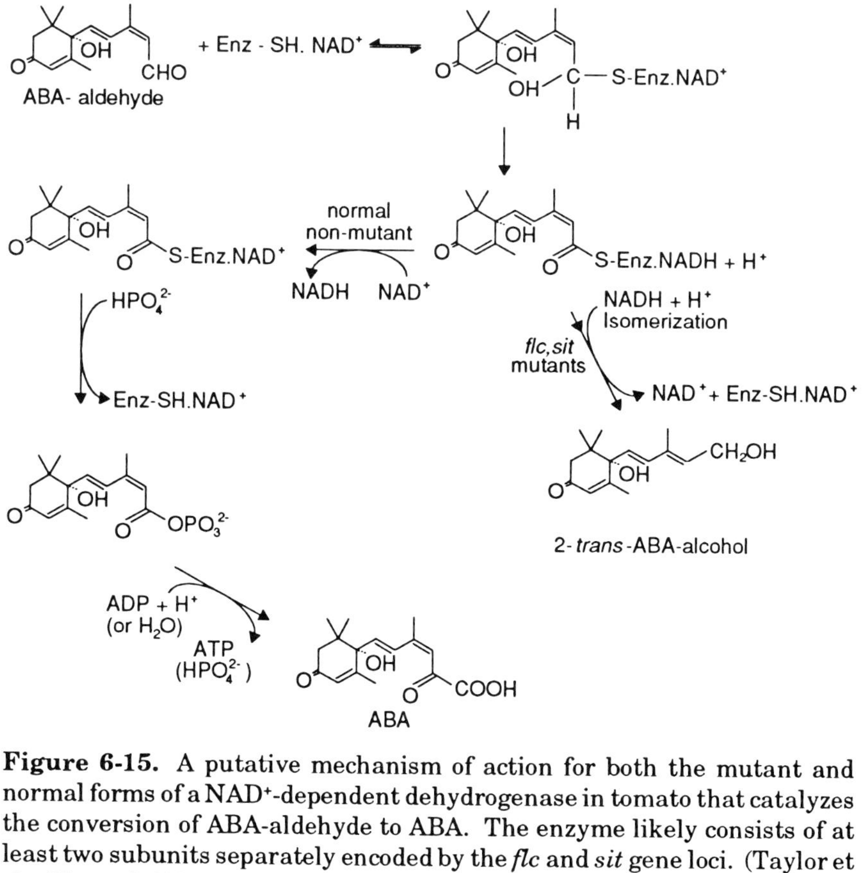

Figure 6-15. A putative mechanism of action for both the mutant and normal forms of a NAD+-dependent dehydrogenase in tomato that catalyzes the conversion of ABA-aldehyde to ABA. The enzyme likely consists of at least two subunits separately encoded by the *flc* and *sit* gene loci. (Taylor et al., *Plant, Cell Environ. 11*: 739-745, 1988.)

6-2.2.3   Further Evidence for ODA as a Byproduct of ABA Biosynthesis and for not as an Upstream Mutant in the ABA Biosynthesis Pathway

When *not* and nonmutant plants were grown for 6 weeks and subjected to water stress, after a 10 percent loss of fresh weight, it was expected that the high stress would induce a high level of ABA biosynthesis and cause maximum accumulation of C-10 byproducts. By comparing the ABA concentrations in Tables 6-8 and 6-10 it can be seen that the apparent severity of ABA deficiency caused by *not* was greater in plants exposed to water stress. In "unstressed" plants grown at 95 percent relative humidity, *not* produced 47.3 percent of the ABA found in nonmutant controls. After water stress the *not* mutants could accumulate only 13.2 percent of control ABA concentration. A similar effect on ODA levels was seen with levels of 33.8 percent and 18.8 percent, respectively. Thus, the mutant *not* would seem to be an upstream mutant. Its ABA deficiency appears to be associated with a low rate of C-40 cleavage. This would lead to a simultaneous fall in the accumulation of both ABA and any byproducts of biosynthesis. It can, therefore, be predicted that *not* plants would be just as efficient as the nonmutant controls in converting labelled xanthoxin to ABA (Taylor, 1987).

It should also be the case that by acting earlier in the pathway, an upstream mutant should be able to prevent the typical increase in byproduct accumulation normally caused by a downstream mutant. It

Table 6-10. Characteristic levels of ABA and the C-10 dicarboxylic acids (ODA and OEA) in *notabilis* and nonmutant control plants of tomato. Plants were glasshouse grown for 6 weeks and subjected to water stress after a 10 percent loss of fresh weight.

Tissue	^{13}C isomer fed	Isotopic composition of extracted ABA (% ^{13}C)		% incorporation of applied Xanthoxin	
		ABA	t-ABA	ABA	t-ABA
wild type	Xan	29.9	53.5	12.0	0.4
	t-Xan	6.0	49.0	1.6	0.3
not	Xan	80.0	n.d.	16.5	-
	t-Xan	26.8	60.9	1.6	1.2
flc	Xan	23.6	n.d.	0.7	-
	t-Xan	4.7	3.9	0.1	0.1
sit	Xan	15.4	60.6	0.4	0.2
	t-Xan	n.d.	32.8	-	2.0

(Taylor, in *Developmental Mutants in Higher Plants*, H. Thomas, D. Grierson (eds.), Cambridge University Press, Cambridge, pp. 197-217, 1987.)

Table 6-11. Abscisic acid content (±SD) of roots of *w 3, vp 5,* and *vp 7* mutants of *Zea mays.*
ND = not detected, with an analysis sensitivity of 6 ng ABA/g FW.

	ABA content (ng g^{-1} FW)	
	Normal	Mutant
*w*3	279±43	ND
*vp*5	237±26	ND
*vp*7	338±61	ND

(Moore and Smith, *Planta 164:* 126-128, 1985.)

can be seen that the ODA concentrations of the low double mutant genotypes containing *not* were lower than that of the nonmutant control (Tables 6-8 and 6-10). These two genotypes have precisely the opposite effect on ODA levels to that of the double mutant homozygous for *flc* and *sit* (Taylor, 1987).

Thus, the complex pattern of ODA concentrations in the whole ABA-deficient genotype series can only be satisfactorily explained on the assumption that ODA is a byproduct of ABA biosynthesis.

6-2.2.4   *ABA and Graviresponsiveness*

The *w 3, vp 5,* and *vp 7* mutants of *Zea mays* have metabolic blocks as illustrated in Figure 6-12. Because ABA was believed by some investigators to be involved in and/or responsible for root gravitropism, the graviresponsiveness of primary roots of these mutants was determined.

Determinations of the ABA content of the mutants showed an absence of the hormone from the roots (Table 6-11; Moore and Smith, 1985).

Root gravicurvatures of normal and mutant seedlings were not significantly different (Table 6-12). Primary roots of normal and mutant

Table 6-12. Gravicurvatures (degrees ± SD) of primary roots of *w 3, vp 5,* and *vp 7* mutants of *Zea mays.*

	Time			
	3 h		6 h	
	Normal	Mutant	Normal	Mutant
*w*3	37±8.0	31±8.2	68±11	61±7.6
*vp*5	29±6.1	27±8.9	51±6.9	47±7.5
*vp*7	35±6.2	29±8.0	64±10	58±9.2

(Moore and Smith, *Planta 164:* 126-128, 1985.)

seedlings were positively gravitropic. This fact indicates that ABA is not necessary for positive gravitropism by primary roots of *Z. mays* and provides a neat and conclusive answer to the point.

Case Study 6-3:   The *seg* Mutants of Barley

The Characteristics of the *seg* Mutants

Several shrunken endosperm (*seg*) mutants of barley (*Hordeum vulgare*) have an unusual inheritance pattern. The genes involved are nuclear but these recessive mutants, unlike most cereal endosperm mutants, do not express xenia. Therefore, the endosperm phenotype depends on the genotype of the maternal plant and is independent of pollen source. This inheritance pattern suggested that defects in the maternal plant were affecting grain filling in some manner to prevent normal endosperm development.

A total of eight *seg* mutants were isolated. A comparison of growth rates revealed that the dry weight increase of Betzes (wild type) and *seg*1 was rapid and not significantly different up to 18 days after anthesis (Fig. 6-16) at which point, the rate of *seg*1 dry weight increase diminished markedly, whereas dry weight of wild type continued to increase at nearly the same rate for 12 more days. This indicated that the smaller *seg*1 grains resulted from a premature termination of grain filling rather than a diminished rate of grain fill throughout maturation (Felker et al., 1983; Peterson et al., 1985).

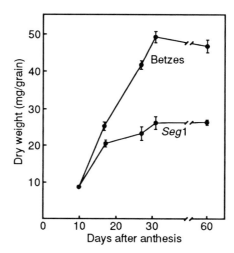

Figure 6-16. Growth of Betzes and seg*1* grains of barley under greenhouse conditions. (Felker et al., *Plant Physiol. 72:* 679-684, 1983; reproduced by permission of the ASPP.)

Excised spikes were cultured in defined media to isolate the spike from its dependency on the maternal plant for its source of assimilates and to determine whether the shrunken characteristic of *seg*1 could, thereby, be relieved. If the spikes were excised at 9 days after anthesis, dry weight increase of grains of both genotypes was comparable for the first 5 days in culture after which the rate of increase of *seg*1 declined relative to Betzes (Fig. 6-17A). If the spikes were not excised until 17 days after anthesis, at which time *seg*1 grains were already smaller than those of Betzes, the dry weight of *seg*1 grains increased very little in culture (Fig. 6-17B). Betzes, by contrast, continued to add dry matter and achieved a grain weight after 15 days that was nearly equal to that of grain maturing on intact plants. This showed that the factors causing early termination of grain filling in *seg*1 may reside in the spike itself, rather than in vegetative tissues such as leaves or roots (Felker et al., 1983).

Microscopic evaluation of the anatomical development of the tissues of maternal origin of Betzes and *seg*1 was conducted and attention focused on those tissues that play a role in assimilate transport. Assimilates from the plant enter the kernel through a single vascular bundle. They are unloaded from the sieve tubes and must cross the chalaza and nucellar projections before they enter the endosperm cavity. Figure 6-18 shows a cross section of the crease region of normal Betzes and *seg*1 25 days after anthesis. In Betzes (Fig. 6-18A), the prominent chalaza lies between the vascular bundle and the nucellar projection. In contrast, the chalaza of *seg*1 (Fig. 6-18B) appeared necrotic and crushed. Individual cells were no longer defined and the entire chalaza was much narrower. The nucellar projection cells also were necrotic and appeared pushed together (Felker et al., 1984a).

It appeared that a premature necrosis of the chalaza, a tissue through which assimilates must pass, was physically blocking transport

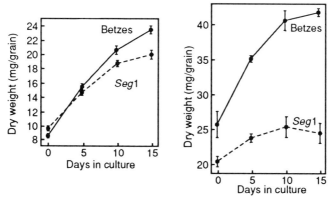

Figure 6-17. Growth of Betzes and *seg*1 barley grains on detached spikes placed in culture 9 days (A) or 17 days (B) after anthesis. (Felker et al., *Plant Physiol.* 72: 679-684, 1983; reproduced by permission of the ASPP.)

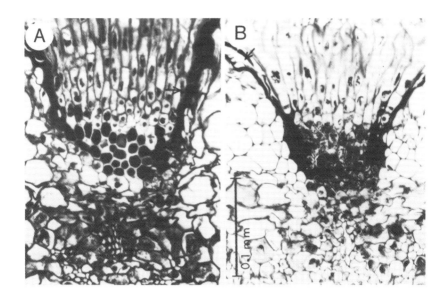

Figure 6-18. Development of crease tissues of normal (Betzes) (A) and *seg*1 mutant (B) barley seeds 25 days after anthesis. Note the crushed nature of the chalazal tissue in the seg1 mutant (arrow in B). (Felker et al., *Planta 161*: 540-549, 1984a.)

and resulting in the shrunken endosperm characteristic of *seg*1. In normal barley chalaza cells small tannin-containing vesicles coalesce into the large central vacuole during development. The cells maintain a parietal cytoplasm and plasmodesmata are abundant. Symplastic transport could, and most likely does, occur through this tissue. In *seg*1, the movement of tannins into the vacuole does not take place normally. Increasing cytoplasmic tannin concentration caused by improper compartmentation or leakage of the small vesicles would precipitate cytoplasmic enzymes leading to cell death. It is not known why tannins are not stored in the central vacuole of *seg*1 chalaza cells. The tannins themselves could be defective or the sequestering process may be prevented by an unknown factor.

*Seg*3, *seg*6, and *seg*7 exhibited the same necrosis and crushing of the chalaza and nucellus projection as did *seg*1. The other four mutants, *seg*2, *seg*4, *seg*5, and *seg*8 showed completely normal development of the pericarp tissues of maternal origin including the chalaza, the seed coat, and the nucellar projection. Endosperm development, however, was abnormal in each mutant and different in each case (Peterson et al., 1985).

It is suggested that some factor of maternal origin is instrumental in programming the normal development of endosperm tissues. Whether the factor is hormonal or nutritional is not known. Its effects may be localized as judged by the development patterns observed. Thus, the eight shrunken endosperm mutants of barley of maternal origin are each separate alleles and each appears to be unique. They can be classified into two groups, however, the chalaza necrosis group and the abnormal endosperm group. The first group can be used to probe the role of the chalaza, and especially the compartmentation of tannins, in assimilate transport. The second group offers a unique opportunity to study ways in which the maternal plant affects early endosperm development. This field has received very little attention and these mutants may provide some new insights (Felker et al., 1984a,b, 1985).

Case Study 6-4:   Mutants Affecting Starch Synthesis in Seeds

Maize endosperm is the major site of starch deposition during kernel development. Many mutants leading to altered levels of starch or its quality are available, many still unmapped. Four mutants, *shrunken-1 (sh-1), shrunken-2 (sh-2), brittle-2 (bt-2),* and *waxy (wx),* however, have been identified with their enzyme products (Chourey, 1981). A recent resurgence of interest in mutants with modified starch metabolism has led to the isolation of a number of them from such sources as *Arabidopsis* (Caspar et al., 1985; Caspar et al., 1989), tobacco (Hanson and McHale, 1988), pea (Smith et al., 1989) and barley (Schulman and Ahokas, 1990).

The *shrunken-1* locus in homozygous form conditions a shrunken or collapsed phenotype due to a reduced level of starch content in the endosperm. No other part of the plant is affected. It would appear that the *Sh-1* locus codes for the enzyme sucrose synthetase and its deficiency leads to a rate-limiting step in starch biosynthesis and, hence, the shrunken phenotype (Chourey, 1981).

Both the *sh-2* and *bt-2* alleles, each in homozygous form, lead to a severe reduction in endosperm starch and a shrunken endosperm phenotype. Both are associated with a complete loss of adenosine diphosphate glucose (ADPG) pyrophosphorylase except for a residual 8 to 10 percent activity that remains. It has been found that both *sh-2* and *bt-2* affect enzyme activity as though each was a structural gene for the enzyme. It has been hypothesized that enzyme subunits are coded by *sh-2* and *bt-2* loci. Very recent *Arabidopsis thaliana* mutants, comparable to those from maize, have confirmed this hypothesis (Chourey, 1981; Preiss et al., 1990).

The *wx* locus influences the type of starch and not the quantity in the affected tissue. Also, the tissue specificity of this mutation is not as stringent as that of several other starch mutants. The *wx* allele is expressed in the endosperm, pollen, and embryo sac; the homozygous *wx* condition is associated with starch made up of 100 percent amylopectin. It has been observed that starch granules from developing endosperms of homozygous *wx* maize did not possess the starch granule-bound enzyme transferring glucose from uridine diphosphate glucose (UDPG) to the nonreducing end of the starch molecule. The *wx* protein is probably the monomeric form of the nucleotide diphosphate - sugar - starch - glucosyl transferase (Chourey, 1981).

Thus, a mutation in each of these four gene systems above can be identified with a rate-limiting step in starch biosynthesis in endosperm cells. As will be seen, it also enables in each case, the uncovering of a second isozyme coded by a gene elsewhere in the genome. Of these two genes,the one encoding the major isozyme is highly specific to the endosperm. Little is known about the tissue specificity of the genes encoding the minor enzymes (Chourey, 1981).

6-4.1 The Isolation, Characterization and Physiological Expression of Mutants Affecting Storage Carbohydrates

6-4.1.1 The sh-1 *Mutation*

Sucrose is one of the predominant sugars in plants. In the case of cereals, the products of photosynthesis are translocated as sucrose to the developing seeds to be stored as starch. Sucrose synthetase (UDPG: D-fructose-2-glucosyltransferase: E.C. 2.4.1.14) catalyzes the reversible reaction:

$$\text{sucrose} + \text{UDP} \longleftrightarrow \text{fructose} + \text{UDPG}$$

This enzyme provides an important link in sucrose-starch conversion reactions in the developing seed (Chourey and Nelson, 1976).

Genetic evidence suggested that the *sh* mutation on chromosome 9 in maize leads to a shrunken or collapsed endosperm due to a reduced amount of starch. Sucrose synthetase activity in the mutant endosperm was reduced to 10 percent of the normal endosperm. The data in Table 6-13 showed that when sucrose synthetase activity was measured as

Table 6-13. Sucrose synthetase activity (sucrose breakdown) in the endosperm preparations from normal and *shrunken-1* maize 22 days after pollination. Values in parenthesis represent percent of normal endosperm.

Stock	Nanomoles reducing sugar/min/mg protein	NanomolesUDPG/ min/mg protein
Normal-W22	112.1	133.8
shrunken -W22	7.5	11.7
	(6.71)	(8.74)

(Chourey and Nelson, *Biochem. Genet. 14*: 1041, 1055, 1976; reproduced by permission of Plenum Publishing Corp.)

Table 6-14. Relative sucrose synthetase activity as a function of *Sh* dosage in the endosperm of w22 inbred maize

[a.] If a linear relationship exists between the number of *Sh* alleles and the amount of sucrose synthetic activity.

	Observed		Expected[a]	
Sh alleles	Nanomoles sucrose/min/mg protein	Percent	Nanomoles sucrose/min/mg protein	Percent
3	945.50	100.00	945.40	100.00
2	735.40	77.78	658.60	69.66
1	394.10	41.68	371.80	39.32
0	84.95	8.98	84.90	8.98

(Chourey and Nelson, *Biochem. Genet. 14*: 1041-1055, 1976; reproduced by permission of Plenum Publishing Corp.)

sucrose breakdown the activity in *sh-1* endosperm was less than 10 percent that in the wild type (Chourey and Nelson, 1976).

Reciprocal hybrids between normal and mutant plants were made to test gene-dose relationship to sucrose synthetase activity. Because maize endosperm is a triploid receiving two genomes from the female and one from the male, it was possible to compare the four genotypes *Sh/Sh/Sh*, *Sh/Sh/sh*, *Sh/sh/sh*, and *sh/sh/sh*. Specific activity measurements in these genotypes showed a correlation between the number of *Sh* alleles and sucrose synthetase activity (Table 6-14; Chourey and Nelson, 1976). Starch content in the homozygous recessive genotype was about 40 percent less than in normal endosperm (Table 6-15). Thus, it might be concluded that a substantial amount of starch

Table 6-15. Starch, sucrose and reducing sugars in endosperms of normal and *shrunken-1* maize endosperms 22 days after pollination.

	Normal				shrunken			
	22-day		Mature		22-day		Mature	
	% dry weight	mg/ endo-sperm	% dry weight	mg/ endo-sperm	%dry weight	mg/ endo-sperm	% dry weight	mg/ endo-sperm
Starch	61.50	45.70	77.00	139.40	42.90	28.20	68.40	82.70
Sucrose	0.86	0.64	0.55	0.99	4.13	2.71	0.44	0.53
Reducing sugars	4.06	3.01	5.25	9.50	11.30	7.43	5.12	6.20

(Chourey and Nelson, *Biochem. Genet. 14:* 1041-1055, 1976; reproduced by permission of Plenum Publishing Corp.)

synthesis is dependent on sucrose mobilization through *Sh*-coded sucrose synthetase (Chourey and Nelson, 1976).

Although it was the case that the sucrose synthetase activity was associated with an "Sh protein" that was absent in the *sh* mutant, it was also found that the homozygous recessive mutant still contained 2 to 6 percent of the synthetase activity of the wild type at maturity. The developmental profile of enzyme activity in the mutant and wild-type endosperm, however was different (Table 6-16; Chourey, 1981). These data suggested that the *sh-1* mutant contained a second sucrose synthetase enzyme that was unrelated to the one associated with the *Sh-1* locus and that supported the synthesis of 60 percent of the normal amount of starch. A small amount of a protein having an electrophoretic mobility on native gels similar, but not identical, to that of the *Sh-1*-encoded sucrose synthetase was present in all endosperm tissue and was shown to have sucrose synthetase activity. It was further suggested that there were marked enzymatic and structural similarities between the two enzymes. They were indistinguishable with respect to substrate specificity and affinity as well as pH optima for both forward and reverse reactions. They had nearly identical monomer sizes (92-kD) and had similar tetrameric native size. The proteins differed slightly in charge and could be separated on polyacrylamide gels. Immunologically, the two proteins were also very similar, all of which suggested that the genes encoding the two activities were related (McCormick et al., 1982).

Biochemical genetic analyses on various *sh-1* mutants indicated that one of the sucrose synthetase isozymes (SS1) was encoded by the *Sh-1* locus located on chromosome 9. The second enzyme, SS2, was

Table 6-16. Developmental profile of sucrose synthetase activity in normal and *shrunken-1* endosperm of maize. The values in parentheses represent percent of wild type.

Days after pollination	Specific activity (nmoles sucrose synthesized/mg protein/min)		
	W22 +/+	W22 *sh-1/sh-1*	
8	38.6	16.8	(43.50)
12	339.6	40.4	(11.89)
16	1034.0	42.0	(4.10)
22	1046.0	52.0	(4.97)
28	1301.0	52.0	(4.00)
32	1512.0	54.7	(3.61)
40	1537.0	48.8	(3.17)
48	619.0	19.0	(3.70)

(Chourey, *Mol. Gen. Genet. 184: 372-376, 1981.*)

encoded by a second gene designated as *Ss-2*, which was also located on chromosome 9 (Gupta et al., 1988).

Different tissues of the maize plant have been analyzed for transcripts of the *Sh-1* gene. Rather high transcription rates have been observed in the etiolated shoot and the primary root of the germinating kernel. If the etiolated seedlings were illuminated, the transcript level dropped by about 95 percent in the greening plant part (first and second leaves) that were active in photosynthesis. A very low transcript level was found in mature green leaves where sucrose was formed from products of photosynthesis by a separate pathway. Upon anaerobic stress of the young seedling, the level of *Sh-1* transcripts increased 10 and 20 times in shoot and root tissue, respectively. Apparently anaerobic induction superseded the negative control that was observed after illumination in the first and second leaves (Table 6-17; Springer et al., 1986).

The expression of the *Sh-1* gene in the endosperm of the maize kernel was an indication that the enzyme catalyzes a rate-limiting step in a series of enzymatic reactions needed for starch biosynthesis, because kernels deficient for this enzyme accumulated less starch and, therefore, showed the shrunken phenotype (Springer et al., 1986). The expression of the gene in etiolated maize seedlings could easily be understood if one assumes that the starch in the kernel endosperm is converted to sucrose upon germination and that sucrose is transported into the growing plant tissues. The shoot tissue is dependent on sucrose translocated from the endosperm until photosynthesis can provide photoassimilates for cell metabolism. Therefore, the enzyme sucrose synthetase, which

Table 6-17. Relative strength of hybridization of polyA⁺ RNA from different wild type tissues of maize to a probe. The probe used for hybridization was the 4.4 kb *Pst1* fragment of the genomic *sh-1* clone.The hybridization signal of the RNA from immature kernels was defined as 100 and the other values were expressed as a percentage of this value.In the *sh-1* deletion mutant, the hybridization in all investigated tissues amounted to about 2 percent

Tissue	Amount of hybridization (%)
Kernels	100
Germinated kernels	10
Shoots	50
Anaerobic shoots	500
Roots	20
Anaerobic roots	400
Leaves	2

(Springer et al., *Mol. Gen. Genet. 205,* 461-468, 1986.)

splits the transport metabolite sucrose, should be shut off in photosynthetically active cells or tissues. Otherwise, the sucrose synthesized from normal photosynthetic pathways would be split by the sucrose synthetase. The absence of the enzyme in leaves is, therefore, not surprising (Table 6-17; Springer et al., 1986). The fact that sucrose synthetase primarily degrades sucrose in maize tissues was recently confirmed through the use of the *sh-1* mutant (Cobb and Hannah, 1988).

It is surprising that homozygous *sh-1* mutations do not have an obvious phenotype in the shoots or roots of maize seedlings as they do in the kernel. The missing phenotype indicates that either other metabolic pathways can bypass the bottleneck caused by the deficient sucrose synthetase A enzyme or the sucrose synthetase B level is sufficient for optimal growth (Springer et al., 1986).

In contrast to the *Sh-1* gene, the sucrose synthetase B gene (*Ss-1*) seems to be transcribed constitutively at a low level in every tissue analyzed. No mutants of this gene have been isolated so far. Physiologically it might be interesting to ask why the sucrose synthetase B gene is expressed at a low level in all tissues, and the sucrose synthetase A gene has acquired all the regulatory signals for tissue-specific expression and the anaerobic response. Does this mean that the two rather similar sucrose synthetase enzymes serve different functions within the cell metabolism (Springer et al., 1986)?

6-4.1.2   *The* sh-2 *Locus*

The *sh-2* mutant in maize is characterized by kernels with poorly developed shrunken endosperms; the kernels are unusually sweet, with a high concentration of sucrose and a low starch content in mature kernels: 16 and 25 percent compared with 1.4 and 65 percent, respectively, in normal maize. Table 6-18 shows the specific activities of ADPG pyrophosphorylase, UDPG pyrophosphorylase, and starch phosphorylase for the *sh-2* mutant and for normal maize.

Table 6-18. Specific activities of ADPG pyrophosphorylase, UDPG pyrophosphorylase, and starch phosphorylase in embryo and endosperm preparations from *sh-2* mutant and normal maize.

| Phosphorylase | Incorporation (count/min) | | | |
| | Embryo | | Endosperm | |
	sh-2	Normal	*sh-2*	Normal
ADPG pyro-	0	10	0	300
UDPG pyro-	364	255	131	127
Starch			6	6

(Tsai and Nelson, *Science 151: 341-343; copyright 1966 by the AAAS.*)

Table 6-19. Reducing sugars, sucrose and starch in endosperms of the *sh-2* mutant and of normal maize 22 days after fertilization.

Endosperm	Concentration (% dry wt)		
	Sugars	Sucrose	Starch
sh-2	3.63	27.34	16.25
Normal	1.60	2.35	67.14

(Tsai and Nelson, *Science 151:* 341-343; copyright 1966 by the AAAS.)

Endosperm preparations from the *sh-2* mutant and from normal maize had about the same UDPG pyrophosphorylase activity; this was also true of embryo preparations. The *sh-2* mutant completely lacked ADPG pyrophosphorylase activity both in embryo and endosperm preparations (Table 6-18). The lack of the enzyme resulted in an accumulation of sucrose and a block in the synthesis of starch (Table 6-19). This indicated that starch synthesis in the normal maize endosperm proceeds largely by way of ADPG as a substrate, and that ADPG is chiefly synthesized through the action of ADPG pyrophosphorylase (Tsai and Nelson, 1966). Thus:

$$ATP + \alpha\text{-glucose } 1\text{-P} = ADPG + PPi$$

$$n(ADPG) + \alpha\text{-1,4 glucan} \text{ ---> } acceptor(\alpha\text{-1,4-glucose})n + n(ADP)$$
$$\text{(acceptor)} \qquad \text{(starch)}$$

Tsai and Nelson (1966) also postulated that the limited amount of starch in the mutants was synthesized by other metabolic routes in the absence of ADPG pyrophosphorylase. Low levels of the enzyme, however, do occur in the *sh-2* endosperm and embryo, as well as in the same tissues in the *bt-2* mutant, which also affects the synthesis of this enzyme in maize (Table 6-20; Dickinson and Preiss, 1969). The *sh-2* mutation seemed to affect only the enzyme in the endosperm, not the embryo, whereas the *bt-2* lesion affected both (Table 6-20).

Table 6-20. The distribution of ADPG pyrophosphorylase in normal (Dekalb 805), *sh-2* and *bt-2* maize.

Maize seed	Specific activity (units/mg protein)		
	Dekalb 805	*Shrunken-2*	*Brittle-2*
Endosperm	3.8	0.28	0.45
Embryo	0.40	0.71	0.16

(Dickinson and Preiss, *Plant Physiol. 44:* 1058-1062, 1969; reproduced by permission of the ASPP.)

These results (Table 6-20) indicated the presence of significant but low amounts of ADPG pyrophosphorylase in the endosperm of *sh-2* maize seeds. Hence, the low level of starch in *sh-2* could have been formed from ADPG and there was no need to postulate any other pathway for starch synthesis. The participation of another pathway in starch synthesis in this mutant, however, could not be ruled out (Dickinson and Preiss, 1969).

The genetic evidence (Tsai and Nelson, 1966; Dickinson and Preiss, 1969) showed that ADPG pyrophosphorylase is a component of the major route of starch biosynthesis. In two maize mutants, *sh-2* and *bt-2*, the levels of ADPG pyrophosphorylase activity in the endosperms was less than 10 percent of that found in normal endosperm. Although the 90 percent reduction in activity was correlated with a 75 percent reduction in the level of starch accumulated in normal endosperm, it was not possible to exclude the involvement of another route for starch biosynthesis in maize amyloplasts using these mutants.

A mutant of *Arabidopsis thaliana* lacking ADPG pyrophosphorylase activity was isolated by screening for the absence of leaf starch. Genetic analysis showed that the deficiency of both starch and ADPG pyrophosphorylase activity were attributable to a single, nuclear, recessive mutation at a locus designated *adg1*. The absence of starch in the mutant demonstrated that starch synthesis in the chloroplast was entirely dependent on a pathway involving ADPG pyrophosphorylase (Table 6-21; Lin et al., 1988a). This evidence effectively excluded the possible participation of both UDPG pyrophosphorylase and starch phosphorylase in net starch synthesis in the chloroplast.

Analysis of leaf extracts using antibodies specific for spinach ADPG pyrophosphorylase showed that two proteins, present in the wild type, were absent in the mutant (TL 25). The heterozygous F_1 progeny of a cross between the mutant and wild type had a specific activity of ADPG pyrophosphorylase indistinguishable from the wild type. These observations suggested that the mutation in the *adg1* gene in TL25 might affect a regulatory locus (Lin et al., 1988a).

Table 6-21. ADPG pyrophosphorylase activity and starch content in leaves of wild type, mutant (TL 25), and F_1 hybrid lines of *Arabidopsis*. Values are the mean $\pm$SE.

Line	Plants tested	ADPG pyrophosphorylase nmol/min/mg protein	Starch content mg/g fresh weight
Wild type (WT)	12	54±5	7.3±0.4
F_1 (WT x TL 25)	12	50±4	6.0±0.5
TL 25	3	0±0	0.0±0.0

(Lin et al., *Plant Physiol. 88:* 1131-1135, 1988a; reproduced by permission of the ASPP.)

Evidence from the use of another mutant of *Arabidopsis, adg2,* suggested that there are two structural genes for ADPG pyrophosphorylase, one coding for a 54-kD polypeptide and the other for a 51-kD polypeptide. The *adg2* mutant, which lacks the 54-kD subunit of the enzyme and has only 5 percent as much activity of the enzyme as the wild type, accumulated 40 percent as much starch as the wild type when grown in continuous light. Thus, there is now the possibility of studying the role in starch synthesis of each subunit of ADPG pyrophosphorylase as well as the control of starch synthesis itself (Lin et al., 1988b). The most recent evidence suggests that *the Adg2 (Sh-2)* locus encodes the large subunit of the enzyme and *Bt-2* the smaller polypeptide (Preiss et al., 1990).

6-4.1.3   The wx *locus*

Since 1943, it has been known that the endosperms of maize seeds homozygous for the *wx* allele contain starch that is entirely amylopectin, that is, contains no amylose. In contrast, seeds that are heterozygous or homozygous for the *nonwaxy (Wx)* allele contain starch with up to 25 percent amylose with the remainder being amylopectin (Nelson and Rines, 1962).

Most starches consist of two types of molecules (1) straight chains that make up the amylose fraction and (2) branched chain molecules that comprise the amylopectin. Ordinary corn starch, for example, contains about 22 percent amylose whereas waxy corn starch has none at all (Table 6-22; Sprague et al., 1943).

It was found that *wx* seeds 16 days after pollination completely lacked a UDPG transferase activity that is present to a substantial degree in similar preparations from *Wx* seeds (Fig. 6-19; Nelson and Rines, 1962). It was subsequently shown that ADPG, not UDPG, was the preferred substrate for enzymes of this type (Nelson et al., 1978).

Reexamination of the ADPG-starch glucosyl transferase activity in waxy and nonwaxy starch granules showed that the relative enzymic

Table 6-22. The relation between maize endosperm genotype and the percentage amylose of the endosperm starch.

Genotype	Amylose in sample, %
WxWxWx	22.0
WxWxwx	20.4
Wxwxwx	18.4
wxwxwx	0

(Sprague et al., *J. Am. Soc. Agron. 35*: 817-822, 1943; reproduced by permission of the American Society of Agronomy, Inc.)

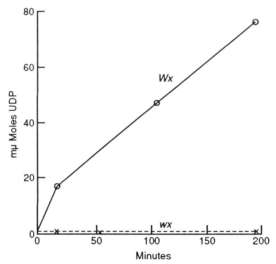

Figure 6-19. The amount of UDP released by starch granule preparations from *wx* and *Wx* maize seeds. (Nelson and Rines, *Biochem. Biophys. Res. Commun. 9*: 297-300, 1962.)

activity of waxy starch granule preparations increases with decreasing substrate (ADPG) concentrations until at the lowest concentration used (0.1 to 1.0 µM), the activities of waxy and nonwaxy starch granules are nearly equal (Nelson et al., 1978). In Table 6-23, the *wx* preparation reached maximum velocity at a lower substrate concentration and hence had a lower K_m. The K_m was estimated as 7.1×10^{-5} M and that for the nonwaxy preparation was estimated as 3×10^{-3} M.

Table 6-23. Incorporation of glucose from ADP-^{14}C-glucose into *wx/wx/wx* and *Wx/Wx/Wx* maize endosperm starch granule preparations.

Glucose donor Concentration	ADP Glucose		UDP Glucose	
	Endosperm Source		Endosperm Source	
	wx/wx/wx	*Wx/Wx/Wx*	*wx/wx/wx*	*Wx/Wx/Wx*
µM	pmol glucose incorporated (percent incorporation)			
0.1	1.7 (43)	1.6 (40)	0 (0)	0.33 (6.9)
1.0	16.7 (42)	16.8 (42)	0.7 (0.2)	2.65 (7.1)
10	107 (27)	163 (41)	0.78 (0.2)	21 (6.4)
100	429 (11)	1520 (38)	7.8 (0.2)	164 (6.6)
1000	881 (2)	10622 (27)	50.3 (0.1)	1641 (6.6)
10000	818 (<1)	51460 (13)	210.0 (0.1)	15230 (3.8)

(Nelson et al., *Plant Physiol. 62*: 383-386, 1978; reproduced by permission of the ASPP.)

Table 6-24. The incorporation of glucose from ADP-^{14}C-glucose into starch granule preparations from *waxy* mutants.

ADPG Concentration	waxy mutant tested		
	wx-B7	*wx-C2*	*wx-R*
μM	pmol glucose incorporated/10 min/3 mg starch		
25	79	85	81
50	101	119	121
100	122	136	176
200	169	168	211
400	202	233	224

(Nelson et al., *Plant Physiol. 62:* 383-386, 1978; reproduced by permission of the ASPP.)

It was also necessary to demonstrate that the second transferase was not encoded by the *wx* locus. To this end, three different *wx* mutants were tested to see if they had different or similar amounts of the second enzyme. Similar activities would indicate that the *wx* allele was not responsible for the second activity. This was demonstrated (Table 6-24; Nelson et al., 1978). This and other observations have been interpreted as indicating the presence in maize endosperm starch granules of a second transferase with a low K_m for ADPG which accounts for the transferase activity observed in *wx* granules.

With the identification of a second starch-granule bound glucosyl transferase having a low apparent K_m for ADPG, the question arose as to whether the two activities increased synchronously during development or whether one preceded the other. Starch granule preparations were, therefore, made from *Wx/Wx/Wx* and *wx/wx/wx* endosperms at various times after pollination (Table 6-25; Nelson et al., 1978). The activity of the *Wx* preparation increased greatly from 12 to

Table 6-25. The activities of *Wx/Wx/Wx* or *wx/wx/wx* starch granules from maize endosperm during development.

Time after Pollination	Source Endosperm Genotype	
	wx/wx/wx	*Wx/Wx/Wx*
Days	pmol glucose incorporated	
12	-	4,200
14	1,725	16,800
16	765	20,550
24	945	19,650

(Nelson et al., *Plant Physiol. 62:* 383-386, 1978; reproduced by permission of the ASPP.)

14 days and somewhat more from 14 to 16 days. By contrast, the activity of the *wx* preparation decreased somewhat from 14 to 16 days (Table 6-25). Because the activity of the low K_m system, which contributed the enzymic activity of the *wx* preparations, dropped somewhat over the period assayed it could be presumed to do so also in the *Wx* preparations. The increase in activity in the nonwaxy starch granules, therefore, resulted from an increase in the high K_m enzyme in the period between 12 and 16 days. This was the period during which a large increase in the activity of enzymes participating in starch synthesis had been found.

The results of the developmental study offered no basis for a decision as to whether one enzyme antedated the other during development. It was clear, however, that the major portion of the high K_m enzyme activity developed after 12 days postpollination, whereas the activity of the low K_m enzyme was not increasing (Nelson et al., 1978).

The analyses of the products of these and other genetic loci have greatly advanced our knowledge of the pathway of starch synthesis in seeds of maize, in particular. Now, it is possible to investigate how several of these genetic lesions together (e.g., *sh-1, sh-2, sh-4, b-1*, and *b-2*) may affect the overall carbohydrate metabolism of mutant maize kernels (Doehlert and Kuo, 1990).

6.4-2   The Use of Other Mutants With Lesions in Starch Metabolism for Physiological Studies

6-4.2.1   *Starchless Mutants Deficient in Phosphoglucomutase Activity*

The accumulation of nonstructural carbohydrate in leaves has been suggested to influence both the rate of photosynthesis and the rate of dark respiration. An effect of leaf carbohydrate concentration on photosynthesis has been repeatedly proposed as a possible explanation for the depression of photosynthetic rate that may result from experimental treatments that increase the source/sink ratio or decrease the rate of translocation of photosynthate from a source leaf. Interpretation of the results of such experiments, however, has been complicated by the possibility that hormonal changes may result from the treatment and from instances in which no correlation has been observed between leaf carbohydrate concentration and photosynthesis rate. An inherent problem in attempting to establish explanatory correlations is that most experimental manipulations do not permit control of the molecular species of carbohydrates that accumulate. Thus, potential regulatory effects of sucrose or other sugars may be obscured by mechanisms that regulate starch/sucrose partitioning and prevent soluble carbohydrate from accumulating (Caspar et al., 1985).

The effect of carbohydrate accumulation on the rate of dark respiration has received less attention but appears to have been consistently observed by a variety of approaches. The essential observation is that the rate of respiration is proportional to carbohydrate content or is stimulated by provision of exogenous carbohydrate. The implication is that the amount of respiration is regulated by substrate supply rather than demand for ATP or reducing equivalents. Indeed, it has been suggested that, under conditions of high carbohydrate supply, a substantial proportion of the reductant generated during carbohydrate catabolism may be consumed by the alternative oxidase without being linked to ATP production. Thus, such respiration may be considered as potentially wasteful and may represent a target for genetic manipulation (Caspar et al., 1985).

M_2 plants descended from ethylmethanesulfonate-mutagenized seed of *Arabidopsis thaliana* L. (Heynh.) were tested with iodine during illumination. The mutant isolation protocol was based on the assumption that the absence of leaf starch would impose no deleterious effect on plants growing in continuous illumination. Two starchless plants were observed from the population of M_2 seeds that remained uniformly starchless for the six generations tested. A quantitative measurement of the starch content of one of the mutant lines and the wild type was obtained by measuring the ethanol-insoluble carbohydrate concentration of plants grown in a 12-hour light/12-hour dark photoperiod. The leaves of the mutant were almost completely lacking in starch under all conditions and in all tissues tested (Fig. 6-20; Caspar et al., 1985). The diversion of carbon from the Calvin cycle to amylose involves only four enzymic steps; phosphoglucomutase (PGM), phosphoglucoisomerase,

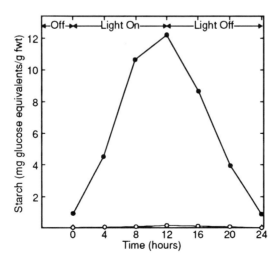

Figure 6-20. Diurnal changes in starch concentration in leaves of wild-type (●) and mutant (○) *A. thaliana*. (Caspar et al., *Plant Physiol.* 79: 11-17, 1985; reproduced by permission of the ASPP.)

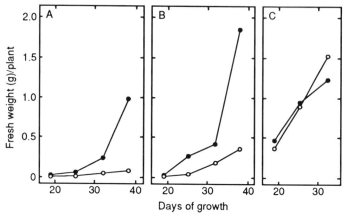

Figure 6-21. Effect of photoperiod on growth of mutant (○) and wild-type (●) *A. thaliana.* The daily period of illumination was: (A), 7 hours, (B), 12 hours, (C), 24 hours. (Caspar et al., *Plant Physiol. 79*: 11-17, 1985; reproduced by permission of the ASPP.)

ADPG pyrophosphorylase, and starch synthase. There was no activity of the chloroplast PGM isozyme in the mutant, which is thus designated as *pgmP.*

To quantitate the effects of the starchless phenotype on growth rate, the rate of increase in fresh weight of mutant and wild-type plants was measured for plants growing in various photoperiods (Fig. 6-21). The results of this experiment showed that when plants were grown in continuous illumination, the growth rate of the mutant was indistinguishable from that of the wild type. As the length of the diurnal dark period was increased, however, growth of the mutant was differentially impaired relative to that of the wild type (Fig. 6-21). The similarity of the growth rate of mutant and wild type in continuous illumination indicated that the *pgmP* mutation was only conditionally deleterious and suggested that the effect was specifically related to the absence of starch (Caspar et al., 1985).

To provide a basis for interpreting the growth characteristics of the mutant, the nonstructural carbohydrate concentrations of leaves were determined at various times during a light/dark cycle. Results showed that the wild type accumulated substantial amounts of starch (Fig. 6-20) but very little soluble carbohydrate (Fig. 6-22). By contrast, quantitative analysis of the carbohydrate concentration of the mutant confirmed that it had barely detectable levels of starch at all times of the day but that photosynthate accumulated in the mutant in the form of sucrose and hexose sugars (Figs. 6-20 and 6-22). The accumulation of abnormally high concentrations of soluble carbohydrate had no obvious

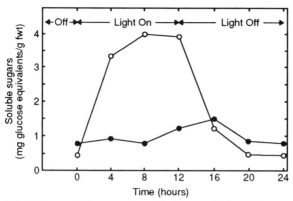

Figure 6-22. Diurnal changes in soluble carbohydrate concentration in leaves of wild-type (●) and mutant (○) *A. thaliana*. (Caspar et al., *Plant Physiol. 79*: 11-17, 1985; reproduced by permission of the ASPP.)

short-term effect on photosynthesis as the photosynthetic rate of mutant plants grown in a 12-hour photoperiod did not change as carbohydrate accumulated during the period of illumination. The photosynthesis rate of the mutant, however, was reduced relative to the wild type when both were grown in short days. Thus, the accumulation of soluble carbohydrate may have triggered a negative long-term regulatory influence on photosynthetic capacity (Caspar et al., 1985).

Comparable investigations to these with *A. thaliana* using a PGM-less mutant of *Nicotiana sylvestris* suggested that when starch cannot be accumulated sucrose storage in the vacuole can be used as a way to store carbohydrate formed in photosynthesis (Hanson and McHale, 1988; see also Hanson, 1990).

6-4.2.2   The Regulation of Amylase Activity

Caspar et al. (1989) have discovered that three classes of mutants of *A. thaliana*, with alterations in starch metabolism, have higher levels of leaf amylase activity than the wild type when grown in 12-hour photoperiods. The effect was largely suppressed during growth in continuous light (Fig. 6-23). The increased amylase activity in the mutants was shown to be due to a 40-fold increase in the activity of extrachloroplastic β-amylase, observations that indicate the existence of a regulatory mechanism that controls the enzyme activity in response to fluctuations in photosynthetic carbohydrate metabolism. The physiological purpose of such a close enzyme regulation is not known (Caspar et al., 1989).

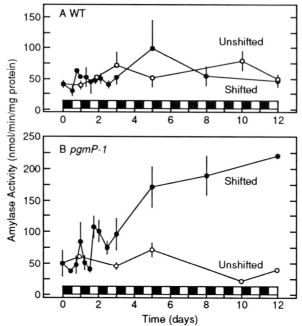

Figure 6-23. Total amylase activity in wild-type (WT) *Arabidopsis* (A) and the *pgmP-1* mutant (B) after a shift from continuous light to a 12-hour photoperiod. Plants were grown for 20 days in continuous light and then shifted to a 12-hour photoperiod (●) or left in continuous light as controls (○). The light conditions for the shifted plants are indicated by the filled (light off) or open (light on) boxes. Values are the means ± SEM, n = 3. (Caspar et al., *Proc. Natl. Acad. Sci. USA 86*: 5830-5833, 1989.)

Reference

Barendse, G.W.M., Kepczynski, J., Karssen, C.M., and Koornneef, M. 1986. The role of endogenous gibberellins during fruit and seed development: Studies on gibberellin-deficient genotypes of *Arabidopsis thaliana*. *Physiol. Plant. 67*: 315-319.

Caspar, T., Huber, S.C., and Somerville, C.R. 1985. Alterations in growth, photosynthesis, and respiration in a starchless mutant of *Arabidopsis thaliana* (L.) deficient in chloroplast phosphoglucomutase activity. *Plant Physiol. 79*: 11-17.

Caspar, T., Lin, T-P., Monroe, J., Bernhard, W., Spilatro, S., Preiss, J., and Somerville, C.R. 1989. Altered regulation of β-amylase activity in mutants of *Arabidopsis* with lesions in starch metabolism. *Proc. Natl. Acad. Sci. USA 86*: 5830-5833.

Chourey, P.S. 1981. Genetic control of sucrose synthetase in maize endosperm. *Mol. Gen. Genet. 184*: 372-376.

Chourey, P.S. and Nelson, O.E. 1976. The enzymatic deficiency conditioned by the *shrunken-1* mutations in maize. *Biochem. Genet. 14*: 1041-1055.

Cobb, B.G. and Hannah, L.C. 1988. *Shrunken-1* encoded sucrose synthase is not required for sucrose synthesis in the maize endosperm. *Plant Physiol. 88*: 1219-1221.

Dickinson, D.B. and Preiss, J. 1969. Presence of ADP-glucose pyrophosphorylase in *shrunken-2* and *brittle-2* mutants of maize endosperm. *Plant Physiol. 44*: 1058-1062.

Doehlert, D.C. and Kuo, T.M. 1990. Sugar metabolism in developing kernels of starch-deficient endosperm mutants of maize. *Plant Physiol. 92*: 990-994.

Dooner, H.K. 1985. *Viviparous-1* mutation in maize conditions pleiotropic enzyme deficiencies in the aleurone. *Plant Physiol. 77*: 486-488.

Duckham, S.C., Taylor, I.B., Linforth, R.S.T., Al-Naieb, R.J., Marples, B.A., and Bowman, W.R. 1989. The metabolism of *cis* -ABA-aldehyde by the wilty mutants of potato, pea and *Arabidopsis thaliana*. *J. Exp. Bot. 40*: 901-905.

Felker, F.C., Peterson, D.M., and Nelson, O.E. 1983. Growth characteristics, grain filling, and assimilate transport in a shrunken endosperm mutant of barley. *Plant Physiol. 72*: 679-684.

Felker, F.C., Peterson, D.M., and Nelson, O.E. 1984a. Development of tannin vacuoles in chalaza and seed coat of barley in relation to early chalazal necrosis in the *seg1* mutant. *Planta 161*: 540-549.

Felker, F.C., Peterson, D.M., and Nelson, O.E. 1984b. [^{14}C]Sucrose uptake and labeling of starch in developing grains of normal and *seg1* barley. *Plant Physiol. 74*: 43-46.

Felker, F.C., Peterson, D.M., and Nelson, O.E. 1985. Anatomy of immature grains of eight maternal effect shrunken endosperm barley mutants. *Am. J. Bot. 72*: 248-256.

Groot, S.P.C., Bruinsma, J., and Karssen, C.M. 1987. The role of endogenous gibberellin in seed and fruit development of tomato: Studies with a gibberellin-deficient mutant. *Physiol. Plant. 71*: 184-190.

Gupta, M., Chourey, P.S., Burr, B., and Still, P.E. 1988. cDNAs of two non-allelic sucrose synthase genes in maize: Cloning, expression, characterization and molecular mapping of the *sucrose synthase-2* gene. *Plant Mol. Biol. 10*: 215-224.

Hanson, J.R. 1990. Steady-state and oscillating photosynthesis by a starchless mutant of *Nicotiana sylvestris*. *Plant Physiol. 93*: 1212-1218.

Hanson, J.R. and McHale, N.A. 1988. A starchless mutant of *Nicotiana sylvestris* containing a modified plastid phosphoglucomutase. *Plant Physiol. 88*: 838-844.

Karssen, C.M., Brinkhorst-van der Swan, D.L.C., Breekland, A.E., and Koornneef, M. 1983. Induction of dormancy during seed development by endogenous abscisic acid: Studies on abscisic acid deficient genotypes of *Arabidopsis thaliana* (L.) Heynh. *Planta 157*: 158-165.

Karssen, C.M. and Lacka, E. 1986. A revision of the hormone-balance theory of seed dormancy: Studies on gibberellin and/or abscisic acid deficient mutants of *Arabidopsis thaliana*. In *Plant Growth Substances*, M. Bopp (ed.), Springer-Verlag, Heidelberg, pp. 315-323.

Karssen, C.M., Groot, S.P.C., and Koornneef, M. 1987. Hormone mutants and seed dormancy in *Arabidopsis* and tomato. In *Developmental*

Mutants in Higher Plants, H. Thomas and D. Grierson (eds.), Cambridge University Press, Cambridge, pp. 119-133.

Koornneef, M. and van der Veen, J.H. 1980. Induction and analysis of gibberellin sensitive mutants in *Arabidopsis thaliana* (L.) Heynh. *Theor. Appl. Genet. 58*: 257-263.

Koornneef, M., Jorna, M.L., Brinkhorst-van der Swan, D.L.C., and Karssen, C.M. 1982. The isolation of abscisic acid (ABA) deficient mutants by selection of induced revertants in nongerminating gibberellin sensitive lines of *Arabidopsis thaliana* (L.) Heynh. *Theor. Appl. Genet. 61*: 385-393.

Koornneef, M. 1986. Genetic aspects of abscisic acid. In *A Genetic Approach to Plant Biochemistry*, A.D. Blonstein and P.J. King (eds.), Springer-Verlag, Wien, chapter 2.

Lin, T-P., Caspar, T., Somerville, C.R., and Preiss, J. 1988a. Isolation and characterization of a starchless mutant of *Arabidopsis thaliana* (L.) Heynh lacking ADP glucose pyrophosphorylase activity. *Plant Physiol. 86*: 1131-1135.

Lin, T-P., Caspar, T., Somerville, C.R., and Preiss, J. 1988b. A starch-deficient mutant of *Arabidopsis thaliana* with low ADP glucose pyrophosphorylase activity lacks one of the two subunits of the enzyme. *Plant Physiol. 88*: 1175-1181.

Linforth, R.S.T., Taylor, I.B., and Hedden, P. 1987. Abscisic acid and C10 dicarboxylic acids in wilty tomato mutants. *J. Exp. Bot. 38*: 1734-1740.

McCormick, S., Mauvais, J., and Fedoroff, N. 1982. Evidence that the two sucrose syntetase genes in maize are related. *Mol. Gen. Genet. 187*: 494-500.

Moore, R. and Smith, J.D. 1985. Graviresponsiveness and abscisic acid content of roots of carotenoid-deficient mutants of *Zea mays* L. *Planta 164*: 126-128.

Nelson, O.E. and Rines, H.W. 1962. The enzymatic deficiency in the *waxy* mutant of maize. *Biochem. Biophys. Res. Commun. 9*: 297-300.

Nelson, O.E., Chourey, P.S., and Chang. M.T. 1978. Nucleotide diphosphate sugar-starch glucosyl transferase activity of *wx* starch granules. *Plant Physiol. 62*: 383-386.

Parry, A.D., Neill, S.J., and Horgan, R. 1988. Xanthoxin levels and metabolism in the wild type and wilty mutants of tomato. *Planta 173*: 397-404.

Peterson, D.M., Felker, F.C., and Nelson, O.E. 1985. *Seg* mutants as probes for regulation of assimilate transport and endosperm development in barley. In *Exploitatation of Physiological and Genetic Variability to Enhance Crop Productivity*, J.E. Harper, L.E. Schrader, and R.W. Howell (eds.), American Society of Plant Physiologists, Rockville, MD, pp. 31-45.

Preiss, J., Danner, S., Summers, P.S., Morell, M., Barton, C.R., Yang, L., and Nieder, M. 1990. Molecular characterization of the *brittle-2* gene effect on maize endosperm ADP glucose pyrophosphorylase subunits. *Plant Physiol. 92*: 881-885.

Quatrano, R.S. 1988. The role of hormones during seed development. In *Plant Hormones and Their Role in Plant Growth and Development*, P.J.

Davies (ed.), Kluwer Academic Publishers, Dortrecht, pp. 494-514.

Robertson, D.S. 1975. Survey of the albino and white endosperm mutants of maize. *J. Hered. 66*: 67-74.

Robichaud, C.S., Wong, J., and Sussex, I.M. 1980. Control of in vitro growth of viviparous embryo mutants of maize by abscisic acid. *Dev. Genet. 1*: 325-330.

Schulman, A.H. and Ahokas, H. 1990. A novel shrunken endosperm mutant of barley. *Physiol. Plant. 78*: 583-589.

Smith, A.M., Bettey, M., and Bedford, I.D. 1989. Evidence that the *rb* locus alters the starch content of developing pea embryos through an effect on ADP glucose pyrophosphorylase. *Plant Physiol. 89*: 1279-1284.

Smith, J.D., McDaniel, S., and Lively, S. 1978. Regulation of embryo growth by abscisic acid in vitro. *Maize Genet. Coop. Newsletter 52*: 107-108.

Sprague, G.F., Brimhall, B., and Hixon, R.M. 1943. Some effects of the *waxy* gene in corn on properties of the endosperm starch. *J. Am. Soc. Agron. 35*: 817-822.

Springer, B., Werr, W., Starlinger, P., Bennett, D.C., Zokolica, M., and Freeling, M. 1986. The *Shrunken* gene on chromosome 9 of *Zea mays* L. is expressed in various plant tissues and encodes an anaerobic protein. *Mol. Gen. Genet. 205*: 461-468.

Taylor, I.B. 1987. ABA deficient tomato mutants. In Developmental Mutants in Higher Plants, H. Thomas, and D. Grierson (eds.), Cambridge University Press, pp. 197-217.

Taylor, I.B., Linforth, R.S.T., Al-Naieb, R.J., Bowman, W.R., and Marples, B.A. 1988. The wilty tomato mutants *flacca* and *sitiens* are impaired in the oxidation of ABA-aldehyde to ABA. *Plant, Cell Environ. 11*: 739-745.

Tsai, C-Y. and Nelson, O.E. 1966. Starch-deficient maize mutant lacking adenosine diphosphate glucose pyrophosphorylase activity. *Science 151*: 341-343.

Wilson, G.F., Rhodes, A.M., and Dickinson, D.B. 1973. Some physiological effects of viviparous genes *vp1* and *vp5* on developing maize kernels. *Plant Physiol. 52*: 350-356.

Author Index

Subject Index

Abscisic acid
 biosynthesis, 364-71
  ABA-aldehyde as intermediate in, 369
  class I and II mutants of, 362-63
  C-10 dicarboxylic acids as intermediates in, 366-67, 370-71
  "downstream" mutants of, 366, 369
  lesions in *flacca* and *sitiens* wilty mutants, site of, 369
  "upstream" mutants of, 366, 369, 370-71
  violaxanthin as precursor in, 364-65
  xanthoxin as intermediate in, 367-70
 deficiency mutants, 6, 9-19, 349, 350-362-72. *See also* Stomate; Seed dormancy; Seed development
 and dormancy, 348, 349, 350-57. *See also* Seed dormancy
 and gravitropism, 276, 285, 286, 371-72
 leaf water potentials and, 9-19
 and salt tolerance, 32-33
 slender barley mutants, effects on, 253-255
 and stomate action, 4, 9-19
 wilty mutants and, 5-6, 9-19, 351, 364-71
Acetolactate synthase (acetohydroxyacid synthase), 82
 cross-resistance to herbicides, 127-30
 feedback sensitivity of, 128-30
 gene amplification and, 130-32

herbicide site of action on, 122-27
Adenosine triphosphate synthetase (coupling factor) *See* Thylakoid
S-Adenosylmethionine decarboxylase. *See* Polyamines
ADPG pyrophosphorylase. *See* Starch synthesis
ADPG-starch glucosyl transferase. *See* Starch synthesis
Alcohol dehydrogenase, 8
 and anaerobiosis, 33-41. *See also* Anaerobiosis; Stress
 gene expression under, 36
 mutants of, 33-41
 seed germination and, 35-37
Alfalfa, 54
Amaranthus hybridus, triazine resistance in, 137, 140-41
Amino acid biosynthesis
 of aromatic amino acids, 106
 aspartate pathway mutants of, 112-20
  amino acid uptake by, 116-117, 118
  aspartokinase in, 113, 114-15, 119-20
  control of, 113-14
  dihydropicolinate synthetase, 113, 114, 115, 120
  homoserine dehydrogenase in, 113, 119
  isolation and characterization, 114-116
  lysine and threonine effects in, 114-116
 auxotrophic mutants of, 75-79, 80-83, 84-86
 ethionine resistance, 117-120

729 84

QK   King, John, 1938-
711.2 The genetic basis of
.K57   plant physiological
1991   processes

AUGUSTANA UNIVERSITY COLLEGE
LIBRARY